Dieter Rehder

Chemistry in Space

Related Titles

Bar-Cohen, Y., Zacny, K. (eds.)

Drilling in Extreme Environments

Penetration and Sampling on Earth and other Planets

2009
ISBN: 978-3-527-40852-8

Grunenberg, J. (ed.)

Computational Spectroscopy

Methods, Experiments and Applications

2010
ISBN: 978-3-527-32649-5

Shaw, A. M.

Astrochemistry

From Astronomy to Astrobiology

2006
ISBN: 978-0-470-09138-8

Dieter Rehder

Chemistry in Space

From Interstellar Matter to the Origin of Life

WILEY-VCH

WILEY-VCH Verlag GmbH & Co. KGaA

The Author

Prof. Dr. Dieter Rehder
Universität Hamburg
Department Chemie
Martin-Luther-King-Platz 6
20146 Hamburg
Germany

Cover
The molecules shown on the cover are formic acid and aminoacetonitrile. Both have recently been discovered in interstellar clouds.

■ All books published by Wiley-VCH are carefully produced. Nevertheless, authors, editors, and publisher do not warrant the information contained in these books, including this book, to be free of errors. Readers are advised to keep in mind that statements, data, illustrations, procedural details or other items may inadvertently be inaccurate.

Library of Congress Card No.: applied for

British Library Cataloguing-in-Publication Data
A catalogue record for this book is available from the British Library.

Bibliographic information published by the Deutsche Nationalbibliothek
The Deutsche Nationalbibliothek lists this publication in the Deutsche Nationalbibliografie; detailed bibliographic data are available on the Internet at http://dnb.d-nb.de.

© 2010 Wiley-VCH Verlag & Co. KGaA, Boschstr. 12, 69469 Weinheim, Germany

All rights reserved (including those of translation into other languages). No part of this book may be reproduced in any form – by photoprinting, microfilm, or any other means – nor transmitted or translated into a machine language without written permission from the publishers. Registered names, trademarks, etc. used in this book, even when not specifically marked as such, are not to be considered unprotected by law.

Cover Design   Grafik-Design Schulz, Fußgönheim
Typesetting   Toppan Best-set Premedia Limited, Hong Kong
Printing and Binding   Fabulous Printers Pte Ltd

Printed in Singapore
Printed on acid-free paper

ISBN: 978-3-527-32689-1

Contents

Preface *IX*

1	**Introduction and Technical Notes** *1*	
	References *5*	
2	**Origin and Development of the Universe** *7*	
2.1	The Big Bang *7*	
2.2	Cosmic Evolution: Dark Matter – the First Stars *10*	
2.3	Cosmo-Chronometry *12*	
	Summary *15*	
	References *15*	
3	**The Evolution of Stars** *17*	
3.1	Formation, Classification, and Evolution of Stars *17*	
3.1.1	General *17*	
3.1.2	Neutron Stars and Black Holes *23*	
3.1.3	Accretion and Hydrogen Burning *25*	
3.1.4	Nuclear Fusion Sequences Involving He, C, O, Ne, and Si *28*	
3.1.5	The *r*-, *s*-, *rp*- and Related Processes *30*	
3.1.5.1	General *30*	
3.1.5.2	Rapid Processes *31*	
3.1.5.3	Slow Processes *34*	
3.2	Chemistry in AGB Stars *35*	
3.3	Galaxies and Clusters *40*	
	Summary *42*	
	References *43*	
4	**The Interstellar Medium** *45*	
4.1	General *45*	
4.2	Chemistry in Interstellar Clouds *50*	
4.2.1	Reaction Types *50*	
4.2.2	Reaction Networks *54*	

Chemistry in Space: From Interstellar Matter to the Origin of Life. Dieter Rehder
© 2010 WILEY-VCH Verlag GmbH & Co. KGaA, Weinheim
ISBN: 978-3-527-32689-1

4.2.3	Detection of Basic Interstellar Species	61
4.2.3.1	Hydrogen	62
4.2.3.2	Other Basic Molecules	68
4.2.4	Complex Molecules	74
4.2.5	Chemistry on Grains	80
4.2.5.1	The Hydrogen Problem	81
4.2.5.2	Grain Structure, Chemical Composition, and Chemical Reactions	82
	Summary	94
	References	95
5	**The Solar System**	**99**
5.1	Overview	99
5.2	Earth's Moon and the Terrestrial Planets: Mercury, Venus, and Mars	107
5.2.1	The Moon	107
5.2.2	Mercury	110
5.2.3	Venus	115
5.2.3.1	General, and Geological and Orbit Features	115
5.2.3.2	Venus' Atmosphere	118
5.2.3.3	Chemical Reactions	121
5.2.4	Mars	126
5.2.4.1	General	126
5.2.4.2	Orbital Features, and the Martian Moons and Trojans	127
5.2.4.3	Geological Features, Surface Chemistry, and Mars Meteorites	129
5.2.4.4	Methane	133
5.2.4.5	Carbonates, Sulfates, and Water	137
5.2.4.6	Chemistry in the Martian Atmosphere	140
	Summary Section 5.2	145
5.3	Ceres, Asteroids, Meteorites, and Interplanetary Dust	146
5.3.1	General and Classification	146
5.3.2	Carbon-Bearing Components in Carbonaceous Chondrites	153
5.3.3	Interplanetary Dust Particles (Presolar Grains)	162
5.4	Comets	167
5.4.1	General	167
5.4.2	Comet Chemistry	171
5.5	Kuiper Belt Objects	176
	Summary Sections 5.3–5.5	179
5.6	The Giant Planets and Their Moons	180
5.6.1	Jupiter, Saturn, Uranus, and Neptune	180
5.6.2	The Galilean Moons	186
5.6.3	The Moons Enceladus, Titan and Triton	191
	Summary Section 5.6	195
	References	196

6 **Exoplanets** *203*
Summary *211*
References *212*

7 **The Origin of Life** *213*
7.1 What is Life? *213*
7.2 Putative Non-Carbon and Nonaqueous Life Forms; the Biological Role of Silicate, Phosphate, and Water *220*
7.3 Life Under Extreme Conditions *230*
Summary Sections 7.1–7.3 *240*
7.4 Scenarios for the Primordial Supply of Basic Life Molecules *241*
7.4.1 The Iron–Sulfur World ("Pioneer Organisms") *242*
7.4.2 The Miller–Urey and Related Experiments *247*
7.4.3 "Clay Organisms" *259*
7.4.4 Extraterrestrial Input *262*
7.5 Extraterrestrial Life? *265*
Summary Sections 7.4 and 7.5 *274*
References *276*

Index *281*

Preface

On 27th December 1984, a team of "meteorite hunters," funded by the National Science Foundation, picked up a rock of 1.93 kg in an Antarctic area known as Alan Hills. Since it was the first one to be collected in 1984, it was labeled ALH84001, ALan Hills 1984 no. 001. Soon it became evident that this meteorite originated from our neighbor planet Mars – a rock that formed 4.1 billion years ago and was blasted off the red planet's crust 15 million years ago by an impacting planetesimal. After roaming about in the Solar System for most of its time, this rock entered into the irresistible force of Earth's attraction, where it landed 13 thousand years ago, in Antarctica and hence in an area where it was protected, at least in part, from weathering. Structural elements detected in this Martian meteorite, considered to represent biomarkers, sparked off a controversial debate on the possibility of early microbial life on our neighbor planet about 4 billion years ago, and shipping of Martian life forms to Earth, a debate which became reignited by recent reinvestigations of the meteoritic inclusions.

Other meteorites, originating from objects in the asteroid belt between Mars and Jupiter, have brought amino acids and nucleobases to Earth, among these amino acids which are essential for terrestrial life forms. Does this hint toward an extraterrestrial origin of at least part of the building blocks necessary for terrestrial life? And if yes – how could amino acids, which are rather complex molecules, have been synthesized and survived under conditions prevailing in space?

The idea of "seeds (*spermata*) of life," from which all organisms derive, goes back to the cosmological theory formulated by the Greek philosopher and mathematician Anaxagoras in the 5th century B.C. Anaxagoras, perhaps better known for his "squaring the circle," thus may be considered the originator of what became established as *panspermia*. Panspermia reached the level of a scientific (and popular) hypothesis in the 19th century through contributions from Berzelius, Pasteur, Richter, Thomson (Lord Kelvin), von Helmholtz, and others, a hypothesis according to which life originated and became distributed somewhere in space, and was transported to the planets from space. In 1903, the Swedish chemist Arrhenius proposed that radiation pressure exerted by stars such as our Sun can spread submicrometer to micrometer-sized "spores of life," a proposal that later (in the 1960s) was quantified by Sagan. The panspermia hypothesis got somewhat disreputable, when Francis H. Crick (who, together with Watson, received the

Chemistry in Space: From Interstellar Matter to the Origin of Life. Dieter Rehder
© 2010 WILEY-VCH Verlag GmbH & Co. KGaA, Weinheim
ISBN: 978-3-527-32689-1

1962 Nobel Prize in Medicine for the discovery of the double-helix structure of desoxyribonucleic acid) and Leslie Orgel published a paper, in 1973, where they suggested that life arrived on Earth through "directed panspermia," where *directed* refers to an extraterrestrial civilization. The likeliness of another civilization somewhere else out in space is even more speculative than the likeliness that Life came into existence at all.

There is no doubt, of course, that life exists on Earth. Whether Earth is the cradle of life (from which it may have been transported elsewhere into our Solar System or even beyond) or whether life has been carried to our planet from outside (exospermia) remains an interesting concern to be addressed. ALH84001 may provide a clue to this question. The discovery of exoplanets (planets orbiting other stars than our Sun in the Milky Way galaxy) is another issue that stimulates imagination as it comes to the possibility of extraterrestrial life. New exoplanets are being discovered at a vertiginous speed, and a few of the about 455 exoplanets known to date, so-called super-Earths, do have features which are reminiscent of our planet.

Hamburg, May 2010   *Dieter Rehder*

1
Introduction and Technical Notes

In the year 1609, Johannes Kepler published a standard work of astronomy, the *Astronomia Nova, sev Physica Coelestis, tradita commentariis de Mortibvs Stellæ Martis*: "The New Astronomy, or Celestial Physics, based on records on the Motions of the Star Mars." In Chapter LIX (59), he summarizes what became known as Kepler's first and second law (Figure 1.1). The heading of this chapter starts as follows: *Demonstratio, qvod orbita Martis ... fiat perfecta ellipsis*: "This is to demonstrate that the Martian orbit ... is a perfect ellipse," or – in today's common phrasing of Kepler's first law: "The planet's orbit is an ellipse, with the Sun at one focus." The second law states that the "line connecting the Sun and the planet sweeps out equal areas in equal time intervals." (The third law was formulated 10 years later: $p_1^2/p_2^2 = r_1^3/r_2^3$, p = revolutionary period, r = semimajor axis; the lower indices 1 and 2 refer to two planets.) Kepler's pioneering mathematical treatise, based on minute observations collected by Tycho Brahe, had been a breakthrough for astronomy, and applications of his laws are still influential in modern astronomy.

A second trailblazing event 400 years ago was the discovery of what is now known as the "Galilean moons," the four large moons of the planet Jupiter. Galileo Galilei announced the discovery of three of the Jovian moons on the 7th of January 1610 (discovery of the fourth moon followed a couple of weeks later) – according to the Gregorian calendar, which corresponds to the 28th of December, 1609, in the Julian calendar. In honor of his mentor Cosimo II de Medici, Galilei named the moons *Cosmica Sidera* (Cosimo's stars), and then *Medicea Sidera* (stars of the Medici). Following a suggestion by Simon Mayr (or Simon *Marius* in the Latinized version) in 1614, the four moons were termed "Io, Europa, Ganymed *atque* (and) Callisto *lascivo nimivm perplacvere Iovi*" (... who greatly pleased lustful Jupiter [Zeus]). Simon Mayr discovered the moons independently of Galilei, but announced his discovery a day later, on the 8th of January 1610. The discovery of the moons, and realization that the moons orbit *Jupiter*, was a final bash against a geocentric worldview of the Universe dominating medieval times.

The two discoveries became duly commemorated in the 2009 International Year of Astronomy, which was also the year for a couple of key discoveries in astronomy, astrophysics, astrochemistry, and astrobiology: (i) detection of the first exoplanets with physical and chemical characteristics approximating those of our home planet;

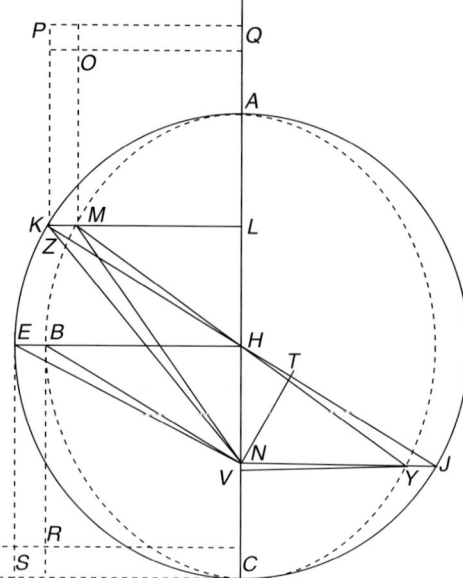

Figure 1.1 Kepler's illustration of his findings on Mars' motions, which became known as Kepler's first and second law of planetary motion; from chapter *LIX* of *Astronomia Nova*, published 1609. The first law states that the planet's orbit is an ellipse – the punctuated line starting with the quadrant AMB, with the Sun (N) at one focus. The second law provides information on the area (BMN) swept by the line (MN, BN) connecting the Sun and the planet.

(ii) reinvestigation of nanosized magnetite crystals, possible biomarkers, in a Martian meteorite recovered in Antarctica in 1984 (see also Preface); (iii) discovery of the glycine precursor aminoacetonitrile (see cover of this book) in the "Large Molecular Heimat," a dense interstellar molecular cloud in the constellation of Sagittarius; (iv) the final proof that our next neighbor in the Cosmos, our Moon, contains sizable reservoirs of water, possibly of cometary origin, deposited in permanently shaded craters; and (v) location of the most distant and oldest object in the Universe, a gamma ray burst associated with a stellar-sized black hole or magnetic neutron star, which formed just 630 million years after the Big Bang, the event which is considered the hour of birth of our Universe, 13.7 billion years ago.

These are just a few selected highlights, supposed to adumbrate the scope of the present treatise, and to be addressed together with other topical and less recent events and discoveries in some detail in this book. The book will focus on aspects in astronomy related to chemistry – in stars and the interstellar medium, in the atmospheres, on the surfaces, and in subsurface areas of planets, planetoidal bodies, moons, asteroids, comets, interplanetary, and interstellar dust grains. A topical point to be covered is the query of the origin of life, either on Earth or somewhere else in our Milky Way galaxy, and the genesis of basic molecules functioning as building blocks for complex molecules associated with life and/or representing life. Along with these chemistry-related issues, general cosmological aspects related to astronomy and astrophysics, and often indispensable for an axiomatic comprehension of chemical processes, will be approached. Some knowledge of the basics of chemical (including bio- and physicochemical) coherency will be afforded to become involved: the book is designed so as to be both an introduction for the interested beginner with some basic knowledge, and a compendium for the more advanced scientist with a background in chemistry and adjacent disciplines.

Several of the crucial points covered in the present book have been treated in book publications by other authors, usually with another target course, that is, less intimately directed toward chemical and biological aspects of astronomical problems. The following glossary (sorted chronologically) is a selection of books and compendia that have animated me during the bygone two decades, and are thus recommended as "Further Reading".

- Duley, W.W., Williams, D.A. (1984) *Interstellar Chemistry*, Academic Press, London.
- Saxena, S.K. (ed.) (1986) *Chemistry and Physics of the Terrestrial Planets* [vol. 6 of Advances in Physical Geochemistry], Springer Verlag, Berlin.
- Lewis, J.S. (1995) *Physics and Chemistry of the Solar System*, Academic Press, San Diego. [2nd Edition (2004): Elsevier/Academic Press]
- Szczerba, R., Górny, S.K. (eds.) (2001) *Post-AGB Objects as a Phase of Stellar Evolution* [vol. 265 of Astrophysics and Space Science Library], Kluwer Academic, Dordrecht.
- Clayton, D.D. (2003) *Handbook of Isotopes in the Cosmos*, Cambridge University Press, Cambridge.
- Green, S.F., Jones, M.H. (eds.) (2003/04) *An Introduction to the Sun and Stars*, Cambridge University Press, Cambridge.
- Thielens, A.G.G.M. (2005) *The Physics and Chemistry of the Interstellar Medium*, Cambridge University Press, Cambridge.
- Shaw, A.M. (2006) *Astrochemistry – From Astronomy to Astrobiology*, John Wiley & Sons, Chichester.
- Plaxco, K.W., Gross, M. (2006) *Astrobiology*, The John Hopkins University Press, Baltimore.
- Kwok, S. (2007) *Physics and Chemistry of the Interstellar Medium*, University Science Books, Sausalito, CA.
- Shapiro, S.L, Teukolsky, S.A (2007) *Black Holes, White Dwarfs, and Neutron Stars*, Wiley VCH, Weinheim.

Scientists enrooted in astronomy do have their subject-specific nomenclature and system of units, which is not always easily accessible to a chemist. As an

Table 1.1 Units for concentration and density, and their conversion into molar units.

Quantity	Description	Unit[a]	Molar unit; conversion factor[b]
Column density, column amount, column abundance N	The number of elementary entities in a vertical column. Column: In atmospheric chemistry the height of the atmosphere;[c] in interstellar chemistry the length of the line of sight between observer and a light-emitting (stellar) object	cm^{-2}	mol m^{-2}; $N \times (6.022 \times 10^{19})^{-1}$
Volume(tric) or number density n	The number of elementary entities per unit volume	cm^{-3}	mol l^{-1}; $n \times (6.022 \times 10^{20})^{-1}$
Fractional or abundance ratio $f(X)^{[d]} = n(X)/n(H_2)$	The number of entities X per number of H_2 molecules	–	–
Molar concentration c	Number of moles per liter of solvent	M ≡ mol L^{-1}	–
Mixing ratio (mole fraction) $c_X = n_X/\Sigma n_i$	The number of moles of a species X in the overall mix (containing i components); $\Sigma c_X = 1$	–	–

a) Number of elementary entities (atoms, ions, molecules, electrons, …) per area (cm^{-2}) or volume (cm^{-3}); the number of entities is a dimensionless quantity.
b) Contains the Avogadro constant $N_A = 6.022 \times 10^{23}$ mol^{-1} elementary entities (i.e., 1 mol).
c) See Eq. (5.9) in Section 5.2.3.2 for additional details.
d) This symbol is also used for mole fraction.

example, if it comes to the term "concentration" (of a specific species X in a mix), chemists use to think in terms of "molarity" (moles of X per liter of the mix) or "molality" (moles of X per kg), where "mole" relates to the amount of substance: 1 mole of *any* substance is equal to 6.022×10^{23} elementary entities. Examples for elementary entities are elementary particles (such as electrons, protons, and neutrons), atoms, ions, molecules, light quanta. In contrast, astronomers commonly refer to concentration in terms of "column density/abundance/amount," "fractional density," and "number/volume density," conceptions so uncommon for chemists that they hardly do associate any perception with these quantifications. From a chemist's point of view, *column amount* quoted in terms of mol m^{-2} (i.e., employing the units of the Système Internationale, the SI system) is "correct" [1] and has been used wherever sensibly applicable–together with the units preferred by astronomers. Table 1.1 provides an overview of conversions of units for "concentration," frequently employed in astronomical and astrophysical articles, into molar units. Conversions will also be provided in the main text wherever this appears to be reasonable.

Most of the units employed in this book are SI units. Where our conceptions from everyday experience are dominated by more classical units, both the SI and the popular units are provided. Examples are temperature (in Kelvin or degrees Celsius), pressure (in Pascal or bar), strength of the magnetic field (the B field; in Tesla or Gauss). Distances in astronomical dimensions, when expressed in meters or 10^3 multiples thereof, are not easily handled by our spatial perception. Astronomical units (AUs), parsecs (pc), and light-years (ly), as defined in Figure 5.2 and Table 5.3, are more easily comprehended and therefore used throughout. Similarly, if it comes to "astronomical ages," years (a, derived from the Latin *annum*) and multiples thereof, such as megayears (Ma = 10^6 a) and gigayears (Ga = 10^9 a) are employed rather than the SI unit "second." Finally, masses (m, SI unit: g) are quoted, were appropriate, in $m_\oplus$ (multiples of Earth; $\oplus$ is the astronomical symbol for Earth), m_J (multiples of Jupiters) and $m_\odot$ (multiples of Suns; $\odot$ is the symbol for the Sun). The lower case letter "m" otherwise stands for magnitude (of a star); the *capital* letter M ($\equiv$ mol l^{-1}) denotes molarity and, in chemical equations, "metal" (all elements beyond helium), while M (in italics) indicates "molecular mass" (g mol^{-1}) [and matrix in reactions on dust particles].

The quantification of "energy" is another point of potential controversy: in chemistry, the (almost exclusive) unit for energy is kilo-Joule per mole (kJ mol^{-1}). In particle physics, this unit is unhandy, and electron volts (eV) are preferred; in spectroscopy, it is common to measure energy in reciprocal centimeters (cm^{-1}) which, strictly speaking, is not energy but energy divided by hc (the product of the Planck constant and the speed of light). Conversions of these units will be provided in the main text wherever appropriate.

References

1 Basher, R.E. (2006) Units for column amounts of ozone and other atmospheric gases. *Quart. J. R. Meteorol. Soc.*, **108**, 460–462.

2
Origin and Development of the Universe

2.1
The Big Bang

The dark sky against which we see stars and galaxies is not completely black. Rather, the Universe is filled with a relic electromagnetic radiation called cosmic microwave background (CMB) radiation, characterized by a frequency of 160.2 GHz, corresponding to a wavelength of 1.9 mm. This radiation represents the cosmologically red-shifted (shifted to longer wavelengths, also termed "Doppler shift") radiation of an incessantly expanding Universe. The intensity to wavelength distribution of the CMB follows an almost perfect black body radiation at a temperature of 2.725 K, and it is almost isotropic, that is, of equal intensity in all directions. Backward extrapolation in time reveals that this background radiation originates from the time where the Universe was 380 000 years old: the time span which elapsed since the Universe started to develop from a singularity in time and space, the starting point of which was termed the "Big Bang." A spacetime (or gravitational) singularity is, according to the general theory of relativity, the initial state of the Universe. 380 000 years after this development started, the Universe was sufficiently cold, about 3000 K (corresponding to energy of 0.25 eV), to allow for the formation of neutral atoms which no longer absorbed photons, making the Universe transparent. Along with the background radiation, the relative abundance of the stable hydrogen isotopes ^{1}H (protium) and ^{2}H (deuterium), and the helium isotopes ^{3}He and ^{4}He in the Universe provide a convincing back-up of the present theory.

What became known as the Big Bang theory for the origin of the Universe was originally proposed by Georges Lemaître (1927–1931), who called this theory "hypothesis of the primeval atom," where "primeval atom" refers to a single point at time $t = 0$ or, rather, to a situation where time and space did not yet exist. The term "Big Bang" goes back to Fred Hoyle (1949) who, incipiently, tried to discredit the hypothesis he was not yet ready to subscribe to. The discovery of the cosmic background radiation in 1964 secured the theory. The "Big Bang event" nowadays is commonly not restricted to the very first fraction of a second where the singularity became resolved, developing into matter, time and space, but to the first few minutes of expansion and evolution of the primordial matter, which includes Big Bang nucleosynthesis. The first about 5 min of the time line, starting 13.73 billion

Chemistry in Space: From Interstellar Matter to the Origin of Life. Dieter Rehder
© 2010 WILEY-VCH Verlag GmbH & Co. KGaA, Weinheim
ISBN: 978-3-527-32689-1

years ago,[1] may have proceeded according to the following succession of epochs; for the hierarchy of fermions addressed below see Scheme 2.1, for characteristics of selected elementary particles see Table 2.1.

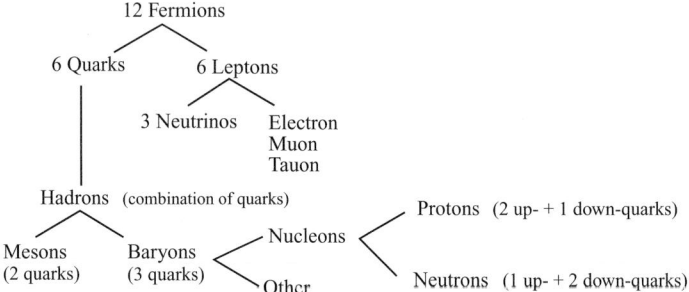

Scheme 2.1 Hierarchy of fermions, the elementary particles of matter. A corresponding set of antifermions also exists. An example for "other" is the *hyperon*, where one of the down-quarks in the neutron is replaced by a heavy strange-quark.

1) The first epoch, the *Planck epoch*, is characterized by the Planck time of 5.4×10^{-44} s. This is the time it takes a photon to travel the Planck length[2] of 1.6×10^{-34} m. In other words: this is the period of uncertainty at the beginning of the Universe. The Planck epoch is further characterized by a temperature of 10^{32} K, and a density of 10^{94} g cm^{-3}. The crucial event within the Planck epoch is decoupling of the gravity off the three other fundamental forces (electromagnetic forces, strong, and weak nuclear forces).

2) Uncoupling of the gravitational force triggered quantum fluctuation (the formation and annihilation of particles of matter and antimatter out of vacuum), followed by inflation, an extremely rapid expansion of the Universe by a factor of 10^{50}, in the time range 10^{-35} to 10^{-33} s, accompanied by a drop in temperature to 10^{27} K.

3) Further cooling to 10^{25} K ended the period of inflation and gave rise to the formation of a quark–gluon plasma, consisting of quarks, antiquarks, and gluons, the building blocks of matter (quarks) and interacting forces ("glues," viz. gluons). Concomitantly, the strong forces separated from the weak and electromagnetic forces.

4) The separation of electromagnetic and weak forces was achieved after 10^{-12} s and a temperature of 10^{16} K. After a time span of 10^{-6} s had elapsed, and the temperature dropped to $T \approx 10^{13}$ K, quarks/antiquarks became glued to form hadrons: mesons formed from two quarks, and baryons formed from three quarks. Protons/antiprotons and neutrons/antineutrons are *baryons*. The

---

1) Compared to the age of the Universe, our Solar System (Section 5.1), 4.57 billion years old, is still in its adolescence.

2) The Planck length is defined by $l_p = (h\,G/c^3)^{1/2}$, where G is the gravitational constant, h the Planck constant, and c the speed of light.

Table 2.1 Properties of standard elementary particles.

Name	Symbol	Charge (e)[a]	Rest mass (rounded) (u)[b]	Rest energy[c] (rounded) (MeV)	Spin	Half-life (s)
Proton	p, $^1_1$H	+1	1.00728	938.272	1/2	Stable
Neutron	n	0	1.00866	939.565	1/2	885.7
Electron	e$^-$, β$^-$	−1	5.486×10^{-4}	0.511	1/2	Stable
Positron (or antielectron)	e$^+$, β$^+$	+1	5.486×10^{-4}	0.511	1/2	Stable
(Electron–) neutrino	ν, ν$_e$	0	$<2\,\text{eV}/c^2$	–	1/2	–
Antineutrino	$\bar{\nu}_e$	0	$<2\,\text{eV}/c^2$	–	1/2	–

a) 1 e (elementary charge) = 1.602 C.
b) 1 u (elementary mass unit) = 1.661×10^{-24} g = $0.9315\,\text{GeV}/c^2$.
c) 1 MeV = 4.29×10^{-15} kJ.

baryogenesis in this so-called *hadron epoch* is believed to have triggered a tiny asymmetry between protons and neutrons (which together represent matter) on the one hand, and antiprotons and antineutrons (antimatter) on the other hand, responsible for today's predominance of matter over antimatter. When the temperature was no longer high enough to create new baryon–antibaryon pairs, baryons and antibaryons started to anneal each other, leaving behind a thinned-out population of baryons (protons and neutrons), and photons, the product of annihilation, the latter in very high energy density. Further, by continuous interconversion of protons and neutrons, neutrinos and antineutrinos were produced, Eq. (2.1):

$$\nu_e + {}^1_0 n \rightleftharpoons {}^1_1 p + e^- \tag{2.1a}$$

$$\bar{\nu}_e + {}^1_1 p \rightleftharpoons {}^1_0 n + e^+ \tag{2.1b}$$

5) After about 10^{-2} s, $T \approx 10^{12}$ K, and a density of 10^{13} g cm^{-3}, leptons (such as electrons e$^-$ and positrons e$^+$) were created (*lepton epoch*) by the collision of photons, again with a tiny excess of electrons over their antimatter equivalent positron. At about 1 s and $T = 10^{10}$ K, an annihilation process similar to that for baryons occurred, leaving behind electrons and photons.

6) Within the next three minutes and a temperature of 10^9 K, nucleosynthesis started, producing deuterium (^{2}H, Eq. (2.2)) and the lighter helium isotope ^{3}He (Eq. (2.3)). Most of the ^{2}H and ^{3}He ended up in the helium isotope ^{4}He (about 25% of the overall amount of gas constituents), Eqs. (2.4) and (2.5). This process, known as primordial or Big Bang nucleosynthesis, stopped after $t \approx 5$ min due to a dramatic loss in density. The remaining almost 75% of matter were represented by protons (p ≡ hydrogen nuclei ^{1}H).

$$^1_0 n + {}^1_1 p \rightarrow {}^2_1 H + \gamma \tag{2.2a}$$

$$2{}_1^1\text{p} \rightarrow {}_1^2\text{H} + \text{e}^+ + \nu_\text{e} \tag{2.2b}$$

$$2{}_1^2\text{H} + \gamma \rightarrow {}_2^3\text{He} + {}_0^1\text{n} \tag{2.3a}$$

$${}_1^2\text{H} + {}_1^1\text{p} \rightarrow {}_2^3\text{He} + \gamma \tag{2.3b}$$

$$\begin{aligned}2{}_1^2\text{H} &\rightarrow {}_1^3\text{H} + {}_1^1\text{p} \\ &\,\bigl|+ {}_1^2\text{H} \\ &\,\longrightarrow {}_2^4\text{He} + {}_0^1\text{n}\end{aligned} \tag{2.4a}$$

$${}_1^3\text{H} \rightarrow {}_2^3\text{He} + \text{e}^- + \bar{\nu}_\text{e} \tag{2.4b}$$

$$2{}_2^3\text{He} \rightarrow {}_2^4\text{He} + 2{}_1^1\text{p} + \gamma \tag{2.5}$$

The hydrogen isotope tritium (^{3}H) intermittently formed in the reaction sequence of Eq. (2.4a) is unstable; its half-life is 12.32 years, the decay products are ^{3}He, an electron and an antineutrino, Eq. (2.4b). All nuclei in today's Universe heavier than ^{4}He have been produced by nucleosynthesis in stars, with the exception of trace amounts of lithium (Eq. (2.6)) and beryllium (Eq. (2.7)), also generated via Big Bang nucleosynthesis:

$${}_1^3\text{H} + {}_2^4\text{He} \rightarrow {}_3^7\text{Li} + \gamma \tag{2.6}$$

$${}_2^3\text{He} + {}_2^4\text{He} \rightarrow {}_4^7\text{Be} + \gamma \tag{2.7}$$

2.2
Cosmic Evolution: Dark Matter – the First Stars

The most distant and oldest object so far discovered in the Universe, a γ-ray burst[3] (associated with a stellar-sized black hole or rapidly rotating magnetic neutron star; cf. Section 3.1.2), dates back 630×10^6 years [1]. The presently accepted scenario for the formation of the first stars about 10^8 years after the Big Bang is described by the cold dark matter (CDM) model of cosmic evolution. The particles making up CDM [2] are interacting only through gravity; they have subrelativistic velocities, that is, they are "slow" and thus "cold" (in terms of low kinetic energy), and they are "dark," that is, beyond detection by electromagnetic radiation. According to present perception, based on, inter alia, the gravitational influence on stars and galaxies, 21% of the contents of our Universe constitute CDM. This corresponds to a current mean density of 3×10^5 atomic mass units per cubic meter (u m^{-3}, 1 u is the approximate mass of a proton and a neutron; Table 2.1), or about three orders of magnitude less than in diffuse interstellar HII regions (thin nebulae essentially consisting of H$^+$; Chapter 4). Of the remaining contents of the Universe, 74% is *dark energy*, and just 5% is common matter, one tenth of which (and just 0.5% of the overall inventory of the Universe) is visible. This "baryonic matter" is not evenly distributed: large galaxies have a higher percentage of baryonic matter than small galaxies. Dark energy has been postulated in order to be

---

3) The red shift is $z = 8.2$, where z is defined by $z = (\lambda_\text{obs.} - \lambda_\text{emit.})/\lambda_\text{emit.}$.

able to explain the "antigravitational" effect, an acceleration of the expansion of the cosmos for the last 5 billion years. The present rate of expansion, defined by the Hubble constant H, is $74.2 \pm 3.6\,\text{km}\,\text{s}^{-1}\,\text{Mpc}^{-1}$ (the Hubble constant relates the speed by which galaxies race apart to their distance). Candidates for CDM particles are neutralinos[4] which, by self-annihilation, produce pions,[5] electron–positron pairs, and high-energy photons (γ rays). Neutralinos, or "weak interacting massive particles",[6] possibly produced in the Big Bang in the course of baryogenesis along with hydrogen and helium (see Section 2.1), are hypothetical "supersymmetric" particles. Supersymmetric refers to a linear combination of partners which differ in spin by ½. CDM neutralinos are the lightest among the neutralinos, typically with a mass of several dozen to several hundred GeV/c^2.

Roughly, the formation of the first stars and galaxies can have proceeded according to the following steps [3]:

1) Fragmentation of the primordial dark matter halo into assemblies of dark matter minihalos: "gas" clouds with an average temperature of ~1000 K, an overall mass of ~$10^6\,m_\odot$ ($m_\odot$ stands for Solar mass), and a mass per minihalo just about that of the Earth, but an extension corresponding to that of the Solar System.

2) Cooling of the primordial gas constituting the minihalo and collapse, primarily leaded to a small protostar and, by further accretion of the surrounding gas, to a massive so-called population III.1 star. Population III stars contain H, D, He and some Li (and Be) only. Just one star per minihalo is formed. Population III.2 stars are formed from gas that has already been processed.

3) Formation of galaxies by feed-back processes. Black holes may attain a central role in these processes. See also Section 3.3 for additional details.

Radiative cooling of the primordial gas, enabling contraction and accretion to stars, requires the presence of small amounts of molecules, H_2 in particular, the formation of which is represented by Eqs. (2.8a) and (2.8b). Contraction and accretion is further accompanied by the formation of a disc-like structure (proto-stellar disc). Accretion stops, mainly due to mass loss driven by photoevaporation, when the mass encompasses ca. 100 $m_\odot$:

$$H + e^- \rightarrow H^- + h\nu \qquad (2.8a)$$

$$H^- + H \rightarrow H_2 + e^- \qquad (2.8b)$$

---

4) To be differentiated from neutrinos (with a rest mass of close to zero) and neutrons (with a rest mass of ca. $1\,\text{GeV}/c^2$ (~1 u).

5) There are three pions (also termed π mesons): neutral (π^0) and charged (π^+ and π^-). The $\pi^\pm$ have a rest mass of $139.6\,\text{MeV}/c^2$ (~0.14 u), a mean life of 2.8×10^{-8} s (decay products are muon [related to electron/positron, but more massive; see also Scheme 2.1] and neutrino), a spin of $I = 1$, and negative parity (ungerade with respect to inversion).

6) Detection of dark matter particles is one of the primary goals of the recently installed Large Hadron Collider at CERN in Geneva.

As the temperature increases on accretion and formation of a massive population III star, secondary (feedback) effects come in, such as photodissociation of H_2 and ionization of H by radiation emitted by the star. This leads to a delay in the formation of additional new stars, but can also subsequently stimulate the formation of molecules within residual HII regions of the population III.1 stars, evolution of population III.1 into population III.2 stars, and star formation in neighboring minihalos. In any case, the population III stars end up as supernovae either by explosion and hence complete disruption, or by collapsing into black holes.

How the first galaxies formed still remains an enigma. Models suggest a crucial role of the feed-back effects, initiating the formation of star assemblies in cold black matter haloes with masses exceeding those of the mini-haloes by orders of magnitude. Chemical enrichment by the first supernovae, that is, supply of "metals" (everything beyond helium in astrochemical terminology) was a precondition for the formation of population II stars with still low but distinct metallicities, enabling a more vivid stellar evolution. Recent large area surveys have identified spheroid dwarf satellite galaxies inside and outside the Milky Way, which are supposed to be survivors of the gravitationally bound systems. The time frame for the formation of the first galaxies supposedly amounts to another 10^8 to 10^9 years.

2.3
Cosmo-Chronometry

As set out in Section 2.2, the chemical composition of a star can be correlated with its cosmological age: The first stars that formed, the population III stars, almost exclusively contain the very lightest elements (hydrogen, deuterium, helium, traces of lithium, and beryllium) only, while the younger stars of populations II and I are characterized by low (population II) and high (population I) "metallicity," that is, increasing amounts of elements heavier than helium formed in the course of various nucleosynthetic processes to be addressed in Chapter 3 (Sections 3.1.3–3.1.5). Our Sun is a representative of the young population I stars.

Along with the relation between age and metallicity, there are correlations between the age of a star and macroscopic physical properties, such as changes of the rotation period with time, and oscillation in brightness with time [4]. Convective stars, like our Sun, develop a permanent magnetic field which, by interaction with the ions constituting the stellar wind, transfers angular momentum to these particles and thus slow down rotation. To what extent there is interaction also depends on the particle density in the stellar wind, and hence the activity and stage of development of the star. The state of development is related to the age. As a star ages, its core composition is the part that changes most. The core composition in turn is related to minor oscillations in brightness.

The oldest stars in the galactic halo, with particularly low metallicities, provide a direct and rather reliable measure to determine the age of a star via the decay of radioactive nuclei with long half-lives, such as the long-living isotopes of thorium and uranium. Radionuclide dating is achieved by comparing observed abundances of the radioactive nuclei with predictions of their initial production rates as based on nucleosynthesis models. The decay routes for ^{232}Th (nuclear charge $z = 90$, half-life $t_{1/2} = 1.4 \times 10^{10}$ a) and ^{238}U ($z = 92$, $t_{1/2} = 4.5 \times 10^9$ a) – both are α emitters – are provided by Eqs. (2.9) and (2.10). As in radionuclide dating of objects on our planet – including objects which have been carried to Earth from various regions of the Solar System – the method relies on the determination of the amount of the radioactive element still present today, against a nonradioactive comparison element. For thorium- and uranium-based cosmo-chronometry, this is usually europium (or, alternatively, osmium or iridium). The initial production rates and modes of formation of Th and Eu, formed almost exclusively by the r-process (Section 3.1.5.2), are well established [5]. By comparing today's Th/Eu ratio with that calculated for the initial production (at $t = 0$), the age of the star or, rather, the time which has elapsed since establishment of the initial Th/Eu ratio, is obtained. Alternatively or in addition to the Th/Eu ratio, the U/Th ratio can also be employed as a chronometer. The age of the red giant star HE 1523-091[7] in the constellation of Libra (a population II star of ca. 0.8 Solar masses, at a distance of 7400 ly) has thus been determined to approximately 13.2×10^9 a [6], which is close to the overall age of the Universe (13.73×10^9 a) as derived from the back-ground microwave radiation:

$$^{232}_{90}\text{Th} \xrightarrow{\alpha} {}^{228}_{88}\text{Ra} \xrightarrow{\beta^-} {}^{228}_{89}\text{Ac} \xrightarrow{\beta^-} {}^{228}_{90}\text{Th} \xrightarrow{\alpha} {}^{224}_{88}\text{Ra} \to \xrightarrow{\alpha,\alpha,\alpha,\beta^-,\beta^-} \to {}^{212}_{84}\text{Po} \xrightarrow{\alpha} {}^{208}_{82}\text{Pb} \quad (2.9)$$

$$^{238}_{92}\text{U} \xrightarrow{\alpha} {}^{234}_{90}\text{Th} \xrightarrow{\beta^-} {}^{234}_{91}\text{Pa} \xrightarrow{\beta^-} {}^{234}_{92}\text{U} \xrightarrow{\alpha} {}^{230}_{90}\text{Th} \to \xrightarrow{4\alpha,2\beta^-,\alpha,2\beta^-} \to {}^{210}_{84}\text{Po} \xrightarrow{\alpha} {}^{206}_{82}\text{Pb} \quad (2.10)$$

The detection of the metals in the stars, and the assessment of their present abundances, relies on the optical emissions of the mono-cations Th$^+$ (ThII, 401.91 nm) and U$^+$ (UII, 385.96 nm). The latter is blended with cyanide features and thus only analyzable with sufficient reliability in carbon-poor stars. Figure 2.1 shows the spectral region around the U-II band.

Other cosmological clocks are in use to determine the age of, for example, meteoritic and Lunar material. The more common ones, Eqs. (2.11) to (2.14), are based on ^{87}Rb/^{87}Sr (β^- emission), ^{40}K/^{40}Ar (β^+ emission or electron capture), ^{129}I/^{129}Xe (β^- emission), and ^{150}Gd/^{142}Nd (via ^{146}Sm; 2α emission). The age determination in meteoritic materials very much works in the same way as the age determination in geological terrestrial samples, that is, the current amount of the radioactive nucleus is determined and compared to the current amount of the daughter nucleus less the amount of the daughter nucleus which has originally been present, the amount of which may be deduced from nucleosynthesis models. The method only works reliably if chemical fractionation, such as effected by melting, did not take place. This can be particularly problematic with meteoritic

---

7) HE stands for Hamburg ESO Survey (ESO = European Southern Observatory).

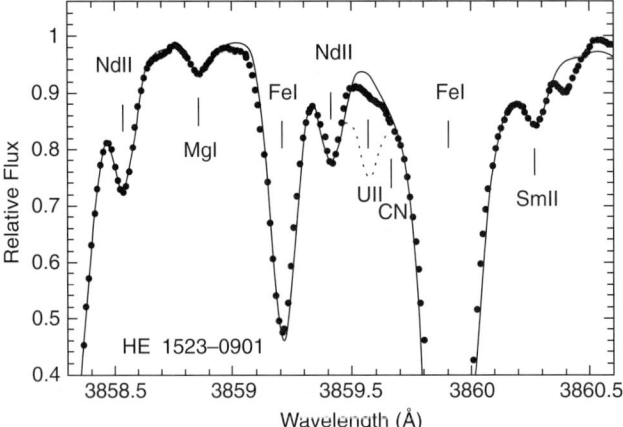

Figure 2.1 Spectral region in the vicinity of the UII (U⁺) line of the star HE 1523-091. The UII line at 385.96 nm appears as a shoulder of the CN/FeI emission. The features above and below this shoulder are synthetic spectra for no uranium (drawn-out line) and uranium without decay (subtly dotted). Reproduced from Ref. [6] with the permission of the AAS journals.

material which had been subjected to shock heating. Were gases are involved, such as ^{40}Ar and ^{129}Xe (and ^{4}He, stemming from α particles), it is necessary that these gases remain locked within the material, which is only warranted if the temperature remains sufficiently low to prevent diffusion of these gases. Since the light helium diffuses more effectively than argon (and argon more effectively than xenon), a mismatch between the relative abundances of found gases with the expected abundances may provide information on processing of the material. Given the short half-life of ^{129}I (1.7×10^7 a), this nuclide should be extinct; here the direct method for age determination only works if there have been processes available to replenish ^{129}I, for example, by fission of ^{235}U and ^{239}Pu. Alternatively, the isotope ratio of the stable isotopes ^{129}Xe/^{127}I is a record of the ^{129}I/^{127}I ratio at the time of isotopic closure. An example for age determination of Lunar material, based on ^{182}Hf/^{182}W chronometry (^{182}Hf also is an "extinct" isotope), will be discussed in more detail in Section 5.2.1.

$$^{87}_{37}\text{Rb} \rightarrow {}^{87}_{38}\text{Sr} + \beta^- + \bar{\nu} \quad t_{1/2} = 4.9 \times 10^{10} \text{ a} \tag{2.11}$$

$$^{40}_{19}\text{K} \rightarrow {}^{40}_{18}\text{Ar} + \beta^+ + \nu \quad t_{1/2} = 1.3 \times 10^9 \text{ a} \tag{2.12}$$

$$^{129}_{53}\text{I} \rightarrow {}^{129}_{54}\text{Xe} + \beta^- + \bar{\nu} \quad t_{1/2} = 1.7 \times 10^7 \text{ a} \tag{2.13}$$

$$^{150}_{64}\text{Gd} \underset{\alpha}{\rightarrow} {}^{146}_{62}\text{Sm} \rightarrow {}^{142}_{60}\text{Nd} + \alpha \tag{2.14}$$

$$t_{1/2} = 1.1 \times 10^{11} \text{a} \quad t_{1/2} = 1.1 \times 10^{8} \text{a}$$

Summary

The hour of birth of our Universe, the Big Bang, dates back to 13.73 billion years. In the first few minutes, Big Bang nucleosynthesis took place, providing hydrogen nuclei (ca. 75%), He-4 (ca. 25%), and traces of Li-7 and Be-7. This initial event left its mark in the form of the 2.725 K cosmic microwave background radiation. Along with matter as we know it (baryonic matter), dark matter (represented by, e.g., neutralinos) and dark energy constitute the Universe, with dark energy making up for about three fourth of the building material of the Universe. The first stars, population III stars, almost exclusively consisted of hydrogen and helium. Successive development lead to the formation of population II stars with somewhat higher metallicities (contents of elements beyond He), and finally to metal-rich, young population I stars such as our Sun. Apart from the metallicity, the age of a star is determined by radionuclide dating based on the long-lived isotopes Th-232 and U-238. The method relies on a comparison of the observed abundances with the initial production rates as derived from nucleosynthesis models.

References

1 Zhang, B. (2009) Most distant cosmic blast seen. *Nature*, **461**, 1221–1223.
2 Calswell, R., and Kamionkowski, M. (2009) Dark matter and dark energy. *Nature*, **458**, 587–589.
3 (a) Bromm, V., Yoshida, N., Hernquist, L., and McKee, C.F. (2009) The formation of the first stars and galaxies. *Nature*, **459**, 49–54; (b) Cattaneo, A., Faber, S.M., Binney, J., Dekel, A., Kormendy, J., Mushotzky, R., Babul, A., Best, P.N., Brüggen, M., Fabian, A.C., Frenk, C.S., Khalatyan, A., Netzer, H., Mahdavi, A., Silk, J., Steinmetz, M., and Wisotzki, L. (2009) The role of black holes in galaxy formation and evolution. *Nature*, **460**, 213–219.
4 Soderblom, D.R. (2009) How old is that star? *Science*, **323**, 45–46.
5 Sneden, C., Cowan, J.J., and Gallino, R. (2008) Neutron-capture elements in the early galaxy. *Annu. Rev. Astron. Astrophys.*, **46**, 241–288.
6 Frebel, A., Christlieb, N., Norris, J.E., Thom, C., Beers, T.C., and Rhee, J. (2007) Discovery of HE 1523-0901, a strongly r-process enhanced metal-poor star with detected uranium. *Astrophys. J.*, **660**, L117–L120.

3
The Evolution of Stars

In Section 2.2, the formation of the very first stars, the population III stars, has been addressed. These stars formed by accretion from minihalos predominantly containing dark matter (neutralinos) and the lightest of the elements, hydrogen (including its heavier isotope deuterium) and helium, plus traces of lithium and beryllium, produced in the first few minutes of the last episodes of the Big Bang event. In this chapter, the development of stars, emphasizing their chemical evolution, will be addressed, along with star clustering such as in globular clusters, open clusters, and galaxies. Clustering of stars is closely related to their formation and evolution. The basis for the rebirth of stars from the interstellar medium provided by evolving and dying stars will be described in Chapter 4.

3.1
Formation, Classification, and Evolution of Stars

3.1.1
General

As noted earlier, the first stars formed in our Universe shortly after the Big Bang. The particularly massive (about 10^6 Solar masses) population III stars would soon have perished in spectacular supernovae, dispersing their material throughout the Universe. From this material, the next generation of stars formed, the population II stars, for which low metallicities are characteristic. "Metallicity" in this context collectively refers to all elements heavier than He. Metallicity is commonly expressed in terms of the Fe/H ratio; iron is the most abundant metal in evolved stars. These old stars predominate in the bulge and the halo of galaxies, and also abound in globular clusters. Subsequent generations of stars were formed from interstellar gas clouds that had become enriched in metals manufactured by previous generations of stars. These young stars with high metallicities, population I stars, are abundant in the disc and the spiral arms of the galaxies. Our Sun is such a population I star.

A relatively dense gas cloud, composed of molecular hydrogen, some helium, and some dust, can collapse, triggered for example, by a shock wave from a

Chemistry in Space: From Interstellar Matter to the Origin of Life. Dieter Rehder
© 2010 WILEY-VCH Verlag GmbH & Co. KGaA, Weinheim
ISBN: 978-3-527-32689-1

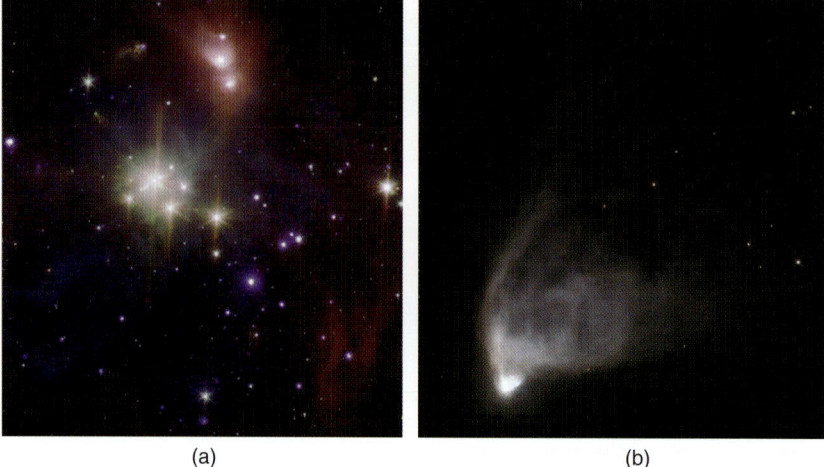

Figure 3.1 (a) The Coronet cluster (at a distance of 420 light-years), in the constellation of Corona Australis, is a center of star formation. In this composite picture, the X-ray patterns (in purple) as taken by the Chandra X-ray observatory are overlaid with the IR pattern (orange, green, and cyan) as obtained by the Spitzer telescope. Both instruments are space-based. Credit: X-ray: NASA/CXC/CfA/J. Forbrich et al.; Infrared: NASA/SSC/CfA/IRAC GTO Team. (b) A T-Tauri variable, NGC 2261 (also known as Hubble's variable nebula) in the constellation of Monoceros, photographed by the Hubble telescope. T-Tauri variables are young protostars; cf. the birth line (track 1) in Figure 3.2. Credit: William Sparks, Sylvia Baggett et al. (STScI) & the Hubble Heritage Team (AURA/STCI/NASA).

supernova, when the gravitational forces within the cloud overcome its internal thermal energy, or gas pressure. This is the case for a critical mass m, known as Jeans' mass, defined by Eq. (3.1), where k and G are the Boltzmann and gravitational constants, respectively, T the temperature measured in K, M the mean molar mass in units g mol^{-1}, and ρ the density:

$$m > (5\,kT/GM)^{3/2} (3/4\pi\rho)^{1/2} \tag{3.1}$$

In the beginning, the collapse will be essentially isothermal by an exchange with the outside Universe. As the collapse continues, the gas will heat to a few hundred Kelvin, and the colliding hydrogen molecules will become rotationally excited and emit their energy in the infrared until the contracting cloud finally becomes opaque. The large cloud may then fragment into smaller cloudlets that further contract and heat up to a few thousand Kelvin, at which point H_2 dissociates to form H atoms, and H atoms become ionized to form protons and electrons. Contraction proceeds with the formation of protostars with core temperatures of several 10^6 K; the temperature at which nuclear fusion of protons starts. The complete scenario of the formation of families of young stars is visualized by stellar nebulae such as the Coronet cluster shown in Figure 3.1a. Figure 3.1b, pictures the situation shortly before a developing young star of about Solar mass becomes established as a main sequence star.

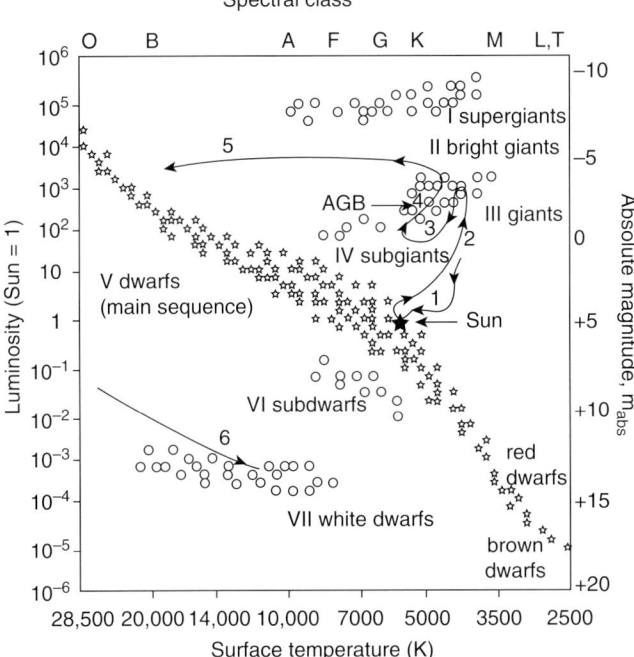

Figure 3.2 The Hertzsprung–Russel diagram. (left abscissa) Luminosity relative to that of the Sun; (right abscissa) absolute magnitude; cf. Eq. (3.5) for the relation between these two quantities. Main sequence stars are represented by the symbol ☆, and other stars by ○. The Roman numbers I to VII refer to classification according to luminosity. The contorted solid black line indicates the evolution of a Sun-like star. Track 1: birth line, including the T-Tauri interim state (Figure 3.1b); track 2: expansion to a red giant by H fusion in the star's shell; track 3: recontraction after He fusion in the core has ceased; track 4: AGB, the second red giant state by He fusion in the shell; track 5: mass loss (PN); track 6: final stage (development toward a white dwarf). See text for additional details.

Once formed, stars evolve, and path and speed of the evolution of a star very much depend on its initial mass. The present situation of the various states of development of visible stars, snapshot in present time, is represented by the Hertzsprung–Russell (HR) diagram in Figure 3.2. In this depiction, the "magnitude" or "luminosity" of a star is plotted against its surface temperature as a measure of the star's overall activity and hence its overall mass. The surface temperature is correlated to the color as it appears in the visible range: high surface temperature stars appear bluish, low surface temperature stars reddish. According to their surface temperature (or color index; see below), the stellar classes O (very hot), B, A, F, G, K, and M (relatively cold) are distinguished, sometimes further extended to L and T. The mnemonic "Oh Be A Fine Girl (or Guy), Kiss Me" may be employed to memorize this sequence. Table 3.1 provides an overview of properties associated with these *spectral classes*. Some chemical characteristics of type M dwarfs and type L and T subdwarfs supposedly have certain

Table 3.1 Properties of O, B, F, G, K, and M class stars.

Class	O	B	A	F	G	K	M
Color	Blue	Blue-white	White	Yellow-white	Yellow	Orange	Red
T (K)	>30 000	30 000–10 000	10 000–7500	7500–6000	6000–5200	5200–3700	<3700
L relative $L_\odot$	10^5	30 000–25	25–5	5–1.5	1.5–0.6	0.6–0.08	<0.08
r relative $r_\odot$	>6.6	6.6–1.8	1.8–1.4	1.4–1.15	1.15–0.96	0.96–0.7	<0.7
m relative $m_\odot$	>16	16–2.1	2.1–1.4	1.4–1.04	1.04–0.8	0.8–0.45	<0.45
Lifetime in 10^6 a	10	100	10^3	5×10^3	10^4	5×10^4	10^5
% of main sequence	3×10^{-5}	0.13	0.6	3	7.6	12.1	76.5
Hydrogen lines	Weak	Medium	Strong	Medium	Weak	Very weak	very weak
Prominent other emissions	He^0, He^+, C^{2+}, N^{2+}, O^{2+}, Si^{4+}	He^0, Mg^+, Si^+	Mg^+, Si^+, Fe^+	Ca^+, Fe^0, Cr^0	Ca^+, M^+, $M^{0a)}$	Si^0, Fe^0, $Mn^{0a)}$	M^0, TiO, VO
Examples	Mintaka (δ Orionis)	Rigel	Sirius	Procyon	Sun, α Cent. A	Aldebaran	Betelgeuse, Proxima Centauri

a) M stands for (any) metal.

properties in common with exoplanets of the categories "hot Jupiters" and "super-Jupiters," and will briefly be dealt with in Chapter 6. The steps between the spectral classes are further subdivided by 10, for example, G0–G9, with G0 being the hottest within the G class. Our Sun is a G2 type star. The "color index" indicates the difference in magnitude at short and long wavelengths, viz. ultraviolet minus blue (U–B index) or blue minus green-yellow (B–V index; V for visible). The smaller the color index, the more pronounced is the contribution of shorter wavelengths.[1] For the B–V index, the following correlations apply:

	B0	A0	F0	G0	K0	M0
B–V	−0.30	−0.00	+0.30	+0.58	+0.81	+1.40

Luminosity is an intrinsic property of a star describing the amount of energy a star radiates per unit time, and may be provided as apparent luminosity

1) Note that a brighter star has a smaller magnitude m.

(visible light only) or bolometric luminosity (total electromagnetic radiation energy). Luminosity is usually given in Solar units; the luminosity of the Sun, $L_\odot$, is 3.839×10^{33} erg s^{-1} (= 3.839×10^{26} W; 1 erg = 0.1 µJ). The luminosity is related to the temperature T and the radius r of the star by Eq. (3.2), and to its brightness b by Eq. (3.3). The *brightness* of a star is quantified by its magnitude m, which may be the apparent (i.e., observed) magnitude m_{obs} (the visible brightness) or the absolute magnitude m_{abs}, which is the brightness corresponding to an interstellar distance of 10 parsec (pc). For the Sun, $m_{obs} = -26.7$, $m_{abs} = +4.8$. The relations between m_{obs} and m_{abs}, and m_{abs} and L are provided by Eqs. (3.4) and (3.5), respectively. Magnitude is a logarithmic measure: a magnitude 3 star is $100^{1/5}$ times (ca. 2.5 times) less bright than a magnitude 2 star, and an $m = 0$ star, such as Vega = α Lyrae, is 100 times brighter than an $m = 5$ star, such as Alcor, the companion of Mizar = ζ Ursae Majoris, at the border of visibility without instruments:

$$L = 4\pi k r^2 T^4 \quad (k = \text{Boltzmann constant} = 5.67 \times 10^{-8} \text{ W m}^{-2} \text{ K}^{-4}) \quad (3.2)$$

$$L = 4\pi d^2 b \quad (d = \text{distance in pc}) \quad (3.3)$$

$$m_{abs} = m_{obs} + 2 - \log_{10} d \quad (d = \text{distance in pc}) \quad (3.4)$$

$$m_{abs} = -5/2 \times \log_{10}(L/L_\odot) + 6 \quad (3.5)$$

The birth lines, or Hayashi tracks, of stars start in the low-temperature range of the HR diagram in the luminosity regimes I to IV (depending on the overall mass) moving toward higher temperatures during evolution until the star reaches "zero age" on the main sequence. Very high-mass stars will start as supergiants (I) or giants (II), and develop into O and B stars within approximately 10^4–10^5 years, where they undergo comparatively rapid development. Red dwarfs, on the lower right branch of the main sequence, which make up about three-fourth of the overall population of stars, have masses too low to allow development off the main sequence. Brown dwarfs with even lower masses cannot sustain nuclear fission. Stars of about the present mass of our Sun (F-, G-, and K-type stars), condensing out of a gas and dust nebula, start as subgiant (luminosity class IV) pre-main sequence stars, arriving at the main sequence after about 100 million years. These Sun-like stars will spend most of their lifetime on the main sequence, and finally end up as white dwarfs. For the Sun, the dwell time on the main sequence is estimated to cover another 10 billion years, in addition to the 4.6 billion years that already have elapsed. Depending on their mass, old stars will evolve through planetary nebulae or supernovae into white dwarfs, neutron stars, or black holes; for planetary nebulae (PN) and supernovae, see Figures 3.3a and b, for a white dwarf "devouring" a giant companion star, see Figure 3.3c.

The typical stages of the development of a star of approximately the mass of the Sun are represented by tracks 1–6 of the evolution line in the HR diagram, Figure 3.2. In short, these stages are as given below:

1) Accretion of a nebula into a protostar that further develops into a T-Tauri variable and finally commences hydrogen fusion to helium (hydrogen burning). This marks the "birth" (or zero age) of the star, which now spends

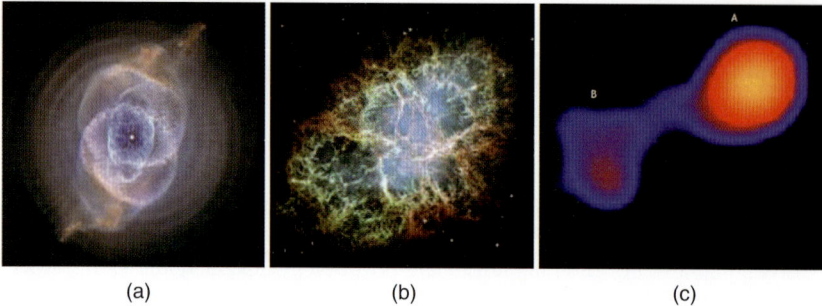

Figure 3.3 Examples for stars at their evolutionary end stages. (a) Cat's Eye Nebula, a PN in the constellation of Draco. Credit: NASA, ESA, J. Hester and A. Loll (Arizona State University). (b) A recent photo of the Crab Nebula in the constellation of Taurus, a remnant of a supernova dating back to 1054. The central star is a pulsar (cf. Section 3.1.2). Credit: NASA, ESA, HEIC and the Hubble Heritage Team. (c) The nova Mira (= o Ceti), a variable binary system: a white dwarf, Mira B, drags and accretes matter from its red giant companion Mira A. Credit: NASA/CXC/SAO/M; Karovska et al.

its main lifetime in an equilibrium situation (balance between gravitational inward pressure and outward radiation pressure) on the main sequence.

2) When H fusion in the stellar core comes to a halt, the core begins to collapse gravitationally, rising its temperature and thus "igniting" the hydrogen shell. At this stage, H fusion in the shell begins, blowing up the star to a red giant. The core is further compressed to the point of electron degeneracy, resulting in an increase in core temperature. This explosive increase in extra energy ignites helium ("helium flash"), thus enabling helium fusion to carbon and oxygen (helium burning at ca. 10^8 K).

3) The accompanying increase in radiation pressure temporarily balances the system, but eventually, when the helium in the core is essentially used up, recontraction takes place.

4) The temperature increase on the contraction of the helium-depleted core initiates carbon burning (provided the mass is at least $4m_\odot$), which ignites helium fusion in the stellar shell, once more blowing up the star and thus leading to a second red giant state. This upward movement in the HRG, also referred to as "asymptotic giant branch," AGB, is of particular interest in the context of the formation of elements beyond iron and the generation of molecular species.

5) The expansion will continue until the outer regions of the stellar atmosphere detach and move outward. The resulting object is referred to as planetary nebula (PN) for historical reasons; this term does not imply formation of planets.

6) Eventually, the star, now represented by the remaining core mass, will end as a white dwarf on the "stellar cemetery." White dwarfs mostly consist of extremely dense, electron degenerate matter. The overall mass compares to

that of the Sun – the volume to that of Earth. Their residual luminosity is due to the radiation of heat.

In general, this is the course for all stars with $m \approx 0.5$–$8m_\odot$. Stars with $m < 0.45m_\odot$ (M-class stars; red dwarfs) have lifetimes longer than the estimated life span of the Universe and hence will never leave the main sequence. When they develop toward the main sequence, these stars often appear as so-called EXors and FUors,[2] eruptive stars varying in luminosity ($\Delta m \approx 5$) with periods of months to years, and thus resembling T-Tauri variables (Section 3.1.3). If the mass drops below $0.08m_\odot$, hydrogen fusion cannot occur, and the star will never reach the main sequence. For these substellar objects, the term "brown dwarfs" has been coined.[3] Stars with $m > 8m_\odot$ (O-class and luminous B-class stars) end up as supernovae, and finally as neutron stars or black holes (see Chapter 4), depending on their mass.

The several stages of development, briefly commented above, are followed up in more detail in Sections 3.1.3–3.1.5. Section 3.1.2 provides a very brief overview on characteristics of neutron stars, pulsars, Wolf-Rayet (WR) stars, black holes, and quasars.

3.1.2
Neutron Stars and Black Holes

Common *neutron stars* are the remnants of core-collapse type II *supernovae* deriving from progenitors of about 8–20 times the mass of the Sun. Most of this mass is blown away in the course of the explosion, leaving a core of $\approx 1.4 m_\odot$ (the Chandrasekhar limit[4]) and a diameter of ≈ 20 km. Under the high pressure involved in the core collapse, protons and electrons combine to form neutrons plus antineutrinos, the latter being emitted and thus carrying off much of the energy. In its final stage, with a temperature of around 10^6 K, the neutron core is degenerate and hence cannot further collapse. The density of a neutron star, about the order of magnitude of that of the nucleus of an atom, is approximately 5×10^{17} g cm^{-3}, that is, a cubic centimeter of this "material" will have a mass of ca. 10^{12} kg. The angular momentum of the original star is essentially preserved as the neutron star forms, leading to rotation periods in the second to millisecond range. The magnetic field of the original star is likewise preserved; its strength is typically 10^8 T. If the axis of the magnetic field is inclined with respect to the axis of rotation, periodic pulses of electromagnetic radiation with an immense intensity are emitted, fed at the expense of the rotational energy. These objects are known as *pulsars*, or *radio pulsars* if, as common, the emitted radiation is in the 0.1–100 GeV regime, or *γ-ray pulsars*, when γ rays are emitted at rotational periods in the

---

2) Named after the prototypes FU Orionis (which is about three degrees NW of Betelgeuse) and EX Lupi.
3) To some extent, brown dwarfs ("infrared dwarfs" would be a more appropriate name; see also footnote 2 in Chapter 6) represent an intermediate status between red dwarfs on the one hand and giant planets on the other.
4) The Chandrasekhar limit defines the maximum mass (of nuclei immersed in a gas of degenerate electrons) that can be supported against gravitational collapse by electron degeneracy pressure.

Figure 3.4 A diffuse emission nebula, formed by the remnants of a supernova. Red features are Hα emission of H atoms (HI) being swept up by the shock of the exploding star shortly before becoming ionized by the hot plasma behind the shock front. Blue is X-band synchrotron radiation emitted by highly energetic electrons. Credit: E. Helder/C. Sharkey, ESO & NASA/Chandra CXC.

millisecond range. If such a gamma ray pulsar is part of a binary sysytem with a normal companion star, the pulsar accretes gas at the expense of this star, consistently speeding up its own rotation.

It is common to distinguish between type II and type I supernovae: These are observationally distinct by the presence or absence, respectively, of the hydrogen Balmer lines in the visible range. Type Ia supernovae, the spectra of which contain the Si^+ (SiII) and S^+ (SII) line, derive from carbon–oxygen white dwarfs which, when approaching the Chandrasekhar limit by accretion of matter (mainly hydrogen) from the surroundings (e.g., a red giant as part of a binary system; Figure 3.3c), are torn apart by a thermonuclear explosion, referred to as explosive hydrogen burning. Tycho Brahe's supernova in the constellation of Cassiopeia, which burst forth in 1572, is a standard type Ia supernova. Type II (strong H lines), Ib (prominent He lines) and Ic (neither H nor He) supernovae are related to more massive, young and thus short-lived stars which eject matter after gravitational collapse (core-collapse SN). Type Ib and Ic may be connected to *Wolf-Rayet* (WR) in that WR stars supposedly are progenitors of type Ib/c supernovae. WR stars are very hot and massive stars with strong emission lines of He and C (or N or O), characterized by particularly fervid stellar winds and hence mass losses. An additional class of novae is the *"luminous red novae,"* stellar explosions which are caused by the merger of two stars.

Supernovae and supernova remnants (Figures 3.3b and 3.4), sometimes organized in agglomerates ("superbubbles"), play a pivotal role in the acceleration of cosmic rays. Intragalactic cosmic rays, mainly protons and helium nuclei plus a small fraction of heavier nuclei, in particular iron, can be accelerated close to the speed of light; they play an important role as inductors of chemical processes in dusty molecular clouds (Section 4.2.5.2). The energy for accelerating cosmic ray particles is provided by the explosion energy "stored" in the expanding envelopes (expanding plasma shells) of the bursting star [1].

Stars exceeding ca. $20 m_\odot$ leave behind, after supernova explosion, a stellar remnant that collapses in on itself to the point where even photons can no longer escape the gravitational field. At this point, where the escape speed is equal to the speed of light, a *black hole* comes into existence. The limiting radius for such an object is the so-called Schwarzschild radius $r = 2Gm/c^2$, where G is the gravitational constant, m the mass, and c the speed of light. The Schwarzschild radius is closely correlated to the event (ereignis) horizon. The event horizon cannot be surpassed by a photon (at the Schwarzschild limit, the escape speed is equal to the speed of light), that is, a black hole is devoid of such an "event."

A massive star eventually can collapse to the point of (almost) zero volume and infinite density – a "singularity." Depending on the inner structure of the collapsing star, the singularity is surrounded by an event horizon and thus invisible, hence a black hole, or there is no such event horizon, and the singularity is consequently referred to as *naked singularity*. A naked singularity is, in principle, visible.

The central region of a galaxy constitutes a super-massive black hole with a mass of up to $10^9 m_\odot$. This region is surrounded by a compact area of matter, emitting – under the gravitational pull of the black hole – extremely strong radiation, including radio waves, and termed *quasar* (quasi-stellar radio source) for this reason. The radiation, and the gravitational influence upon by-passing light, can be employed to identify black holes. Mergers of galaxies produce quasars, which are first hidden by gas and dust (obscured quasars) and later become visible over the complete electromagnetic spectrum.

3.1.3
Accretion and Hydrogen Burning

In the pre-main sequence phase, that is, while developing from a voluminous protostar into a star, the accreting stellar objects are powered by gravitational energy, their central temperature still being too low for hydrogen fusion. This situation of stellar evolution is represented by very young protostars (10^5–10^8 years) observed in nebulae where star formation is going on (Figure 3.1), so-called T-Tauri stars (named after their prototype in the constellation of Taurus), which, in many cases, are embedded in protoplanetary nebulae and discs (Figure 3.1b). T-Tauri stars are variable stars; their erratic brightness changes due to instabilities of the accretion disc, violent activity ("stellar winds") in the thin stellar atmosphere (characterized by strong emission lines, mainly Hα of the Balmer series, and Ca$^+$), and obscuration by parts of the inhomogeneous surrounding cloud. A very characteristic feature of the T-Tauri stars is the higher abundance, relative to the main sequence stars, of lithium, detectable by its 670.7 nm line. The lithium isotope ^{7}Li is formed in the pp (proton–proton) chain (see below) according to Eq. (3.6), but subsequently eliminated, at temperatures $>2.5 \times 10^6$ K by lithium burning (Eq. (3.7)), during the last highly convective stages when the star enters the main sequence. The rapid rotation of the T-Tauri stars, typically between 1 and 12 days, enforces the transport of lithium into the hot stellar core and eventually, lithium depletion:

$$^7_4\text{Be} + e^- \rightarrow {}^7_3\text{Li} + \nu_e \tag{3.6}$$

$$p + {}_3^7Li \rightarrow {}_4^8Be + h\nu$$
$$\hookrightarrow 2{}_2^4He \quad (3.7)$$

Gravitational contraction in the protostar leads to a constant increase in temperature until hydrogen burning (hydrogen fusion) commences. Hydrogen fusion is coincident with the star entering the main sequence. The first step is the fusion of two protons to form deuterium, a positron and a neutrino (Eq. (3.8a)); followed by annihilation of the positron through an electron (Eq. (3.8b)). The next step is the production of helium-3 (Eq. (3.9)):

$${}_1^1H + {}_1^1H \rightarrow {}_1^2H + e^+ + \nu_e \quad (3.8a)$$

$$e^+ + e^- \rightarrow 2\gamma \quad (3.8b)$$

$${}_1^2H + {}_1^1H \rightarrow {}_2^3He + \gamma \quad (3.9)$$

The reaction then branches in three possible paths to produce helium-4, summarized in Scheme 3.1:

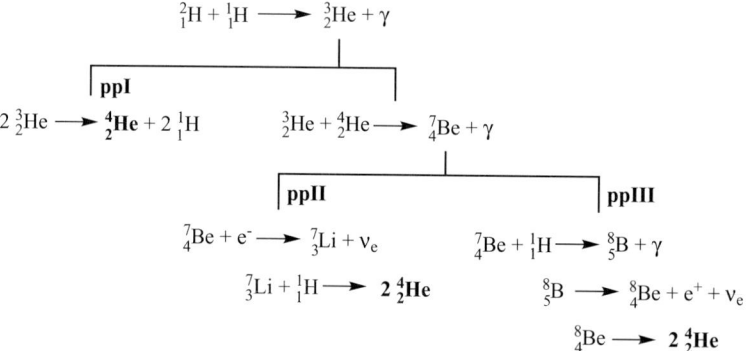

Scheme 3.1 The three pp branches for the formation of helium-4 by hydrogen fusion.

- Branch I is the fusion of two helium-3 nuclei to form helium-4 and two protons. This process dominates in the temperature range 10–14 × 10⁶ K and hence in the Sun (core temperature 15.6 × 10⁶ K). The main part of He, 86%, forms in the Sun along this path.

- In branch II, beryllium-7 is produced, which further forms lithium-7 by electron capture. Lithium-7 then combines with a proton, ending up in two helium-4 nuclei; cf. Eqs. (3.6) and (3.7). Branch II, with an optimum temperature of 14–23 × 10⁶ K, accounts for 14% of the Solar helium production.

- Branch III also starts with the formation of beryllium-7 that further incorporates a proton to form boron-8, which decays to beryllium-8 by positron emission, a process which is accompanied by the generation of particularly high-energetic neutrinos, up to 14.06 MeV in the Sun, as compared to 0.42 MeV for the neutrinos generated according to Eq. (3.8a). Since most of the neutrinos escape, branch III is an important source of energy loss. Beryllium-8 finally falls apart, leaving behind two helium-4 nuclei.

In addition to the formation of deuterium according to Eq. (3.8a), that is, by positron emission, deuterium can also be produced by electron capture (Eq. (3.10)). The neutrino formed in this marginal so-called pep (proton–electron–proton) process carries energy of 1.44 MeV:

$$^1_1H + e^- + ^1_1H \rightarrow ^2_1H + \nu_e \qquad (3.10)$$

The net reaction of the pp-process is represented by Eq. (3.11). The energy released, 26.73 MeV per elementary process in the form of γ rays and carried by neutrinos, corresponds to the mass defect $\Delta m = 4m_p - m_\alpha = 0.03037$ u (α stands for the helium-4 nucleus, u for atomic mass unit); for conversion to energy, the Einstein equation $E = \Delta m c^2$ applies. A total of 26.7 MeV energy for the elementary process corresponds to 2.7×10^9 kJ mol^{-1} for molar turnover, which is 5–6 orders of magnitude more than that for a common chemical reaction![5)]

$$4 ^1_1H \rightarrow ^4_2He + 2e^+ + 2\nu_e \qquad (3.11)$$

The pp chain starts at a temperature of 4×10^6 K and hence is the dominant process also in stars with masses lower than the Sun, including red dwarfs. The minimum mass for attaining a sufficiently high temperature to enable H fusion is $0.075 m_\odot$. An alternative route for the formation of helium from hydrogen proceeds via the carbon–nitrogen–oxygen (CNO) process, also known as Bethe-Weizsäcker cycle. C, N, and O essentially act as "catalysts"; the net reaction, that is, the formation of ^{4}He, two positrons, and two neutrinos by the fusion of four protons (Eq. (3.11)), is the same as in the pp-process. In addition to helium, some ^{14}N is also left over in this process. For the CNO process to occur, a minimum temperature of 13×10^6 K is required; some of the helium formed in the Sun, about 1.7%, is thus produced in the CNO cycle. The CNO cycle becomes the dominant process at temperatures above 17×10^6 K, hence in high-mass stars. Another requirement is, of course, the presence of sufficient amounts of C, N, and O, excluding the CNO process in stars with low metallicities, hence old stars. The nuclear processes going on in the CNO cycle are shown in Scheme 3.2. Branch I, the classical Bethe-Weizsäcker cycle, is the more prominent one. Branch II, of minor importance, includes the formation of ^{17}F. For massive stars, there is an additional sideline, diverting from ^{17}O in branch II, and involving ^{18}F and ^{19}F (Eq. (3.12)):

$$^{17}_8O(p,\gamma)^{18}_9F(e^+\nu_e)^{18}_8O(p,\gamma)^{19}_9F(p,^4_2He)^{16}_8O(p,\gamma)^{17}_9F(e^+\nu_e)^{17}_8O \qquad (3.12)$$

The neutrinos generated in all of these processes are neutral particles with a rest mass close to zero (cf. Table 2.1); they thus have an exceedingly low effective cross section. The cross section improves for so-called charged current interactions, in which the neutrino transforms into its partner lepton (an electron in the case of

---

5) Cf., for example, the oxyhydrogen (knallgas) reaction, the formation of (liquid) water from hydrogen and oxygen, with a reaction enthalpy of 286 kJ mol^{-1}.

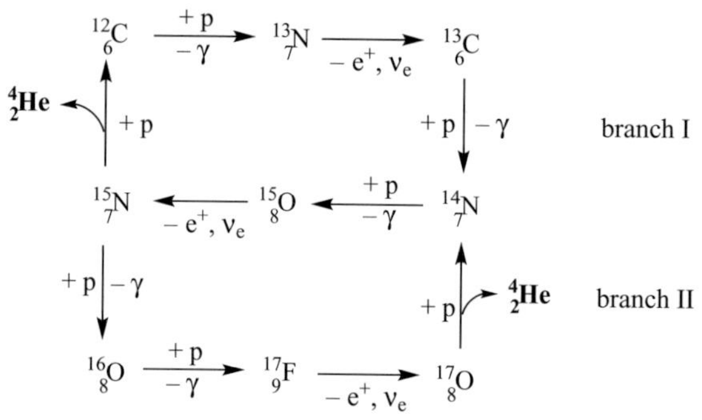

3 The Evolution of Stars

Scheme 3.2 The CNO process of helium formation, showing branch I (the Bethe-Weizsäcker cycle) and, of minor importance, branch II.

medium-energy Solar neutrinos). The basis of such a transformation is the conversion of a neutron into a proton and an electron by neutrino capture. Equations (3.13)–(3.15) provide typical reactions for the "chemical" detection of neutrinos:

$$^{37}_{17}Cl + \nu_e \rightarrow {}^{37}_{18}Ar + e^-$$
chlorine as C_2Cl_4 (3.13)

$$^{71}_{31}Ga + \nu_e \rightarrow {}^{71}_{32}Ge + e^-$$
gallium as $GaCl_3$ in water (3.14)

$$^{2}_{1}H + \nu_e \rightarrow 2{}^{1}_{1}H + e^-$$
deuterium as D_2O (3.15)

3.1.4
Nuclear Fusion Sequences Involving He, C, O, Ne, and Si

When the majority of hydrogen is consumed in helium formation, the stellar core contracts, increasing its temperature through liberation of gravitational energy, until the temperature ($\approx 10^8$ K) allows for He burning. In stars of Solar mass and less (down to $0.45 m_\odot$), a temperature sufficiently high to initiate helium fusion requires backup by electron degeneracy. Electrons are present as the counterparts of the positively charged nuclei. As the density in the stellar core becomes sufficiently high, the electrons become packed together extremely close, forcing electrons into the same otherwise unoccupied high-energy states: the electrons become degenerate. In accordance with the Heisenberg uncertainty principle, these electrons gain high momentum, or "electron degeneracy pressure", which is unaffected by temperature and is characterized by high conductivity for heat. The degeneracy pressure supports ignition and spreading of helium burning.[6]

There are two classes of nuclear fusion processes by which stars convert helium into heavier elements with mass numbers that are integer multiples of 4, viz. (i) the alpha process and (ii) the triple-alpha process. In the triple-alpha process,

---
6) Electron degeneracy pressure also supports white dwarfs against gravity collapse.

three helium nuclei are transformed into carbon via ^{8}Be (Eqs. (3.16) and (3.17)). The beryllium isotope ^{8}Be is unstable (its half-life is only 10^{-16} s) and thus readily decays back to form two ^{4}He; Eq. (3.16) therefore is formulated as a forward–backward reaction. As the temperature in the fusion region rises to about 10^8 K, fusion of ^{8}Be with an additional ^{4}He to form ^{12}C (Eq. (3.17)) is faster than its decay back to helium. The net energy release of the overall process is 7.275 MeV:

$$_2^4He + {}_2^4He \rightleftarrows {}_4^8Be + \gamma \qquad (3.16)$$

$$_4^8Be + {}_2^4He \rightarrow {}_6^{12}C + \gamma \qquad (3.17)$$

Once carbon is present, other reactions consuming helium are possible. The element formed from carbon by the α process ("α capture") is oxygen (Eq. (3.18)), and the reaction chain can proceed up to the formation of ^{56}Fe. Most of the isotopes formed in these α capture processes are intrinsically very stable. Notably, ^{8}Be is an exception, but ^{44}Ti and ^{44}Sc also are short-lived isotopes. For the stable final isotopes in this chain, there is a steady increase in stability up to ^{56}Fe, which has the highest stability (lowest mass per nucleon):

$$_6^{12}C + {}_2^4He \rightarrow {}_8^{16}O + \gamma$$
$$\xrightarrow{(\alpha,\gamma)} {}_{10}^{20}Ne \xrightarrow{(\alpha,\gamma)} \rightarrow {}_{22}^{44}Ti \xrightarrow{+e^-} {}_{21}^{44}Sc \rightarrow {}_{20}^{44}Ca + e^+ \qquad (3.18)$$
$$\xrightarrow{(\alpha,\gamma)} \rightarrow \rightarrow {}_{26}^{56}Fe$$

In stars with a mass $m > 4m_\odot$, carbon atoms can fuse as the temperature reaches 6×10^8 K. Equations (3.19)–(3.23) are typical examples for element formation by carbon burning. Two of these reactions – the formation of ^{23}Mg, Eq. (3.20) and ^{16}O, Eq. (3.23) – are endothermic. Carbon burning is one of the processes by which nuclei other than multiples of ^{4}He can be generated:

$$_6^{12}C + {}_6^{12}C \rightarrow {}_{12}^{24}Mg + \gamma \qquad (3.19)$$

$$_6^{12}C + {}_6^{12}C \rightarrow {}_{12}^{23}Mg + n \qquad (3.20)$$

$$_6^{12}C + {}_6^{12}C \rightarrow {}_{11}^{23}Na + p \qquad (3.21)$$

$$_6^{12}C + {}_6^{12}C \rightarrow {}_{10}^{20}Ne + {}_2^4He \qquad (3.22)$$

$$_6^{12}C + {}_6^{12}C \rightarrow {}_8^{16}O + 2{}_2^4He \qquad (3.23)$$

Hydrogen, helium, and carbon fusion are the main processes by which elements up to iron are formed, and which determine the overall abundance of these elements in population I stars (and, more or less, the overall cosmic abundance). Table 3.2 provides an overview of element abundances for the 10 most abundant elements in the Solar System (the Sun is a population I star) and in our galaxy. In white dwarfs, oxygen, neon, and magnesium (formed by carbon burning) are the most abundant elements. For more massive stars ($m > 8m_\odot$, i.e., O class and high-mass B class stars) and at high core temperatures around 1.2×10^9 K, processes involving neon become effective when the carbon is used up. At an even later stage, oxygen burning (1.5×10^9 K) comes in. Selected examples for O burning and Ne-dependent processes are provided by Eqs. (3.24)–(3.28). Equation (3.27a)

Table 3.2 The abundances (in % of the overall amount of nuclei) of the 10 most abundant elements in the Solar System (representative for population I stars) and in the Milky Way galaxy.

Isotope	^{1}H	^{4}He	^{16}O	^{12}C	^{20}Ne	^{56}Fe	^{14}N	^{28}Si	^{24}Mg	^{32}S
Solar System	70.57	27.52	0.592	0.303	0.155	0.117	0.111	0.065	0.051	0.040
Milky Way	73.9	24.0	0.104	0.460	0.134	0.109	0.096	0.065	0.058	0.044

represents photodisintegration of neon, and Eq. (3.28a) represents neutron capture by neon:

$$^{16}_{8}O + ^{16}_{8}O \rightarrow ^{28}_{14}Si + ^{4}_{2}He \tag{3.24}$$

$$^{16}_{8}O + ^{16}_{8}O \rightarrow ^{31}_{15}P + ^{1}_{1}H \tag{3.25}$$

$$^{16}_{8}O + ^{16}_{8}O \rightarrow ^{32}_{16}S + \gamma \tag{3.26}$$

$$^{20}_{10}Ne(\gamma, \alpha)^{16}_{8}O \tag{3.27a}$$

$$^{20}_{10}Ne(\alpha, \gamma)^{24}_{12}Mg \tag{3.27b}$$

$$^{20}_{10}Ne(n, \gamma)^{21}_{10}Ne \tag{3.28a}$$

$$^{20}_{10}Ne(\alpha, n)^{23}_{12}Mg \tag{3.28b}$$

The final stage in a massive star, at around 3×10^9 K, is α capture by silicon, starting with the formation of ^{32}S, and proceeding to ^{56}Ni (Eq. (3.29)). The nickel isotope ^{56}Ni is liable to β^+ decay (half-life 6.02 days), as is the resulting cobalt isotope ^{56}Co (half-life 77.3 days), the β^+ decay of which produces stable ^{56}Fe (Eq. (3.30)). Hence, as in the sequence (3.18), we again end up with particularly stable ^{56}Fe:

$$^{28}_{14}Si + ^{4}_{2}He \rightarrow ^{32}_{16}S + \gamma \xrightarrow{(\alpha,\gamma)} \rightarrow \rightarrow ^{56}_{28}Ni \tag{3.29}$$

$$^{56}_{28}Ni \rightarrow ^{56}_{27}Co + e^+ + \nu_e \rightarrow ^{56}_{26}Fe + e^+ + \nu_e \tag{3.30}$$

When the Si burning phase is completed, the core further collapses until it is crushed into a neutron star or black hole, whereas the outer layers are blown off in an explosion, visible as a type II supernova (Section 3.1.2).

3.1.5
The r-, s-, rp- and Related Processes

3.1.5.1 General
These are processes by which elements heavier than iron are formed by neutron capture (r- and s-processes) or proton capture (rp-process). The r- (for rapid) and

rp-processes take place in core-collapse supernovae, the s-process (s for slow) in AGB stars, that is, in low- or intermediate-mass stars ($0.8m_\odot$ to $8m_\odot$) in their second red giant phase or close to the final state of this phase. *Rapid* and *slow* refer to the timescale of neutron capture as related to that of the β^- decay of the resulting nucleus. The identification of the two types of neutron capture processes goes back to pioneering work carried out by Fred Hoyle and colleagues half a century ago [2], who also proposed the rp-process to explain the presence, and the abundance pattern, of proton-rich heavy nuclei. Related to the r-process is the vp-process, by which a proton combines with an antineutrino to form a neutron and a positron, with concomitant incorporation of the neutron into a nucleus. A complementary process is the γ-process (sometimes confusingly also referred to as p-process) – a photodisintegration process – by which neutrons, α-particles, or protons are knocked out of a nucleus by γ-rays. Equations (3.31)–(3.35) provide an overview. Here, M and M' represent elements ("metals") beyond iron, z is the atomic number (number of protons, i.e., nuclear charge), and m is the mass number (sum of protons and neutrons).

$$r\text{-process }(n,\gamma) \quad {}^{m}_{Z}M + xn \xrightarrow{rapid} \to \to {}^{m+x}_{Z}M + \gamma \tag{3.31}$$

$$s\text{-process }(n,\gamma) \quad {}^{m}_{Z}M + n \xrightarrow[\gamma]{slow} {}^{m+1}_{Z}M \; (\to {}^{m+1}_{Z+1}M' + e^-) \tag{3.32}$$

$$rp\text{-process }(p,\gamma) \quad {}^{m}_{Z}M + p \to {}^{m+1}_{Z+1}M' + \gamma \tag{3.33}$$

$$\gamma\text{-process }(\gamma,p) \quad {}^{m}_{Z}M + \gamma \to {}^{m-1}_{Z-1}M' + p \tag{3.34}$$

$$vp\text{-process }(\bar{\nu},\beta^+) \quad p + \bar{\nu}_e \to n + e^+; \quad n + {}^{m}_{Z}M \to {}^{m+1}_{Z}M \tag{3.35}$$

The r-process (which affords a particularly high neutron density) and the s-process (which goes on at lower neutron densities and intermediate temperatures) contribute about equally to the nucleosynthesis of heavy nuclei, but there are specific nuclei that are (almost) exclusively produced by either the r- or the s-process. Typical r-process elements are Ge, Zr, Te, Xe, Eu, Os, Pt, and Au, and typical s-process elements are Sr and Ba. Nuclei such as ^{232}Th, ^{235}U, and ^{238}U can only be produced in the r-process. Further, since the s-process occurs later in galactic history than the r-process, elements produced by the s-process are delivered to the interstellar medium at a later stage. So-called halo stars, which circle our galaxy in randomly oriented eccentric orbits, and which are among the very oldest stars (evidenced by their very low iron abundance), have a clear signature of elements made in the r-process. Figure 3.5 is a graphical representation of typical paths for an r- and s-processes.

3.1.5.2 Rapid Processes

The r-process [3], associated with very large neutron densities, requires a few seconds only. The timescale for neutron capture is ca. 10^{-4}s. Up to 10 and more neutrons can be picked up before β^- decay occurs when the neutron flux runs out. For neutron numbers per nucleus exceeding 184, neutron-induced *nuclear fission* processes also come in [4], contributing to the distribution of the lighter r-process

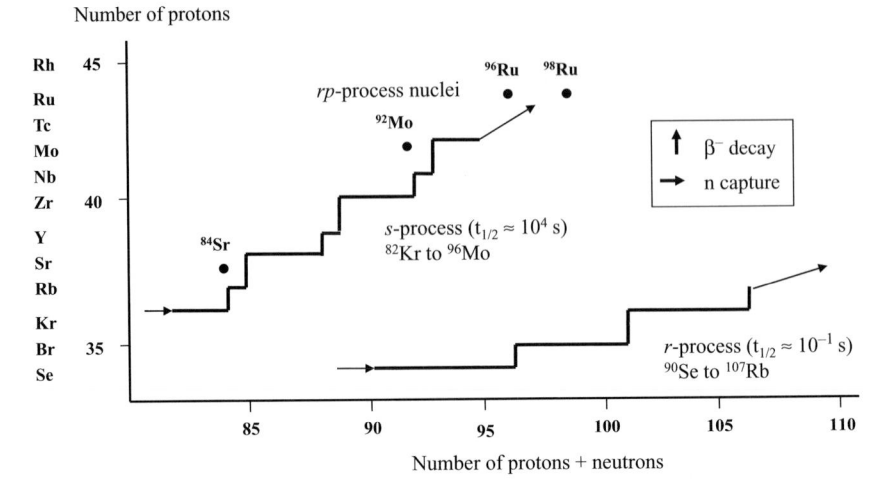

Figure 3.5 Sections of the tracks for the formation of heavy (trans-iron) elements along the r-process (rapid neutron capture, lower trace) and the s-process (slow neutron capture, upper trace). In both tracks, neutron capture is followed by β⁻ decay (vertical lines); $t_{1/2}$ = half-life. The dots represent nuclei that are formed by rapid proton capture (rp-process).

nuclei. Neutron capture at high temperature produces high-energy γ-rays; this leads to (γ,n) photodisintegration competing with n-capture. The overall situation is reflected by Eqs. (3.36)–(3.38), where E represents any heavy nucleus: Eq. (3.36) describes the balance between n-capture and photodisintegration, and Eq. (3.37) describes the successive built-in of neutrons. Equation (3.38a) describes, in a generalized form, the first β⁻ decay, and reaction (3.38b) exemplifies this for the terminal step in the formation of ^{198}Pt (z = 78) by β⁻ decay of ^{198}Ir (z = 77). The high neutrino flux released in a supernova explosion also gives rise to the interaction between neutrinos and neutrons (Eq. (3.39)), thus increasing the number of protons (z) at the expense of the number of neutrons (m − z); cf. also Eqs. (3.13) and (3.14) in Section 3.1.3:

$$^{m}_{z}E + n \rightleftharpoons {}^{m+1}_{z}E + \gamma \tag{3.36}$$

$$^{m+1}_{z}E \xrightarrow{+\,xn} \rightarrow \rightarrow {}^{m+1+x}_{z}E \tag{3.37}$$

$$^{m+1+x}_{z}E \rightarrow {}^{m+1+x}_{z+1}E + e^- + \bar{\nu}_e \tag{3.38a}$$

$$^{198}_{77}Ir \rightarrow {}^{198}_{78}Pt + e^- + \bar{\nu}_e \tag{3.38b}$$

$$^{0}_{0}\nu_e + {}^{1}_{0}n \rightarrow {}^{1}_{1}H + {}^{0}_{-1}e^- \tag{3.39}$$

But where do the neutrons consumed in many of these processes come from? Equations (3.40)–(3.42) describe neutron sources in the r-process. The nuclei ^{13}C and ^{14}C required in these reactions are provided in the CNO cycle (Scheme 3.2). The processes described by Eqs. (3.40) and (3.41) also deliver neutrons for the s-process (vide infra), whereas the neutron production by fusion of magnesium-25 with helium (Eq. (3.42)) is restricted to the high temperatures accompanying the r-process. The reaction sequence (3.41) starting with the consumption of ^{14}N (the most abundant of the CNO nuclides) also contributes to the relative depletion (relative to ^{15}N) of ^{14}N in silicon nitride (Si$_3$N$_4$) in so-called X-type grains picked up by asteroids from the interplanetary dust, and found in meteorites (see also Section 5.3.3):

$$^{13}_{6}C + ^{4}_{2}He \rightarrow ^{16}_{8}O + n \tag{3.40}$$

$$^{14}_{7}N(\alpha,\gamma)^{18}_{9}F(e^+\gamma)^{18}_{8}O(\alpha,\gamma)\boxed{^{22}_{10}Ne(\alpha,n)^{25}_{12}Mg} \tag{3.41}$$

$$^{25}_{12}Mg + ^{4}_{2}He \rightarrow ^{28}_{14}Si + n \tag{3.42}$$

The explosive collapse of a pre-supernova leaves behind a neutron core with an initial temperature of several 10^{11} K that produces an extensive flux of neutrinos and antineutrinos with energies typically between 10 and 20 MeV. In proton-rich, cooler ejecta of the supernovae, protons can be converted into neutrons by the absorption of antineutrinos (left-hand side of Eq. (3.35)). In principle, this is possible for free protons as well as for protons confined to nuclei. Since protons are sufficiently more abundant in the stellar envelopes than heavy nuclei, antineutrino capture occurs predominantly on free protons, leading to a drastic increase of neutron density for a few seconds, and subsequent neutron capture, in particular by nuclei that are comparatively proton-rich, that is, where the number of neutrons is about that of the number of protons. The net result of this process, termed *vp*-process (right-hand side of Eq. (3.35)) [5], compares to that of the *r*-process. Along with the absorption of neutrons by neutron-deficient nuclei, neutron capture can also be coupled to the disposal of a proton. This (n,p) reaction delivers nuclei with a higher effective cross section for proton capture, producing nuclei such as ^{92}Mo, ^{94}Mo, ^{96}Ru, ^{98}Ru, and ^{102}Pd, which are otherwise not easily accessible.

A hydrogen-rich environment is a prerequisite for the *rp*-process to occur, along with a sufficiently high temperature, $>1 \times 10^9$ K, to overcome the Coulomb barrier for the fusion of a proton with the positively charged nucleus. These conditions are fulfilled on the surface of accreting neutron stars as remnants of supernovae [6]. The *rp*-process provides, along with the *s*-process, nuclei that are comparatively rich in protons. The *rp*-process is, to some extent, complementary to the *r*-process (the rapid neutron capture): In the *r*-process, β^- emission occurs after a series of neutron uptakes; in the *rp*-process, proton uptake is followed by β^+ emission. Nuclei beyond ^{56}Ni (the last nucleus arising from the α process, Eq. (3.29)) can be synthesized by the *rp*-process; the upper limit is ^{107}Te, formed according to Eq. (3.43a). This tellurium isotope decays by α-emission to form ^{103}Sn (Eq. (3.43b)), which propagates, via ^{103}In, to ^{105}Sn (Eq. (3.43c)). The *rp*-process hence can apparently not proceed beyond this Sn–Sb–Te cycle. Among the nuclei ranging from ^{56}Ni to ^{107}Te mainly produced by the *rp*-process are ^{98}Ru and ^{102}Pd:

$$^{105}_{50}Sn(p,\gamma)^{106}_{51}Sb(p,\gamma)^{107}_{52}Te \tag{3.43a}$$

$$^{107}_{52}\text{Te} \rightarrow ^{103}_{50}\text{Sn} + \alpha \tag{3.43b}$$

$$^{103}_{50}\text{Sn}(\beta^+\nu_e)^{103}_{49}\text{In}(p,\gamma)^{104}_{50}\text{Sn}(\beta^+\nu_e)^{104}_{49}\text{In}(p,\gamma)^{105}_{50}\text{Sn} \tag{3.43c}$$

3.1.5.3 Slow Processes

In contrast to the rapid processes so far discussed, and typical of supernova scenarios, a slow neutron capture process is a characteristic event for the formation of trans-iron elements in the AGB phase of a star. In this phase (see track 4 in the HR diagram, Figure 3.2), an extremely dense core primarily composed of carbon and oxygen is surrounded by a helium burning shell as the main source of energy and, slightly farther out, by a hydrogen burning shell, followed by a tenuous hydrogen-rich envelope. When most of the helium has been consumed, the giant passes through a brief period of flickering (thermal pulses by alternately switching on and off He and H fusion), whereby material from the inner regions is mixed into the outer layers (dredge-up), and vice versa, accompanied by mass loss through violent "stellar winds." The mass loss gives rise to a circumstellar envelope extending over several light-years.

In the s-process (Eq. (3.32)), for which the neutron capture time is slow relative to the β^- decay rates of the daughter nucleus, lower neutron densities and temperatures (around 3×10^8 K) than in the r-process are required. Neutrons are delivered by the (α,n) process as depicted in Eq. (3.40) – the dominating process in $m < 4m_\odot$ stars – and by the final (framed) step in Eq. (3.41) – the dominating process for $m = 4$–$8m_\odot$ stars. The neutron densities are $<10^7$ neutrons cm^{-3} for the ^{13}C$(\alpha,n)^{16}$O source (Eq. (3.40)) and $>10^{10}$ neutrons cm^{-3} for the ^{22}Ne$(\alpha,n)^{25}$Mg source (Eq. (3.41)). Neutrons provided by this latter source yield particularly high abundances of the long-lived rubidium isotope ^{87}Rb ($t_{1/2} = 4.7 \times 10^{10}$ a) [7a], detected in mass-rich AGB stars in our galaxy [7b]. The final steps of this process, involving n capture and β^- decay, are illustrated in Eq. (3.44):

$$\begin{array}{c} ^{88}_{38}\text{Sr} \rightarrow \\ \nwarrow \beta^- \\ ^{85}_{37}\text{Rb} \xrightarrow{n} ^{86}_{37}\text{Rb} \xrightarrow{n} ^{87}_{37}\text{Rb} \\ \nwarrow \beta^- \qquad \beta^- \searrow \\ \rightarrow ^{83}_{36}\text{Kr} \xrightarrow{n} ^{84}_{36}\text{Kr} \xrightarrow{n} ^{85}_{36}\text{Kr} \xrightarrow{n} ^{86}_{36}\text{Kr} \end{array} \tag{3.44}$$

Nuclei heavier than ^{209}Bi are not formed in the s-process, which terminates by the cycle depicted in Eq. (3.45). The net reaction for this Bi–Po–Pb cycle is the formation of ^{4}He (plus two electrons plus two antineutrinos) out of four neutrons:

$$\begin{array}{c} ^{209}_{83}\text{Bi} \xrightarrow{n} ^{210}_{83}\text{Bi} \searrow ^{210}_{84}\text{Po} \searrow ^{206}_{82}\text{Pb} \xrightarrow{3n} ^{206}_{82}\text{Pb} \\ \uparrow \gamma \qquad e^-,\bar{\nu}_e \qquad \alpha \\ \swarrow \\ e^-,\bar{\nu}_e \end{array} \tag{3.45}$$

3.2
Chemistry in AGB Stars

The evolutionary period of a star passing through the AGB (Figure 3.2) encompasses about 10^6 years. Of particular interest in terms of chemistry is the late AGB phase, and the succeeding phases of mass loss, viz., the preplanetary nebula (PPN)[7] and the planetary nebula (PN), all of which represent scenarios for the generation of a variety of neutral and ionic organic and inorganic molecular species, plus the formation of submicrometer dust particles by condensation out of the gaseous phase as the temperature in the circumstellar environment of the original star drops [8]. Near the stellar photosphere, the exclusively gaseous envelope material has a temperature of ca. 1500 K and a density n of ca. 10^{10} atoms/molecules cm^{-3}. As the matter flows away from the star, it cools with $T \propto r^{-1}$ and expands with $n \propto r^{-2}$, and dust begins to form. At the outer edges of the envelope, $T \approx 25$ K and $n \approx 10^5$ cm^{-3} ([9]; see also Figure 4.1 for visualization of the densities). The chemical species produced in these "laboratories" will eventually "feed" interstellar clouds, and thus also become pristine material of new stellar systems as these interstellar clouds collapse. Interstellar grains as constituents in meteorites found on Earth or picked up in the stratosphere, or brought home from encounters with comets, are witnesses of the pre-solar formation of, for example, carbonaceous matter and silicates. This issue will be addressed and discussed in some detail in Sections 5.3 (on asteroids and meteorites) and 5.4 (on comets).

Stars with a high metallicity, viz., population I stars like our Sun, begin their AGB phase with more oxygen than carbon–as a consequence of α capture on ^{12}C in the stellar core (Eq. (3.18)) and carbon burning, Eqs. (3.19)–(3.23). As more and more carbon becomes available via the 3α process in the core (Eqs. (3.16) and (3.17)) and is dredged up to the shell and surface, the abundance of carbon in the stellar photosphere will eventually exceed that of oxygen. Most of the oxygen will react with carbon to form the stable molecule carbon monoxide CO, leaving an excess of carbon for the formation of a variety of carbon-based molecules: the originally oxygen-rich AGB star has developed into a carbon star.

Characteristics of the three stages of a star evolving off the AGB branch are briefly summarized:

- The late AGB phase lasts for 10^4–10^5 years. A variety of molecules is present in the stellar atmosphere. Stellar winds and radiation pressure finally give rise to mass loss and the formation of a circumstellar envelope.

- The stage where the circumstellar envelope develops and evolves (representing an intermediate stage between the late AGB phase and the actual planetary nebula) is termed PPN. These PPNs, as the late AGB stars, contain predominantly

---

7) More commonly termed protoplanetary nebula in the literature. Since "proto- or pro-planetary nebula" implies confusion with the nebula from which the planets of Solar Systems evolve, the term "preplanetary nebula" is employed here and throughout for the envelopes of (former) AGB stars.

simple neutral molecules. Solid particles condensing out of the gas phase provide opacity and a comparatively low temperature: the dust grains absorb high-energy radiation from the stellar core and re-emit the energy in the infrared regime out into space. In this stage, the first IR emission features typical of aromatic compounds appear. The PPN phase lasts for a maximum of 10^3 years.

- Mass loss, at a rate of ca. $10^{-4} m_\odot$ per year, finally thins out the circumstellar envelope of the star: A PN forms, in which the remaining gas and dust envelope are subjected to the radiation of the central hot core until the temperature reaches 25 000–30 000 K, giving rise to photoionization. The PN is characterized by the presence of molecular ions along with increasingly complex molecules and dust. Aromatic molecules now dominate over aliphatic ones. The PN phase ends after 10^3–10^4 years, and the star becomes a white dwarf.

The persistence of molecules in the PN despite of the destructive photodissociation originates from what is known as "Parker instability" or "magnetic buoyancy instability" in the late AGB and PPN stages, creating stable vortices (self-shielding knots) that continue to stay on as the medium constituting the stellar envelope disperses into interstellar space [9].

In order to promote the rich chemistry observed in the cosmologically very short period where an AGB star develops into a PN, several conditions have to be fulfilled. These include (i) a high density, allowing for short collision times for reactive species in the order of seconds, (ii) reduction of the UV flux by grain extinction to allow for the generation of reactive radicals, (iii) a stationary or just slowly expanding molecular region, and (iv) shielding of complex molecules against destructive radiation by absorption to grains.

A selection of molecules and molecular ions detected by their rotational and vibrational modes in the microwave and infrared regimes is collated in Scheme 3.3; typical IR-spectral features for neutral molecules (at a temperature of ca. 200 K)

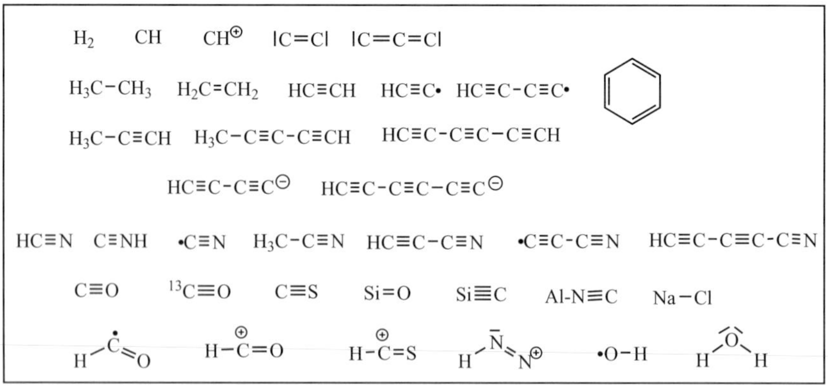

Scheme 3.3 A selection of organic and inorganic, neutral, and ionic molecular species detected in the late phase of AGB stars, in PPNs, and in PNs. Molecular *cations* are essentially confined to PNs.

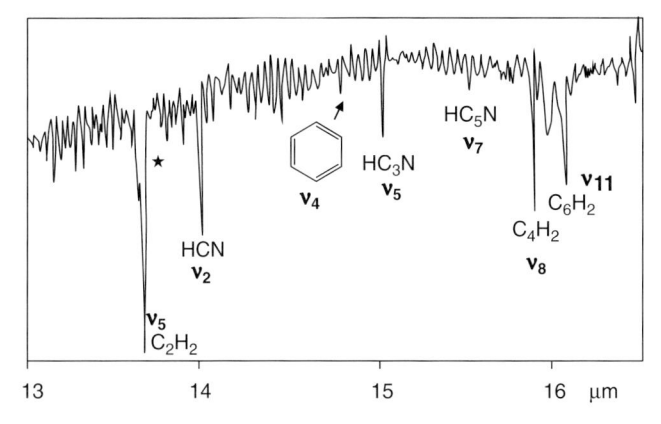

Figure 3.6 IR absorption features (bending modes) of carbon-based molecules in carbon-rich PPNs (constellation of Auriga) in the range 13–16.5 μm (~770–610 cm^{-1}). The low-energy shoulder of the C_2H_2 fundamental indicated by an asterisk corresponds to H^{13}C^{12}CH. The ν_4 for benzene represents the situation where all H atoms synchronously move off the C_6 plane in the same direction. The picture is based on refs. [8b] and [10]. © Ref. [8b] Figure 4 (Elsevier).

are shown in Figure 3.6. Column densities N for key molecules such as HCN and C_2H_2 in PPNs are ca. 2×10^{17} cm^{-2} (corresponding to column amounts of ca. 3×10^{-3} mol m^{-2}), which is about eight orders of magnitude less than the density of nitrogen in Earth's atmosphere. The ratio $^{12}C/^{13}C$ derived from the ratio ($^{12}C)_2H_2/^{13}C^{12}CH_2$ in a specific PPN (CRL 618 in the constellation of Auriga; see also Figure 3.6) is 36, that is, ^{13}C is roughly three times more abundant than on Earth [10]. Vital components in PNs are aromatic compounds and compounds containing aromatic moieties, that is, cyclic planar molecules with sp^2-hybridized carbons and a conjugated system of π electrons. Cyclization of alkynes to aromatic compounds is an exergonic process. Two scenarios for their formation can be anticipated:

- The cyclo-oligomerization of "primitive" building blocks such as that represented by the ethynyl radical C_2H, which is among the first molecules formed in a carbon-rich stellar envelope. The appearance of aromatic species, for example, benzene, in pre-planetary nebulae is thus explained straightforwardly.

- Pre-planetary nebulae are otherwise rich in aliphatic molecules, in particular ethane. The prominent presence of aromatic compounds in the more evolved planetary nebulae may then be traced back to their processing by progressive UV-induced dehydrogenation of ethane to ethene and ethyne, and further of ethyne to the ethynyl radical, as shown in the second row of Scheme 3.3, followed by cyclotrimerization (in the case of benzene) of the latter.

The growth of carbon chains and hence the formation of polyynes and cyanopolyynes shown in Scheme 3.3 (rows 2–5) involves atom–molecule and molecule–molecule reactions of the general type $A + B \rightleftarrows C + D$, according to semiquantitative models for the neutral layers of the circumstellar envelopes of PPNs [11]. In these models, chemistry begins with the photodissociation (but not ionization) of the most abundant simple molecules (H_2, CO, C_2H_2, and HCN). Representative reactions are provided for (i) the formation of tricarbon (allenediyl, C_3) from carbon atoms and ethynyl (C_2H) or ethyne (C_2H_2) (Eq. (3.46)), (ii) the involvement of molecular hydrogen in the generation of the ethynyl radical from dicarbon (Eq. (3.47)), and of alkynes from alkynyls (Eq. (3.48)), (iii) chain propagation by reaction between alkynes and alkyne radicals (Eq. (3.49)), and (iv) the synthesis of cyanopolyynes (Eq. (3.50)). The reaction rate for these reactions, assuming a temperature of 300 K, is typically ca. $2 \times 10^{-10}\,cm^3\,s^{-1}$:

$$C + C_2H \rightarrow C_3 + H \quad (3.46a)$$

$$C + C_2H_2 \rightarrow C_3 + H_2 \quad (3.46b)$$

$$H_2 + C_2 \rightarrow C_2H + H \quad (3.47)$$

$$H_2 + C_nH \rightarrow C_nH_2 + H \quad (n > 2) \quad (3.48)$$

$$C_2H + C_2H_2 \rightarrow C_4H_2 + H \quad (3.49a)$$

$$\text{generally: } C_nH + C_mH_2 \rightarrow C_{n+m}H_2 + H \quad (3.49b)$$

$$CN + C_2H_2 \rightarrow HC_3N + H \quad (3.50a)$$

$$\text{generally: } C_nN + C_mH_2 \rightarrow HC_{n+m}N + H \quad (3.50b)$$

Anions such as C_nH^- (fourth row in Scheme 3.3) form by radiative attachment, that is, uptake of an electron by the neutral precursor radical and radiation of the impact energy as shown by Eq. (3.51). Cyanides and isocyanides of metals such as Na, Mg, and Al can be formed by radiative association between the metal cation M^+ and cyanopolyynes, followed by dissociative recombination of the cationic adduct thus generated. These reactions are represented by Eqs. (3.52) and (3.53). The uptake of a metal ion, for example, Mg^+, by a cyanopolyyne is particularly efficient for large cyanopolyynes ($n = 5$ and 7) [12]. The isocyanide MNC primarily formed after recombination of the cation with an electron (Eq. (3.53)), easily isomerizes to provide an about equal amount of the metal cyanide MCN. For $M = $ magnesium, the bonding energies are 314.7 (Mg–CN) and 320.8 kJ mol^{-1} (Mg–NC); the activation barrier is 19 kJ mol^{-1}. For a systematic treatment of the reaction types in stellar envelopes and in interstellar clouds, see Section 4.2:

$$C_nH + e^- \rightarrow C_nH^- + h\nu \quad (3.51)$$

$$M^+ + HC_nN \rightarrow MNC_nH^+ + h\nu \quad (3.52)$$

$$MNC_nH^+ + e^- \rightarrow MNC + C_{n-1}H; \quad MNC \rightarrow MCN \quad (3.53)$$

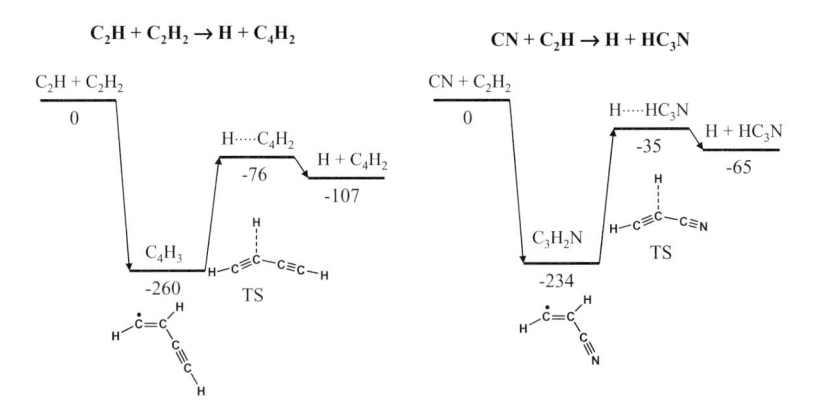

Figure 3.7 Energy potential pathways (showing relative energies in kJ mol^{-1}) for the formation of butadiyne (diacetylene C_4H_2) and cyanoethyne (cyanoacetylene HC_3N), the starting points for the propagation of polyynes and cyanopolyynes, as described by Eqs. (3.49) and (3.50). TS = transition state. Redrawn from Ref. [13].

Chain propagations such as those starting with Eqs. (3.49a) and (3.50a) are two-step processes that commence by long-range electrostatic interaction and finally collision and chemical bond formation (step 1) between the two reaction partners. From the reaction complex (a radical) thus formed in the "entry channel," a hydrogen atom dissociates via a transition state (TS) located in the "exit channel" to form the reaction product (step 2). The energy pathways for reactions (3.49a) and (3.50a), as obtained by *ab initio* calculations [13], are illustrated in Figure 3.7 together with the calculated structures of the reaction complex and the TS. The potential energy term for the long-range interaction is, somewhat simplified, described by Eq. (3.54), where V is the (isotropic) potential, r the distance between the two reactants, μ the dipole moment of the radical (C_2H and CN, respectively), and α the polarizability of ethyne (C_2H_2):

$$V \propto \mu^2 \frac{\alpha}{r^6} \tag{3.54}$$

The chemical bond formation (step 1) is initiated by interaction of the incoming radical (C_2H or CN) with the π electron cloud of ethyne. The final reaction products (H + butadiyne C_4H_2, or cyanoethyne HC_3N) are energetically more favorable than the reactants. The exit channel, however, involves a minor energy barrier (the TS), reflecting, in the case of Eq. (3.50a), the rather unusual temperature behavior of the reaction rate for the overall reaction in the temperature range 0–300 K: While the reaction coefficient increases, as expected, with increasing temperature up to 50 K, there is a slight decreases as the temperature goes further up to 300 K.

More complex organic compounds containing hundreds or even thousands of carbon atoms aggregate (together with siliciumcarbide α-SiC and condensed inorganic compounds) to form dust particles, which resemble carbonaceous materials also found in meteorites. The high-molecular organics are typically polycondensed aromatic compounds, with the aromatic building blocks interlaced by aliphatic (sp^3 carbon) linkers, and furnished with O-, N-, and S-functional groups. For a representative model of these substances, see Figure 5.20 in Section 5.3.2.

If (excess) oxygen is available, a variety of oxygen-containing molecules (such as OH, H_2O, CO_2, and formaldehyde H_2CO) are generated (Scheme 3.3, bottom lines). A reaction by which formaldehyde can be formed is depicted in Eq. (3.55):

$$O + CH_3 \rightarrow H_2CO + H \tag{3.55}$$

Along with the molecular oxygen species, condensed, mineralized oxides accrue, among these silicates, mostly amorphous, but also crystalline ones, indicating heat processing of the originally amorphous forms. A representative selection of oxidic materials is provided below. A prominent secondary oxygen source in carbon-rich (and thus CO-rich) AGB stars is the dissociation of carbon monoxide. The energy for the dissociation of CO can be provided by UV, and by shock waves of a velocity of ~200 km s^{-1} associated with the outflow of protons adjacent to the hot (ca. 30 000 K) core regions:

SiO_2, $(Mg_{1-x},Fe_x)_2SiO_4$ (olivines), $(Mg_{1-x},Fe_x)SiO_3$ (pyroxenes), $\alpha\text{-}Al_2O_3$ (corundum), $MgAl_2O_4$ (spinel), $Mg_xFe_{1-x}O$ TiO_2 (rutile)

3.3
Galaxies and Clusters

In Section 2.2, the question of the formation of the first stars and galaxies had briefly been addressed. Stars do not occur as single objects, but are assembled in large aggregations, commonly galaxies, which in turn are organized in clusters of galaxies (see Figure 3.8a). According to the classical picture, these galaxies form when gas collects at the centers of collapsing haloes[8] of dark cold matter. They further evolve to form larger galaxies by mergers, or by supply with cold gas flowing along rapid streams that penetrate the dark matter haloes [14]. These early galaxies are about 10 billion years old, and they expand in size from an originally more-compact to a less-compact system. A typical galaxy comprises an overall stellar and gas mass of ca. 10^{11} $m_\odot$, residing in dark matter haloes of ca. 10^{13} $m_\odot$. The presence of the dark matter can be inferred by its gravitational influence. Typical star formation rates are ca. 150 per year, which compares to 4 per year in younger galaxies, such as the Milky Way.

There are three main types of galaxies, easily distinct by their external appearance: (1) elliptical galaxies, (2) spiral galaxies, and (3) irregular (plus peculiar) galaxies. The Milky Way is an example of the category of spiral galaxies. Its overall mass is 6×10^{11} $m_\odot$ (approximately 10^{11} stars), its diameter 10^5 light-years, and the thickness is 10^3 light-years. Next to the Milky Way are the Great Andromeda Nebula

---

8) A somewhat misleading term in this context because these "haloes" do not wrap anything.

Figure 3.8 (a) A cluster of galaxies, clearly showing a spiral and (top left) an elliptical galaxy. The various spots also represent galaxies; Credit: FORS1, ESO. (b) Globular cluster M4 in the constellation of Scorpius; Credit: T2KA. NOAO, AURA, NSF.

(Andromeda galaxy, M31), the small Magellanic Cloud, and the large Magellanic Cloud. The Andromeda galaxy, at a distance of 2.5×10^6 light-years, has characteristics very similar to the Milky Way galaxy. The small Magellanic cloud (distance 2×10^5 light-years, diameter 10 light-years, $2 \times 10^9 m_\odot$) and the large Magellanic Cloud (distance 1.6×10^5 light-years, diameter 25 light-years, $10^{10} m_\odot$) are examples for small irregular galaxies. The irregularity likely is caused by gravitational influence exerted by the Milky Way.

Elliptical galaxies, or ellipticals for short, fall within two categories [15], (i) and (ii): Type (i): ellipticals are luminous, slowly rotating galaxies with a triaxial shape. They contain very old stars and have a core. Type (ii) ellipticals are characterized by low luminosity, rapid rotation, and an oblate spheroidal shape. They also contain younger stellar populations, and are coreless. Generally, ellipticals contain little cold gas and a few young (blue) stars only, that is, they are dominated by old red stars. They are considered to be at the end-point of star formation, and to have evolved from spirals and irregulars when star formation in the latter became quenched. Spirals on the other hand contain plenty cold gas and thus abound from vivid regions of star formation. Spirals have central bulges that structurally resemble ellipticals. Both ellipticals and bulges of the spirals contain a supermassive black hole at their center, making up about 1/1000 of the galaxy's overall stellar mass.

The bulges of spirals, on the one hand, and dwarf elliptical galaxies of low ellipticity, on the other hand [16], share common features with *globular clusters* (Figure 3.8b). These globulars, which are situated in the halo[9] of spiral galaxies and orbit

---

9) "Halo" in this context refers to a spherical shell of stars (also accommodating the globular clusters) about the central part of the galaxy, extending beyond the galactic disk.

the galactic core, contain hundreds of thousands of old stars clustering in a comparatively small volume. The diameter typically ranges from 20 to 100 pc; the average star density is around 0.4 stars per cubic pc, with dramatically increased densities toward the core region. In most of these globulars, which also can house a black hole, the stars are all of approximately the same age (population II stars, i.e., stars of low metallicity), and there is no gas and dust, and hence no star formation. A few globulars deviate from this conception, that is, they also contain young (population III) stars and a high abundance of calcium and other heavy elements, which can only have been supplied by supernova explosions. Among the ca. 150 globular clusters in the Milky Way galaxy, ω Centauri represents such an exception. This specific globular probably represents the core of a dwarf galaxy that became swallowed and disrupted by the Milky Way galaxy [17].

In contrast to globular clusters with their high density of stars tightly tied by gravity, *open clusters* contain a few hundred to thousand stars only. In addition, there are *associations*, which are groups of typically 10–100 stars. The stars in open clusters and associations are just loosely gravitationally bound to each other. All of these stars formed from the material making up a giant molecular cloud that became unstable through, for example, a shock event caused by a supernova explosion. The stars in open clusters thus are young stars with high metallicities (population I stars), embedded in the remnants of the cloud, and giving rise to radiative ionization of hydrogen to form HII regions. Unlike globulars, open clusters are confined to the galactic plane, and they are associated with those parts of the spiral arms that are particularly rich in gas and dust. Elliptical galaxies, which are devoid of gas, do not contain open clusters. The Pleiades (Seven Sisters; M45) are a typical example for an open cluster in our galaxy. Hardly any stellar ensemble in our sky has provoked more imagination, reflected in mythological accounts, than the Pleiades. In the Greek mythology, the Pleiades are the seven daughters of Atlas and Pleione; only six of the stars shine brightly: the seventh, Merope, is shamed (and shines dully) because she had an affair with a mortal. In the mythology of the Australian Aborigines, the Seven Sisters are seven women who, after having endured every ordeal, every exceptional torture and fear to show they were ready for womanhood, were snatched from the midst of their friends by the gods and spirits of the high heavens, and taken up to the sky to encourage other women to follow their example.

Summary

Young, metal-rich population I stars formed by accretion in gravity-instable gas and dust clouds, the remainders of disintegrated massive earlier stars. Different stellar masses give rise to different surface temperatures T and luminosities (magnitudes m). A suitable classification system is the Hertzsprung–Russel diagram (T vs m), in which evolving stars in an equilibrium situation occupy the main sequence. The birth, or zero-age, of a star of about the mass of the Sun ($m_\odot$) is indicated by the start of hydrogen fusion in the star's million degrees hot core.

Stars of masses >4$m_\odot$ develop off the main sequence as AGB stars and further as PNs. These objects are of particular interest for the interstellar inventory of molecules. Stars exceeding 8$m_\odot$ finally explode, forming supernovae and ending up as neutron stars, pulsars, quasars, or singularities, the latter either embedded in an event horizon (and thus a black hole) or as a naked singularity.

Hydrogen burning, the fusion of four protons, produces helium-4 (α, along with two positrons and two neutrinos). The net energy release is 2.7×10^9 kJ mol^{-1}. In high-mass stars, He can also be produced in the CNO cycle. Heavier elements, up to iron-56, are predominantly formed by a fusion followed by a successive capture and – for elements with nuclei other than multiples of He-4 – by carbon burning. These processes afford temperatures around 5×10^8 K. At temperatures exceeding 1.5×10^9 K, oxygen burning, and finally silicon burning, come in. Nuclei beyond Fe-56 are mainly formed by rapid (*r*) or slow (*s*) neutron capture cascades, accompanied by β^- emission, or by *rp* capture. The *r*-process is typical for supernovae, whereas *s*-processes dominate in AGB stars. The high neutron flux necessary for efficient neutron capture is provided by, inter alia, the conversion of protons into neutrons plus positrons.

AGB stars develop into PPNs and further into PNs; the envelopes of which exhibit rich chemistry. Molecular species detected include C_5-alkynes and nitriles (cyanopolyynes), aromatic compounds, HCO, HCS$^+$, HN$_2^+$, H$_2$O, NaCl, SiC, and SiO. Condensed high-molecular mass organic compounds, together with inorganic oxidic materials (such as silicates), accrue to form dust particles.

References

1 Helder, E.A., Vink, J., Bassa, C.G., Bamba, A., Bleeker, J.A.M., Funk, S., Ghavamian, P., van der Heyden, K.J., Verbunt, F., and Yamazaki, R. (2009) Measuring the cosmic-ray acceleration efficiency of a supernova remnant. *Science*, **325**, 719–722.

2 Burbidge, E.M., Burbidge, G.R., Fowler, W.A., and Hoyle, F. (1957) Synthesis of the elements in stars. *Rev. Mod. Phys.*, **29**, 547–650.

3 Cowan, J.J., and Thielemann, F.-K. (2004) R-process nucleosynthesis in supernovae. *Phys. Today*, **57**, 47–53.

4 Martínez-Pinedo, G., Kelić, A., Langanke, K., Schmidt, K.-H., Mocelj, D., Fröhlich, C., Thielemann, F.-K., Panov, I., Rauscher, T., Liebendörfer, M., Zinner, N.T., Pfeiffer, B., Buras, R., and Janka, H.-Th (2007) Nucleosynthesis in neutrino heated matter: the *vp*-process and the *r*-process. *Proc. Inst. Nucl. Theory*, **15**, 163–173. (arXiv:astro-ph/0608490v1).

5 Fröhlich, C., Martínez-Pinedo, G., Liebendörfer, M., Thielemann, F.-K., Bravo, E., Hix, W.R., Langanke, K., and Zinner, N.T. (2006) Neutrino-induced nucleosynthesis of *A* >64 nuclei: the np-process. *Phys. Rev. Lett.*, **96**, 142502-1–142502-4.

6 Schatz, H., Aprahamian, A., Barnard, V., Bildsten, L., Cummings, A., Ouellette, M., Rauscher, T., Thielemann, F.-K., and Wiescher, M. (2001) End point of the *rp*-process on accreting neutron stars. *Phys. Rev. Lett.*, **86**, 3471–3474.

7 (a) Beer, H., and Macklin, R.L. (1989) Measurement of the ^{85}Rb and ^{87}Rb capture cross section for *s*-process studies. *Astrophys. J.*, **339**, 962–971.; (b) García-Hernández, D.A., García-Lario, P., Plez, B., D'Antona, F., Manchado, A., and Trigo-Rodríguez,

J.M. (2006) Rubidium-rich asymptotic giant branch stars. *Science*, **314**, 1751–1754.

8 (a) Kwok, S. (2004) The synthesis of organic and inorganic compounds in evolved stars. *Nature*, **430**, 985–991; (b) Kwok, S. (2007) Molecules and solids in planetary nebulae and proto-planetary nebulae. *Adv. Space Res.*, **40**, 655–658.

9 Ziurys, L.M. (2006) The chemistry in circumstellar envelopes of evolved stars: following the origin of the elements to the origin of life. *Proc. Natl. Acad. Sci. USA*, **103**, 12274–12279.

10 Cernicharo, J., Heras, A.M., Tielens, A.G.G., Pardo, J.R., Herpin, F., Guélin, M., and Waters, L.B.F.M. (2001) Infrared space observatory's discovery of C_4H_2, C_6H_2 and benzene in CRL 618. *Astrophys. J.*, **546**, L123–L126.

11 Cernicharo, J. (2004) The polymerization of acetylene, hydrogen cyanide, and carbon chains in the neutral layers of carbon-rich proto-planetary nebulae. *Astrophys. J.*, **608**, L41–L44.

12 Petrie, S. (2003) Deep space organometallic chemistry. *Aust. J. Chem.*, **56**, 259–262.

13 (a) Woon, D.E., and Herbst, E. (1997) The rate of the reaction between CN and C_2H_2 at interstellar temperatures. *Astrophys. J.*, **477**, 204–208; (b) Herbst, E., and Woon, D.E. (1997) *Astrophys. J.*, **489**, 109–112.

14 Dekel, A., Birnboim, Y., Engel, G., Freundlich, J., Goerdt, T., Mumcuoglu, M., Neistein, E., Pichon, C., Teyssier, R., and Zinger, E. (2009) Cold stream in early massive hot haloes as the main mode of galaxy formation. *Nature*, **457**, 451–453.

15 Ciotti, L. (2009) Anatomy of elliptical galaxies. *Nature*, **460**, 333–334.

16 van den Bergh, S. (2008) Globular clusters and dwarf spheroidal galaxies. *Mon. Not. R. Astron. Soc.*, **385**, L20–L22.

17 Lee, J.-W., Kang, Y.-W., Lee, J., and Lee, Y.-W. (2009) Enrichment by supernovae in globular clusters with multiple populations. *Nature*, **462**, 480–483.

4
The Interstellar Medium

4.1
General

About 50% of the baryonic matter of the Universe is represented by the warm-hot intergalatic medium (WHIM), mainly protons, with an average density of $1 \, m^{-3}$ and a temperature of ca. 10^6 K. This matter is not uniformly distributed; rather, it forms filamenteous structures between the galaxies. These structures have been detected mainly by absorptions of highly ionized oxygen, O^{5+} (OVI) and O^{6+} (OVII). In addition, the interstellar space within the galaxies accommodates matter. This interstellar medium is not distributed uniformly, but rather forms clouds of variable size, mass, structure, density, and temperature; see Table 4.1 for an overview, and Figure 4.1 for particle densities as compared to other systems. The main types of clouds are (i) essentially ionized diffuse clouds of low-to-medium density, (ii) neutral so-called reflection nebulae with low particle densities, and (iii) dark clouds of comparatively high density containing molecular materials and dust grains. An intermediate type of clouds, termed "translucent clouds," possibly transitional between diffuse and dense clouds, is responsible for various absorption phenomena in the line of sight of bright stars. Parts of dense molecular clouds can develop into compact regions ("hot molecular cores") with number densities exceeding $10^6 \, cm^{-3}$ and temperatures above 150 K, believed to be markers of the earliest phase of star formation. In many cases, there is smooth transition and mixing between interstellar clouds and circumstellar matter of evolved stars ejected into space on the one hand and contracting clouds delivering matter to clusters of star formation on the other hand. In Figure 4.2, examples of clouds are shown. Clouds can incorporate material corresponding to several thousand Solar masses; linear extensions of interstellar clouds span 1–500 light years.

The first interstellar species detected in 1904 was Ca (by its 393.4 nm emission), followed, in 1919, by Na (by its doublet emission at 589.0 and 589.6 nm). The first molecules were discovered in the early fifties (CH^+, CN) and sixties (OH, CO). Particularly rich in chemistry are the comparatively dense molecular clouds, because molecules formed here, or supplied by mass loss from envelopes of evolved stars, are protected to some extent from the destructive forces of radiation. In addition, the dust particles present in these clouds act as a promoter of chemical

Table 4.1 Typical characteristics of interstellar matter[a].

Type	Number density $n^{[b],[c]}$ (cm^{-3})	Temperature (K)	Major constituents	Main detection technique[d]
Intergalactic coronal gas	≈10^{-3}	10^6–10^7	H$^+$, C^{3+}, N^{4+}, O^{5+}	UV absorption, X-ray emission
Neutral and ionized (warm intercloud) medium	0.2–0.5	6–10 × 10^3	H, H$^+$, He, He$^+$	21 cm emission of H, Hα emission
Ionized diffuse clouds (Emission nebulae, HII regions)	10^2–10^4	8 × 10^3	H$^+$, He$^+$, O$^+$, C$^+$, N$^+$, CH$^+$, H$_3^+$, HCO$^+$	Hα emission
Neutral diffuse clouds (reflection nebulae, HI regions)	20–50	50–100	H, He, H$_2$, H$_3^+$, C$^+$, CO, CN, OH, NH, C$_2$, dust	21 cm emission of H atoms
Molecular clouds (dark nebulae)	up to 10^6	10–20	H$_2$, many molecules, ice-coated dust grains	MW and IR emissions and absorptions

a) H, H$^+$, C^{3+}, etc., correspond to HI, HII, CIV, etc., the more commonly used notations in the astrochemical and astrophysical literature.
b) Divide by 6 × 10^{20} to transform to the unit mol l^{-1}.
c) See also Figure 4.1.
d) See Section 4.2.3 for details.

reactions and exert a protective function by absorbing chemical species. These clouds (or nebulae) are not stable in space and time, and they actually represent an interim situation between the death and the birth of condensed stars. Evolved stars, such as supernovae, asymptotic giant branch (AGB) stars, and the planetary nebulae derived thereof deliver matter into interstellar space by stellar outbursts and stellar winds. In addition, red supergiants are steadily supplying matter to the interstellar space. Red supergiants are luminosity class-I stars of spectral type K or M (see the Hertzsprung–Russel diagram, Figure 3.2); examples are Betelgeuse (α Orionis, the left shoulder star of Orion) and Antares (α Scorpii). These extremely voluminous giant stars have comparatively low surface temperatures of ca. 4000 K, allowing for chemistry in the circumstellar shells similar to that in AGB stars (Section 3.2).

AGB stars are mainly responsible for the supply of carbon and oxygen (and molecules derived thereof), whereas the hotter supernovae make available heavier elements – along with hydrogen and helium, which are by far the dominant constituents in the interstellar medium. The mean lifetime of an interstellar cloud amounts to 10^6–10^8 years. Shock waves provided by supernova events destabilize a cloud, which then begins to contract, finally resulting in the formation of protosuns with protoplanetary discs. Figures 3.1a and 4.2a illustrate such star nurseries. Species such as HCO$^+$, HCN, and C$_2$H formed in preplanetary nebulae may be

4.1 General

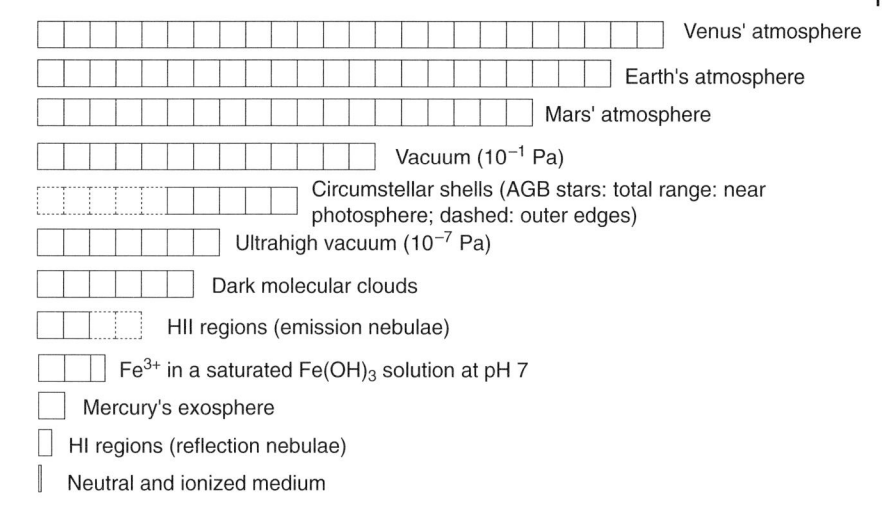

	Venus' atmosphere
	Earth's atmosphere
	Mars' atmosphere
	Vacuum (10^{-1} Pa)
	Circumstellar shells (AGB stars: total range: near photosphere; dashed: outer edges)
	Ultrahigh vacuum (10^{-7} Pa)
	Dark molecular clouds
	HII regions (emission nebulae)
	Fe^{3+} in a saturated $Fe(OH)_3$ solution at pH 7
	Mercury's exosphere
	HI regions (reflection nebulae)
	Neutral and ionized medium

Figure 4.1 Densities, quoted as particles (molecules, atoms, and ions) cm^{-3}, of interstellar media (Table 4.1) compared to the atmospheres of the four inner planets, circumstellar envelopes, and a saturated aqueous solution of ferric hydroxide. Fe^{3+} in neutral media has been chosen here for its key role (limiting factor) in life. The scale is logarithmic ($\log_{10} 10^x$); extending from $x = 24$ (Venus) to $10^{-1} cm^{-3}$ (neutral and ionized intercloud medium).

linked, as survivor molecular material, to the variety of compounds found in interstellar clouds.

Hydrogen and helium constitute more than 99% of the overall interstellar material: Of 1 million atoms, 878 thousand are hydrogen and 121 thousand are helium, followed by oxygen (606), carbon (264), nitrogen (79), magnesium (29), silicon (26), iron (23), and sulfur (13). The very hot and thin intergalactic coronal gas, so termed because its temperature is comparable to that of the gas found in the Sun's corona, permeates through about 98% of the interstellar space. It consists of H^+ and highly ionized ions of the heavier elements; from a chemistry point of view, it is thus less interesting. More intriguing are the neutral diffuse clouds and the dark clouds. Neutral diffuse clouds do not emit light by themselves (as do the ionized diffuse clouds), but are visible because they reflect light of adjacent stars. Dark clouds are so dense that they obscure that part of the sky where they are located. Dark clouds can easily be observed as irregular, rift-like structures dimming out parts of the band of the Milky Way. It is these dark clouds, along with "hot cores" in protostellar regions (Section 4.2.4), where a rich chemistry takes place. Chemistry in dark clouds is not initiated by electromagnetic radiation (which cannot penetrate into the depths of these clouds) but by cosmic rays. Cosmic rays are highly energetic particles, mainly protons (87%), followed by α particles (12%) and about 1% of completely ionized heavier nuclei, plus electrons, neutrinos, and γ quanta.

The low temperature in these molecular clouds, 20 K and less, presupposes that chemistry here is kinetically rather than thermodynamically controlled. An illustrative example is the abundance of hydrogen cyanide (H—C≡N) and hydrogen

Figure 4.2 (a) This section of the Orion cloud complex, at a distance of about 1500 lightyears, shows the three main types of interstellar clouds: The horsehead nebula (Barnard 33, spanning ca. 5 lightyears) is a dark cloud, silhouetted here against the red emission nebula IC 434. The red color is due to the 656.3 nm Hα emission. The blue nebula in the lower left is the reflection nebula NGC 2023, reflecting the light of a nearby bluish O-type star. Reproduced with permission by Daniel Verschatse, Observatorio Antilhue, Chile; link: http://www.verschatse.cl/nebulae/ic434-b33-ngc2023/maximum.jpg. (b) This false-color X-ray image shows a stellar nursery in the Great Orion Nebula (M 42) below Orion's belt. The bright, young stars in the center belong to the Trapezium cluster; the stars shown in blue and orange are young sun-like stars. Credit: NASA/CXC/Penn State/E. Feigelson & K. Getman *et al.*

isocyanide (H—N≡C): These two isomers are observed in about equal abundances (see also Eq. (4.41b) in Section 4.2.2) despite of the fact that the iso-form HNC, which is unstable under terrestrial conditions, is less stable by 65.5 kJ mol^{-1}. Many of the possible interstellar chemical reactions have been modeled in laboratories. The model reactions have provided insight into possible mechanisms for the formation of interstellar molecules, including those that might be expected to be present but have not yet been detected. A major problem in these laboratory simulations is the time factor: There are hundreds of thousands of years available in space, but certainly not in an Earth-bound laboratory.

To date (as of May 2010), ca. 160 molecules have been found in interstellar clouds and circumstellar envelopes in our galaxy, about 40 of which, containing up to 7 atoms, have also been detected in the intergalactic space. The number of molecules adding to this list presently amounts to an average of 8 per year. In addition, a couple of heavy isotopomers have been detected such as ^{13}CO and the deuterium (^{2}H) isotopomers HD, H$_2$D$^+$, HDO, D$_2$O, DCN, DNC, DCO, N$_2$D$^+$, NH$_2$D, NHD$_2$, ND$_3$, HDCO, D$_2$CO, CH$_2$DC≡CH, and CH$_3$C≡CD. Only a restricted number of elements are involved in the built-up of the interstellar inventory: H, C, N, O, F, Si, P, S, Cl, Na, K, Mg, Al, and possibly Fe (in FeO). Carbon is a predominant constituent in most of the molecules, and in all of the more complex ones (molecules with 6–13 atoms). The only purely inorganic species, containing four to five atoms, are ammonia NH_3, the isoelectronic hydroxonium ion H_3O^+, and silane SiH_4.

The largest molecule that has so far been found is cyanodecapentayne $HC_{10}CN$. Laboratory experiments carried out in 1985 by Curl, Kroto, and Smalley (Nobel prize 1996) to explain the interstellar formation of polyynes have resulted in the discovery of fullerenes such as C_{60}. Fullerenes, or "bucky balls" in popular language, constitute spheroidal cage structures composed of five- and six-membered polygons (exactly like a soccer ball in the case of C_{60}) reminiscent of the geodesic domes designed by the architect Buckminster Fuller around the forties of the last century. In Kroto's experiments, carbon was vaporized by a special laser-supersonic beam. Fullerenes (see Figure 5.19 in Section 5.3.2) and "buckyonions" (multishell fullerenes) are believed to be constituents of the carbonaceous matter that, together with silicates, build up interstellar dust particles, and have also been found in meteorites. Typical sizes of these dust grains vary from 10 nm to 1 μm; a "standard size" of 100 nm is often assumed. The grains are coated by ice, predominantly water, but also contain CO, CO_2, formaldehyde, methanol, NH_3, and CH_4. Heavier elements, not found in the gas fraction of the interstellar medium, are thought to be confined to the dust particles. Dust grains play an important role in interstellar chemistry, and will thus be addressed in Section 4.2.5.

The exotic conditions in interstellar clouds – low temperature and, in particular, low density – allow for the presence of many exotic (and otherwise reactive and/or unstable or metastable) molecules such as radicals, molecular ions, including negative ones, and subvalent (carbon) species. A high degree of unsaturation is typical of interstellar molecules, manifest by polyynes, cyanopolyynes, and (poly) aromatic compounds. A few saturated molecules, some of which are predominantly products of surface chemistry on grains rather than of chemical reactions in the gas phase, have also been detected. Examples for the latter are ethanol (C_2H_5OH), glycol (CH_2OHCH_2OH), acetic acid (CH_3CO_2H), and methylamine (CH_3NH_2). On the other hand, the saturated molecules dimethyl ether ($O(CH_3)_2$), diethyl ether ($O(C_2H_5)_2$), methylformate ($HCOOCH_3$), and acetone ($OC(CH_3)_2$) are more likely products of gas-phase reactions. Scheme 4.1 contains a selection of a few rather unusual molecules from the inventory of molecular clouds. Currently maintained websites for databases providing the current stand on inter- and circumstellar molecules are:

Scheme 4.1 A selection of unusual molecules from interstellar molecular clouds. For c-propenylidene and c-propenone, see also Eqs. (4.53) and (4.54) in Section 4.2.4.

http://www.ph1.uni-koeln.de/node/470
http://www.cv.nrao.edu/~awootten/allmols.html

In his novel "The Black Cloud" [1], Fred Hoyle describes a scenario where the complex molecules of an interstellar cloud are organized and cooperating in such a manner that an intelligent superorganism emerges, seriously threatening and finally, by obscuring the Sun, harming Earth on its approach to our planet. Fortunately, such a scenario has so far remained fiction. Which fact does not exclude an important if not essential (and, blessedly, beneficial) role of the molecular inventory of interstellar clouds for life on Earth: The bulk of carbon on Earth, and thus of the central element of life, was brought from interstellar space via meteorites, comets, and dust grains. "This organic material, with its ultimate origin coupled to carbon-rich chemistry of preplanetary nebulae, may have been the basis of development of life on Earth" [2]. Note again (Section 3.2) that the term *prepl*anetary nebula is used here (and will be used throughout) for the transient phase between the late AGB stage of a star and the planetary nebula formed toward the end of the star's evolution, to have it distinct from the *proto*planetary nebula (also referred to as protoplanetary disc) from which planets form in the course of, or rather after, contraction and accretion of a cloud to a protosun.

4.2
Chemistry in Interstellar Clouds

4.2.1
Reaction Types

Among the many reaction types describing chemical networks, only a few are pertinent to chemical processes occurring in interstellar clouds. Starting point in many cases is the ionization of an atom, such as H, He, or C, Eq. (4.1), or a molecule, predominantly H_2, which forms by recombination on the surface of dust particles, M, Eq. (4.2). Carbon in interstellar clouds is ionized by UV radiation stemming from close-by or embedded (new-born) stars. The necessary minimum energy for the ionization of a C atom is 11.26 eV, which corresponds to ultrashort (extreme or far) UV, 110.2 nm. Virtually all elemental carbon present in the clouds is in the form of C^+. Hydrogen and helium cannot be ionized by stellar UV because of their high ionization potentials; they are ionized by impact with cosmic rays, Eqs. (4.3a) and (4.4), which transfer part of their kinetic energy into the ionization process. Along with H_2^+, both H and H^+, Eqs. (4.3b) and (4.3c), can also be generated by cosmic rays. Ionization by cosmic rays is also the only way for a primary ionization process in the inner parts of a dense cloud, shielded from any electromagnetic radiation. He^+ is stable in a medium dominated by H_2, because the reaction between He^+ and H_2 to form He and H_2^+ is strongly forbidden [3]. The reaction between He^+ and CO is an alternative source for the production of C^+, Eq. (4.5). The molecular cation H_2^+, formed in molecular clouds, is rapidly converted to H_3^+, Eq. (4.6). Ionization rates for $H_2 \to H_2^+ + e^-$ are typically $10^{-17}\,\text{s}^{-1}$ ± one order of magnitude, from which a density of $n(H_3^+) \approx 5\times10^{-5}\,\text{cm}^{-3}$ [4] (corresponding to $\approx 10^{-25}\,\text{mol}\,\text{l}^{-1}$) derives. This is ca. nine orders of magnitude less than the density of neutral hydrogen (H plus H_2), but sufficient to initiate a rich sequential chemistry. The estimated overall fractional ionization in *dense* prestellar cores originating from contracting dense clouds is even less, viz. $\approx 10^{-7}\,\text{cm}^{-3}$.

$$C \xrightarrow{\text{UV}} C^+ + e^- \tag{4.1}$$

$$H + H + M \to H_2 + M \tag{4.2}$$

$$H_2 \xrightarrow{\text{cosmic rays}} \begin{cases} H_2^+ + e^- & (88\%) & (4.3\text{a}) \\ 2H & (10\%) & (4.3\text{b}) \\ H + H^+ + e^- & (8\%) & (4.3\text{c}) \end{cases}$$

$$He \xrightarrow{\text{cosmic rays}} He^+ + e^- \tag{4.4}$$

$$He^+ + CO \to He + C^+ + O \tag{4.5}$$

$$H_2^+ + H_2 \to H_3^+ + H \tag{4.6}$$

Modeling interstellar chemistry has to take into account the very low temperature, a maximum of 20 K, and the long free mean path even in dense clouds, leading to just one encounter of two potential reactants per day at the best. In order for such an encounter to be "successful" (in the sense that a new species forms), the activation barrier for the chemical reaction has to be low to nonexistent, the reaction should be exergonic, and the kinetic energy of the impacting particles has to be carried away to prevent the new-born molecule from being instantly torn apart. This latter condition is fulfilled, if the energy is taken up by a second reaction product or a matrix. The lifetime of a cloud, typically several million years, helps to accumulate even comparatively complex species, as does the protection of molecules by absorption on grain surfaces. Nonetheless, for an efficient chemistry to take place, the reaction barrier should be pretty low, and this is secured by ion–molecule reactions, although a couple of reactions involving neutral reactants also fulfill this condition. Equations (4.5) and (4.6) are examples of *ion–molecule reactions*, or, more precise (but less commonly used), *ion–neutral reactions*, and thus the more common reaction type under interstellar conditions. Interstellar components such as CO, O, and N_2 undergo ion–neutral reactions with the comparatively abundant ions C^+ and H_3^+ (H_3^+ does, however, not react with H_2) recycling H and H_2; for examples see Eqs. (4.7)–(4.10). Commonly, cations are involved in ion–neutral reactions, but as exemplified by Eq. (4.11), anions such as the hydride ion H^- may also come in. For additional details on ion–neutral reactions, see Section 4.2.2.

Ion–neutral reactions:

$$H_3^+ + N \to NH_2^+ + H \tag{4.7}$$

$$H_3^+ + CO \to HCO^+ + H_2 \tag{4.8}$$

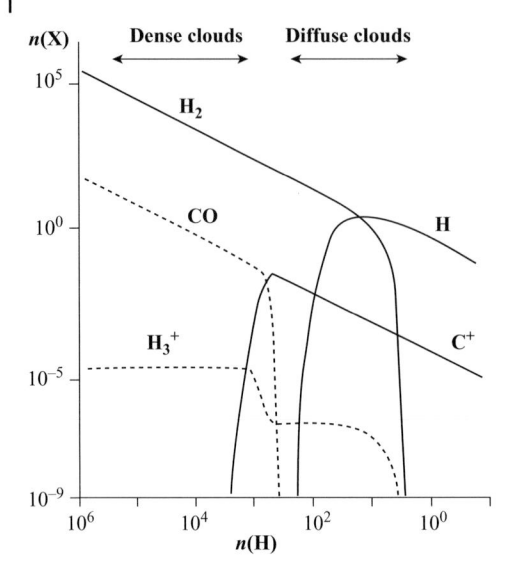

Figure 4.3 Number densities $n(X)$ for X = H, H_2, H_3^+, C^+, and CO versus the cloud density expressed as $n(H)$, in cm^{-3}. Adapted from Ref. [5a]; © (2009) National Academy of Sciences, U.S.A.

$$H_3^+ + O \rightarrow OH^+ + H_2 \tag{4.9}$$

$$H_3^+ + HD \rightarrow H_2D^+ + H_2 \tag{4.10}$$

$$H^- + C_2H_2 \rightarrow C_2H^- + H_2 \tag{4.11}$$

The proton donor H_3^+ holds a prominent role in these ion–neutral reactions because it is (i) comparatively abundant in dense *and* diffuse clouds and (ii) readily formed and readily reacting. Its destruction in dense clouds is mainly via reaction (4.8), whereas in diffuse clouds, it is mainly destroyed by electron capture, Eq. (4.12a). In clouds with very high densities ($>10^6\,cm^{-3}$) and temperatures as low as 10 K, where most molecules may be frozen on grains, the main destruction process likely is the slightly exothermic reaction with HD, Eq. (4.10) [5]. The central importance of the most abundant particles in dense and diffuse interstellar clouds, viz. H, H_2, H_3^+, C^+, and CO, is reflected in Figure 4.3, where the number densities $n(X)$ of these particles are related to the overall density of the cloud.

Electron capture by an ion generally leads to the formation of an atom and a molecule (or two molecules). Typical examples are the annihilation of H_3^+, Eq. (4.12a), and the reformation of carbon monoxide from the formyl cation, Eq. (4.12b). This type of reaction is termed *dissociative recombination*.

Dissociative recombination:

$$H_3^+ + e^- \rightarrow H_2 + H \tag{4.12a}$$

$$HCO^+ + e^- \rightarrow H + CO \tag{4.12b}$$

Dissociative electron attachment, Eq. (4.13), is one of the major paths to the recently discovered anions in dark clouds and envelopes of evolved stars. For more details on this fascinating issue, neglected for a long time, see Section 4.2.4. Other electron capture processes result in the release of the kinetic energies of the combining cation + electron in the form of electromagnetic radiation (*radiative recombination*, Eq. (4.14)) or, if the electron is captured by an atom (or molecule) to form an anion, *radiative association*, Eq. (4.15). In cases, where the molecule is sufficiently large to distribute and finally to dissipate the impact energy, simple *electron attachment* occurs. This is the case for polycyclic aromatic hydrocarbons PAHs (Eq. (4.16)), which are believed to be present in the interstellar medium and are an integral constituent of dust grains. The reverse of the radiative association, by which an anion is formed, is the *photodetachment* of an electron from an anion, Eq. (4.17):

Dissociative attachment: $HNC_3 + e^- \rightarrow C_3N^- + H$ (4.13)

Radiative recombination: $C^+ + e^- \rightarrow C + h\nu$ (4.14)

Radiative association (a): $C_4H + e^- \rightarrow C_4H^- + h\nu$ (4.15)

Electron attachment: $PAH + e^- \rightarrow PAH^-$ (4.16)

Photodetachment: $C_2H^- + h\nu \rightarrow C_2H + e^-$ (4.17)

Radiative associations are not restricted to electron capture processes. An example where a molecular cation and H_2 are involved in such a process is depicted in Eq. (4.18). If a (positive) charge is transferred from an ion to a molecule (or atom), the respective reaction is referred to as a *charge transfer reaction*, Eqs. (4.19) and (4.20). If the ionization potentials of the involved particles are almost alike, as in the case of equilibrium Eq. (4.20), the charge transfer is also referred to as *resonance charge exchange*.

Radiative association (b):

$$CH_3^+ + H_2 \rightarrow CH_5^+ + h\nu \quad (4.18)$$

Charge transfer reaction:

$$P^+ + H_2S \rightarrow P + H_2S^+ \quad (4.19)$$

$$H^+ + O \rightleftarrows H + O^+ \quad (4.20)$$

The effective capture cross section for a reaction between two neutral particles (*neutral–neutral reactions*), commonly two molecules or a molecule and an atom, is lower than that of an ion–molecule reaction. Nonetheless, these so-called *insertion reactions*, resulting in chain propagation, can be rapid at low temperatures if a radical is involved [6]. Two examples – the insertion of a carbon atom into ethyne and of the cyano radical into ethyne – are shown in Eqs. (4.21) and (4.22), respectively. For the latter reaction and the energy paths involved in this reaction, see also Eq. (3.50a) and Figure 3.7 (right) in Section 3.2. Another type of reaction where only uncharged particles are involved is the *neutral exchange* (or *neutral–neutral*) reaction, Eq. (4.23).

Insertion reactions:

$$C + C_2H_2 \rightarrow C_3H + H \tag{4.21}$$

$$CN + C_2H_2 \rightarrow NC_2CN + H \tag{4.22}$$

Neutral exchange: $CH + O \rightarrow CO + H$ (4.23)

Finally, chemical bonds may be broken by photons, to some extent, the reverse to the radiative association as depicted in Eq. (4.18). In such a photochemically induced reaction, or *photodissociation*, the excess energy of the photon is shared, in the form of excitation and kinetic energy, by the two fragments. The minimum photon energy necessary to achieve a bond breakage is the bond energy E_B, which is related to the energy of the photon $h\nu$ by Eq. (4.24), where N_A is Avogadro's number (6.022×10^{23} mol^{-1}), h the Planck constant (6.63×10^{-37} kJ s), and c (3×10^{17} nm s^{-1}) and λ the speed of light and the wavelength, respectively. For the photochemical cleavage of the C≡O bond according to Eq. (4.25), for which the bond energy is 1075 kJ mol^{-1}, a minimum wavelength of $\lambda = 105$ nm is necessary, that is, far-UV, while middle-UV suffices for the rupture of the C–H bond (bond energy 436 kJ mol^{-1}, $\lambda = 274$ nm):

$$E_B = N_A \times h\nu = N_A hc/\lambda = 119625/\lambda \text{ kJ mol}^{-1} \tag{4.24}$$

Photodissociation: $CO + h\nu \rightarrow C + O$ (4.25)

4.2.2
Reaction Networks

The various reaction types exemplified in Section 4.2.1 for basic initiating reactions can be intimately interlaced, leading to complex reaction networks, a selection of which is introduced in this chapter. For elusive and more complex molecules, also of interest in the frame of precursor molecules for building blocks of life, see Section 4.2.4. In Section 4.2.3, methods of detection of interstellar and circumstellar species are visualized.

Basic reactions leading to the three main families (the carbon, nitrogen, and oxygen family) of interstellar molecules via the generation of key active species are assorted below:

Carbon family:

$$C + h\nu \rightarrow C^+ + e^- \tag{4.26}$$

$$CO + He^+ \rightarrow C^+ + O + He \tag{4.27}$$

$$CO + H_3^+ \rightarrow HCO^+ + H_2 \tag{4.28}$$

Nitrogen family:

$$N_2 + He^+ \rightarrow N + N^+ + He^+ + e^- \tag{4.29}$$

$$N + H_3^+ \rightarrow NH_2^+ + H \tag{4.30}$$

Oxygen family:

$$H_3^+ + O \rightarrow OH^+ + H_2 \tag{4.31}$$

$$H^+ + O \rightleftarrows H + O^+ \tag{4.32}$$

As already mentioned, almost all of the carbon is present in the form of C^+, and this should presuppose its reaction with the by far predominant species, H_2. The reaction between C^+ and H_2, Eq. (4.33), is, however, one of the rare ion–molecule reactions that are endothermic and hence not likely to occur at 10–20 K. In contrast, radiative association, Eq. (4.34), can channel C^+ into the reaction networks. The ion CH_2^+ thus formed can further react with H_2 to generate the methyl cation CH_3^+, Eq. (4.35), which is another key ion in a variety of successive reactions. The formyl cation, the formation of which is exemplified by Eq. (4.28), is surprisingly stable and thus particularly abundant. HCO^+ thus is a central molecular cation in all reaction networks. The only effective reaction, by which it becomes destroyed, is by dissociative recombination through electron capture; cf. Eq. (4.12b):

$$C^+ + H_2 \rightarrow CH + H^+ \tag{4.33a}$$

$$C^+ + H_2 \rightarrow CH^+ + H \tag{4.33b}$$

$$C^+ + H_2 \rightarrow \{CH_2^+\}^* \rightarrow CH_2^+ + h\nu \tag{4.34}$$

$$CH_2^+ + H_2 \rightarrow CH_3^+ + H \tag{4.35}$$

Rate constants for exothermic ion–molecule reactions are of the order of magnitude 10^{-10}–10^{-9}, and they are essentially temperature-independent. What are the conditions for a reaction between an ion and a molecule to occur under interstellar conditions? Let us consider, as a typical example, the formation of OH^+ by collision between O^+ and H_2; Eq. (4.36):

$$O^+ + H_2 \rightarrow OH^+ + H \tag{4.36}$$

The two particles supposed to undergo a reaction, here O^+ and H_2, can interact only if they are at a critical minimum distance b_0, termed the impact parameter. The ion induces a dipole moment in the molecule, the size of which depends on the molecule's polarizability α. The interaction energy E is described by

$$E = -\alpha e^2 / 2r^4 \tag{4.37}$$

where e is the elementary charge and r the distance between the ion and the induced molecular dipole. For $r > b_0$, there will just be a deflection; for $r = b_0$, the ion will orbit the induced dipole with an orbital radius of $b_0/\sqrt{2}$, from which a capture cross section of $1/2 \pi b_0^2$ derives. For $r < b_0$, the ion spirals into the molecule. The impact parameter b_0 is related to the polarizability α, the reduced mass of the colliding particles $\mu = m_{ion} \times m_{mol}/(m_{ion} + m_{mol})$, and their velocity v by

$$b_0 = (4e^2 \alpha / \mu v^2)^{1/4} \tag{4.38}$$

For the rate constant k = (capture cross section) × (speed of collision), the expression

$$k = 1/2\pi(4e^2\alpha/\mu v^2)^{1/2} \times v = \pi e(\alpha/\mu)^{1/2} \qquad (4.39)$$

is obtained. The rate constant thus is independent of the velocity of the interacting particles and hence independent of the temperature. For reaction (4.36) and $\alpha(H_2) = 0.79 \times 10^{-24}\,cm^3$, the rate constant calculated from Eq. (4.39) is $k = 1.6 \times 10^{-9}\,cm^3\,s^{-1}$, which comforts with the experimental value. For systems with nonspherical symmetry, and for polar molecules, a more complex situation applies.

A compilation of selected reactions and particles playing a central role in reaction networks is provided in Scheme 4.2. A fractional reaction network involving O^+ and HCO^+, that is, the ions initiating chemistry leading into the interstellar oxygen family is depicted in Scheme 4.3.

Scheme 4.2 Summary of important initiating reactions and pivotal ions/molecules (framed) in interstellar chemistry.

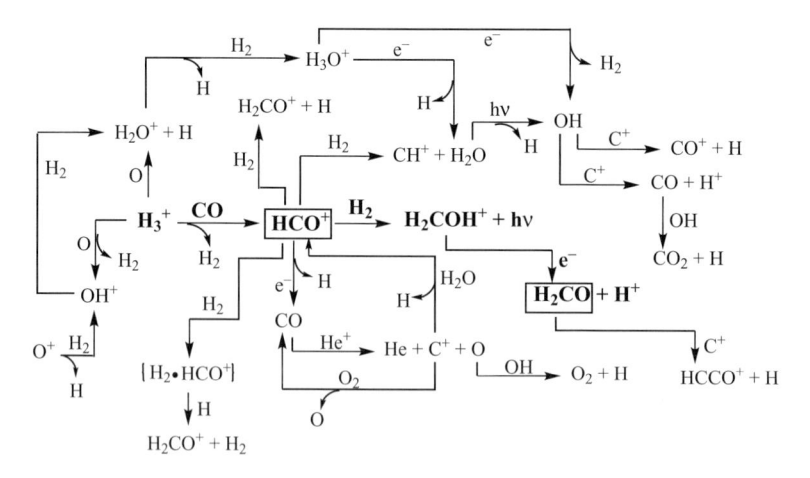

Scheme 4.3 Reaction network for the generation of simple representatives of the oxygen family, based, in part, on information in Refs. [4, 6]. The reaction path leading to the formyl cation, and from the formyl cation to formaldehyde is highlighted (see text for details).

Accentuated in Scheme 4.3 is the reaction path leading—via the formyl cation formed by the ion–neutral reaction between H_3^+ and CO—to formaldehyde. An intermediate in this reaction sequence is protonated formaldehyde H_2COH^+ (formed from HCO^+ and H_2 by radiative association) which, by electron capture, delivers formaldehyde. Formaldehyde can be a key molecule in the formation of sugars (by self-condensation) or amino acids (by condensation with ammonia or hydrocyanic acid), and also for the extension of carbon chains, as shown for the formation of the ketenyl cation $HCCO^+$ by carbon insertion (lower right in Scheme 4.3). There are alternate paths to formaldehyde such as the formation of H_2CO^+ via the adduct $H_2 \cdot HCO^+$ (lower left in Scheme 4.3) or the reaction sequence (4.40):

$$C^+ + H_2 \rightarrow CH^+ + H$$
$$\xrightarrow{H_2} CH_2^+ + H$$
$$\xrightarrow{e^-} CH_2 + H \qquad (4.40)$$
$$\xrightarrow{OH} H-C\overset{H}{\underset{O}{\lessgtr}} + H$$

In addition to formaldehyde (one of the molecular species shown on this book's cover), ammonia (NH_3) and hydrogen cyanide (HCN) are potential building blocks for more complex molecules that adopt a role in life. A straightforward reaction path leading to the formation of these two molecules is provided by Eq. (4.41). Accordingly, a sequence of ion–neutral reactions involving H_2 and starting with N^+ delivers NH_3^+ that transfers its charge to magnesium atoms. The resulting ammonia can further react with a carbon ion to produce protonated hydrogen cyanide ($HCNH^+$), which undergoes dissociative recombination, Eq. (4.41b). Hydrogen cyanide and hydrogen isocyanide are formed here in equal amounts:

$$N^+ + H_2 \rightarrow NH^+ + H$$
$$\xrightarrow{H_2} NH_2^+ + H$$
$$\xrightarrow{H_2} NH_3^+ + H \qquad (4.41a)$$
$$\xrightarrow{Mg} \underset{H}{\overset{\overline{N}\cdots}{\nwarrow}}\overset{H}{\underset{H}{|}} + Mg^+$$
$$\xrightarrow{C^+} HCNH^+ + H \qquad (4.41b)$$
$$\downarrow e^-$$
$$(H-C\equiv N \text{ and } C\equiv N-H) + H$$

A more complex network representing reactions and species of the nitrogen and carbon families (Scheme 4.4) has been arranged around the methyl cation (CH_3^+) as the central species. The methyl cation can be delivered by the reaction sequence (4.42). More recently, it has been shown that dinitrogen N_2 likely is the major nitrogen-bearing molecule in the interstellar medium [7]; for its formation, see the dashed frame in Scheme 4.4. This simple diatomic molecule is the major

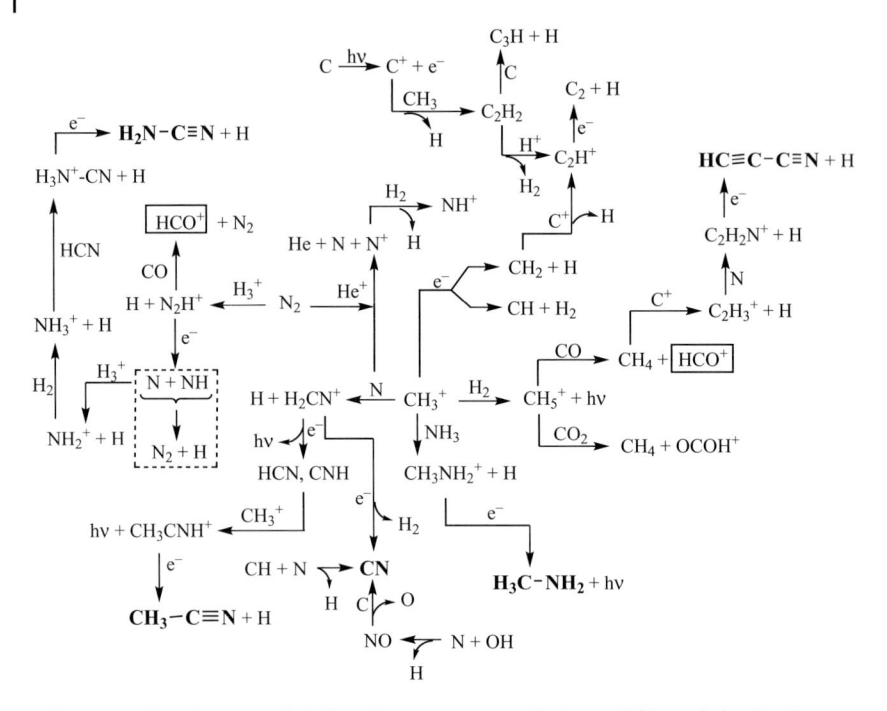

Scheme 4.4 Reaction network (selected reactions) for the generation of particles belonging to the carbon and nitrogen families. The reactions shown here partially are based on Refs. [7, 8]. The highlighted molecules are acetonitrile (lower left), cyanamide (upper left), methylamine (bottom center), and cyanoethyne (upper right). Links to the oxygen family (HCO⁺, Scheme 4.3) are framed; the formation of N_2 is accentuated by a dashed frame.

constituent in the atmospheres of Earth and Saturn's moon Titan, and is also present in trace amounts in the atmospheres of Venus and Mars. On the basis of gas-phase models, an $N_2:H_2$ ratio of 10^{-5} (dense clouds) to 10^{-8} (diffuse clouds) has been predicted; the predicted abundance increases if grain-surface chemistry for the formation of N_2 is included. The detection of interstellar dinitrogen, a symmetric diatomic molecule without a dipole moment, has been particularly problematic; detection in fact relies on electronic transitions in the far-UV (Section 4.2.3), which is inaccessible for Earth-based telescopes:

$$C + h\nu \rightarrow C^+ + e^-$$
$$\xrightarrow{H_2} CH^+ + H$$
$$\xrightarrow{H_2} CH_2^+ + H$$
$$\xrightarrow{H_2} CH_3^+ + H \qquad (4.42)$$

Of particular interest in the context of the formation of unsaturated carbon chains as precursors for polyaromatic compounds present in the carbonaceous fraction of dust grains are the chain propagation reactions, upper right in Scheme 4.4, leading to polyynes (acetylenes) and cyanopolyynes. The second predominant fraction in grains, silica (SiO_2), and silicates, relies on the presence of silicon and gas-phase chemistry providing silicon monoxide as the precursor of silica and silicates. Scheme 4.5 is a reaction network providing some insight into the interstellar silicon family. Silicon, as carbon, is present in its ionized form (Si^+) down to large optical depths, where it is rapidly converted to SiO by the sequence shown in bold in Scheme 4.5, right. This scheme further emphasizes (center and bottom) the formation of silicon carbide, also abundant in interstellar dust.

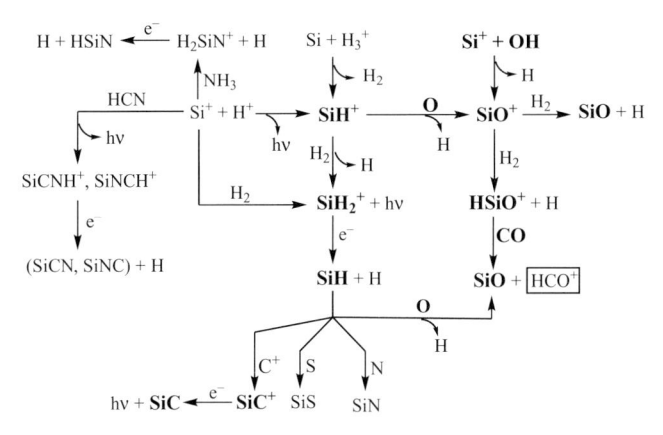

Scheme 4.5 Reactions within the silicon family. The tracks leading to SiO and SiC are highlighted in bold; a gateway to the oxygen family (Scheme 4.3) is framed.

The isomers SiCN and SiNC have recently been detected in a stellar envelope [9]. A possible mode of their generation is radiative association of Si^+ and HCN (and/or HNC), followed by dissociative recombination, as shown in Scheme 4.5, left-hand margin. The isomer SiCN is more stable by 6.3 kJ mol^{-1} than the isomer SiNC; interconversion between the two isomers is, however, energetically unfavorable: Although, in principle, isomerization can proceed via two energetically slightly different transition states, the energy levels of these states, about 90 kJ mol^{-1} above ground state, is not easily surmounted [10] (Figure 4.4). Since the reaction path depicted in Scheme 4.5 delivers both isomers in equal amount, isomerization is not necessary to explain the mutual abundances.

To some extent, interstellar sulfur chemistry is related to oxygen chemistry (if it comes to chemical similarities) and silicon chemistry (as far as the ease of ionization to form Si^+ and S^+, respectively, is concerned). A reaction scheme summarizing reactions leading to several of the sulfur compounds detected in interstellar matter is provided in Scheme 4.6.

Which of the reactions in the networks provided here actually occur, is highly dependent on the elemental abundance ratios of the reactants (in particular the

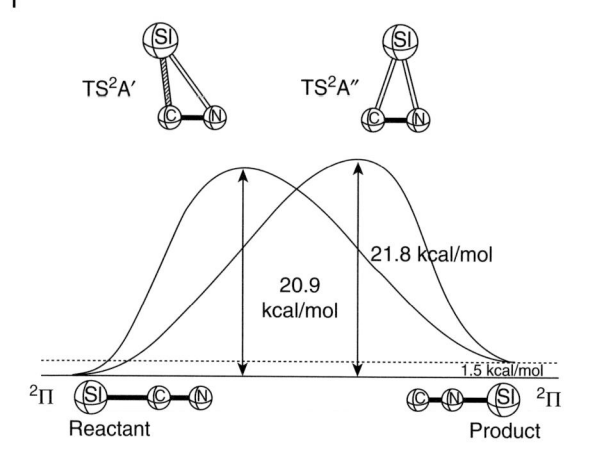

Figure 4.4 Relative energy levels of the ground states of the two isomers SiCN and SiNC ($\Delta E = 6.3$ kJ mol^{-1}), and the two transition states (TS^{2}A' and TS^{2}A", 87.5 and 91.3 kJ mol^{-1}, respectively) for isomerization. Reprinted with permission from J. Chem. Phys. (Ref. [10]).

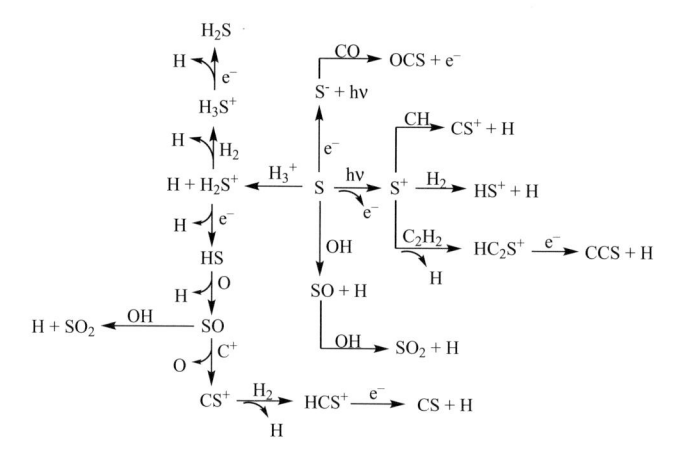

Scheme 4.6 Some reactions within the sulfur family, based, in part, on Ref. [11].

O/C ratio), the importance of competing reactions and of destructive reactions, the temperature and other energy sources available, and, in molecular clouds, gas–grain coupling.

Along with "normal" hydrogenated species of the various families of interstellar compounds, deuterated variants have been found in many instances. The reservoir of deuterium in dense clouds is HD, the fractional abundance of which is about 10^{-5} that of H$_2$ [6]. The relative abundance of singly and multiply deuterated isotopomers of the various hydrogenated species in interstellar clouds is much higher, and this effect, known as fractionation, roots in the equilibrium reaction (4.43), which is slightly exothermic for the formation of H$_2$D$^+$. At temperatures

Table 4.2 Survey of atmospheric windows.

	Wavelength band	Available information
Ultraviolet and visible	300–900 nm (3000–9000 Å)	Electronic transitions (neutral and charged atoms and molecules)
Infrared	1–4 µm (10 000–2 500 cm^{-1})[a] 8–13 µm (1 250–770 cm^{-1})	Vibrational transitions (functional groups in molecules and solids)
Microwave (radar)	1–5 mm (300–60 GHz) 2.5 cm to 3.8 m (12–0.3 GHz)	Rotational transitions Atomic hyperfine splitting[b]

a) Near-infrared (NIR). There are five narrow windows within this band.
b) Such as the 21 cm line of HI regions, coming from the inversion of the spin of the electron relative to that of the proton.

close to 10 K, the forward reaction therefore dominates. The reaction is further driven to the right by continuous depletion of H_2D^+ through successive reactions of this ion with a variety of neutral molecules, reactions by which deuterium becomes transferred and incorporated into newly formed species such as HDO, D_2O, DCN, HDCO, NH_xD_{3-x}, CH_2CCCD, CH_3CCD, and others:

$$H_3^+ + HD \rightleftarrows H_2D^+ + H_2 \tag{4.43}$$

4.2.3
Detection of Basic Interstellar Species

In principle, almost the complete electromagnetic spectrum can be exploited to gather information on the chemical nature, the abundance, and the physical properties (such as temperature and motion) of interstellar species. For Earth-based telescopes, the information is restricted to those bands in the overall spectrum, for which the terrestrial atmosphere is transparent, that is, provides "windows" (Table 4.2). This is the case for the near (long wavelength) UV and visible regions, several windows in the near and thermal IR, and the micro- to meter-range of radio waves. Dust and water droplets in the atmosphere as well as atmospheric turbulences caused by density fluctuations further deteriorate the information. These problems can be bypassed, in part, by placing telescopes onto the summits of high mountains. All of these restrictions are essentially irrelevant, of course, for space-based telescopes, commonly mounted on satellites, and these have opened a much broader range of information, including far (short-wave) UV, X-ray, and γ-ray spectroscopy. Examples of powerful telescopes in this respect are the Spitzer Infrared Telescope (launched in 2003), the Chandra X-Ray Observatory (launched 1999), and the Hubble Space Telescope (for UV–vis and IR; launched in 1990 and reequipped for the last time in 2009).[1]

1) The eponyms for these satellite-based telescopes are astrophysicist Lyman Spitzer, and the astronomers Subrahmanyan Chandrasekar and Edvin Hubble.

Generally, information on the nature of a species is available from emission and/or (mostly) absorption of electromagnetic radiation by the target species. The information encompasses characteristic lines[2] stemming from electronic, vibrational, rotational, and spin transitions (Table 4.2). The synergy between laboratory data and those observed in interstellar space is (after correction for relative motion) a prominent factor for the reliable identification of a specific species, but predictions based on sophisticated theoretical calculations become increasingly important. In this chapter, selected examples will be addressed in some detail. For additional information, see also:

- Figure 3.6 in Section 3.2 (IR of carbon-based molecules)
- Figure 5.5 in Section 5.2.2 (mass spectrum of Mercury's exosphere)
- Figure 5.8 in Section 5.2.3 (IR features of O_2 and H_2O in Venus' atmosphere)
- Figure 5.13 in Section 5.2.4 (IR features of CH_4 and H_2O in Mars' atmosphere)
- Figure 5.15 in Section 5.2.4 (UV + vibrational fine structure of NO on Mars)
- Figure 5.16 in Section 5.3.1 (γ-ray pattern of Eros' surface regolith)
- Figure 5.21 in Section 5.3.3 (energy-dispersive X-ray pattern of the meteorite Acfer 094)
- Figure 5.26 in Section 5.4.2 (IR stretching pattern for CH_n functional groups; IR features of silicate)
- Figure 5.28 in Section 5.6.1 (IR emissions in the upper stratosphere of Jupiter)
- Figure 6.2 in Chapter 6 (optical and IR features of an M8 subdwarf).

4.2.3.1 Hydrogen

There are five hydrogen species present in interstellar clouds: H (HI), H^+ (HII), H^-, H_2, and H_3^+, plus the deuterium analogs and partially deuterated isotopomers. Protons (H^+) are the main constituents of diffuse clouds. They are formed by ionization through photons stemming from near-by or embedded hot stars, with photon energies exceeding the ionization potential for H atoms (13.6 eV), corresponding to a wavelength of 91 nm – hence in the far-UV – or a molar photon energy of 1312 kJ mol^{-1}. For the conversion of wavelengths λ (nm) into energy E (in electron volts or kilo-Joule per mole), see Eq. (4.44):

$$\lambda \text{ (nm)} = 1240/E \text{ (eV)} \tag{4.44}$$

$$E \text{ (kJ mol}^{-1}) = E(\text{eV}) \times 1.602 \times 10^{-22} \times 6.022 \times 10^{23}$$

(1 eV = 1.602×10^{-22} kJ; 6.022×10^{23} mol^{-1} is the Avogadro constant)

Recombination of H^+ and an electron produces an excited hydrogen atom H^*, Eq. (4.45), which moves back to its ground state, emitting several cascades of photons, corresponding to the energy difference between the 1s ground state ($^2\Sigma$) and the various excited electronic levels of the H atom. These cascades are known as Lyman (in the UV), Balmer (visible), Paschen, Bracket, and Pfundt series (the latter

---

[2] The term "line" reflects the classical mode of imaging a spectrum via the *slit* of the spectrometer.

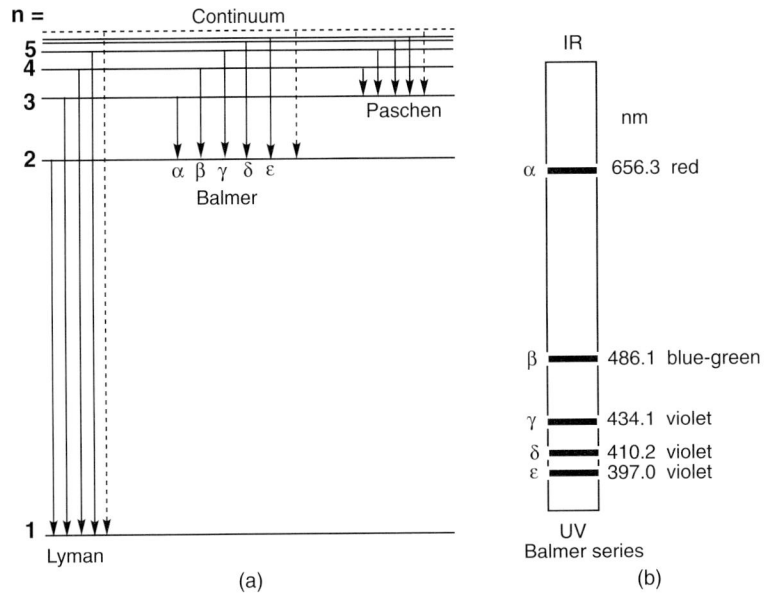

Figure 4.5 (a) Graphical representation of the electronic transitions between excited states and the ground state ($n = 1$, Lyman series) of the hydrogen atom, and electronic transitions ending on the levels $n = 2$ (Balmer series) and $n = 3$ (Paschen series). (b) Projection of the Balmer series into an optical spectrum.

three all in the IR). Three of these series are sketched in Figure 4.5. Of particular interest is the Balmer series, corresponding to electronic transitions that temporarily end on the $n = 2$ level (L shell). The most prominent of these is the Hα transition at 656.3 nm, in the red part of the visible spectrum, and responsible for the reddish glow of diffuse clouds such as the emission nebula IC 434 in the constellation of Orion (Figure 4.2) and emission nebula originating from a supernova remnants (Figure 3.4):

$$H^+ + e^- \rightarrow \{H\}^* \rightarrow H + h\nu \tag{4.45}$$

The negatively charged hydrogen, or hydride ion H^-, formed by radiative association of H atoms with an electron (line 3 in Scheme 4.2) is detected by its absorption pattern in the range 0.75–0.4 eV, corresponding to 1653–3100 nm (6050–3280 cm^{-1}), hence in the near-infrared (NIR). For the detection of neutral hydrogen atoms present in diffuse reflection clouds and dense clouds, microwave spectroscopy (the 21 cm line) has been particularly helpful and allowed to "map" a considerable part of our galaxies with respect to hydrogen occurrence and abundance in so-called HI regions. Under ambient conditions, the single electron of the hydrogen atom is residing in the ground state, characterized by the main quantum number $n = 1$, the orbital quantum number $l = 0$ (or s) and a spin of 1/2. The ground state is thus represented by $^2\Sigma$ (doublet sigma). Spin–orbit coupling gives rise to the total

4 The Interstellar Medium

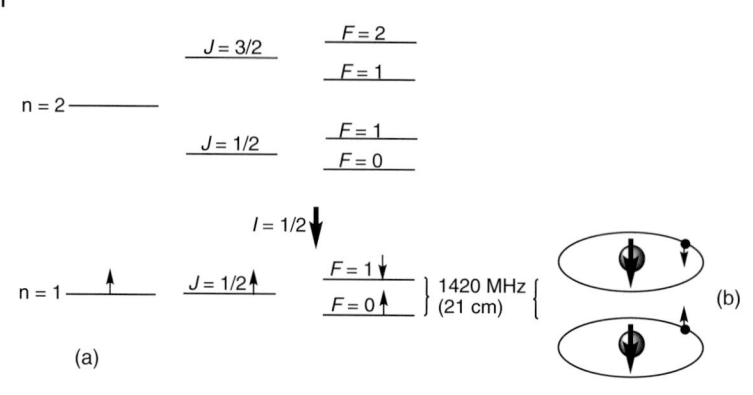

Figure 4.6 (a) Fine structure splitting of the atomic ground level ($n = 1$) and the first excited level ($n = 2$) by spin–orbit coupling ($J = s + l$) of the electron spin (s) and the electron angular momentum (l). The electronic levels, characterized by the J quantum number, are further fine-structured by F coupling ($F = J + I$), that is, coupling between the overall electron momentum J and the nuclear spin I (bold arrow). (b) Visualization of the $F = 0 \rightarrow F = 1$ transition, corresponding to a spin-flip of the electron, with respect to the nuclear spin, from the antiparallel to the parallel orientation.

angular momentum quantum number $J = 1/2^{3)}$ (Figure 4.6). The first excited state is the $n = 2$ state, subdivided into s ($l = 0$) and p ($l = 1$) subshells. If the single electron is in the 2p sublevel, this corresponds to an excited $^2\Pi$ (doublet pi) term that splits, by spin–orbit coupling, into $J = 1/2$ and $J = 3/2$. Coupling of the total electron angular momentum J with the nuclear spin $I = 1/2$ leads to hyperfine splitting F as shown in Figure 4.6. The ground-state transition $F = 0 \rightarrow F = 1$ is the source of the 21 cm line. This transition, depicted by a flip of the spin of the electron from the antiparallel (with respect to the proton spin) to the parallel orientation, is highly forbidden. The transition probability is in fact just $3 \times 10^{-15}\,\text{s}^{-1}$. Given the abundance of interstellar HI, this suffices for its detection. The lifetime of the excited state amounts to 11×10^6 years. Lifetime t of an excited state and energy E are related by the Heisenberg uncertainty principle:

$$\Delta E \times \Delta t \geq h/2\pi \tag{4.46}$$

where h is the Planck constant and ΔE and Δt refer to the uncertainties of the energy and the lifetime, respectively. With a large Δt, the term ΔE becomes particularly small, and the line is thus extremely sharp. In reality, some broadening by various effects occurs, such as Doppler broadening by the range of velocities of the H atoms in the line of sight.

---

3) The letter "J" is used for different designations: (1) In NMR spectroscopy, "J-coupling" indicates the coupling between the nuclear spins of two or more magnetic nuclei in a molecule; (2) J is employed as the "rotational quantum number"; (3) J is used to designate the "total angular momentum quantum number" resulting from the coupling of the orbital momentum and the spin momentum of an electron: $J = S + L$, where $S = \Sigma(s_i)$ and $L = \Sigma(l_i)$.

This situation becomes sufficiently more complex, when *molecules* are considered, even in the case of the simplest perceivable neutral molecule, molecular hydrogen H_2, which even lacks a permanent dipole moment. This molecule, which forms from two H atoms by surface chemistry (Eq. (4.2) and Section 4.2.5.1), is by far the dominant molecular constituent in dense and neutral diffuse clouds; its direct observation is, however, largely restricted to diffuse clouds. Irradiation by far-UV from nearby stars in the sight-line results in electronic excitations, Eq. (4.47), that can be observed in the absorption mode and, in rare cases, in the emission mode as well. Emissions from $\{H_2\}^*$ is actually one of the principle means by which the interstellar gas cools. About 15% of the $\{H_2\}^*$ undergoes radiative decay, as indicated, in parentheses, in Eq. (4.47):[4]

$$H_2 + UV \rightarrow \{H_2\}^*(\rightarrow H + H + h\nu) \tag{4.47}$$

Molecular hydrogen absorbs in the range 110.8 and 91.2 nm (11.2–13.6 eV), the so-called Lyman band ($\lambda < 110.8$ nm) and Werner band ($\lambda < 100.8$ nm). The Lyman band reflects an excitation from the ground state $^1\Sigma_g^+$ into the antibonding σ-type molecular orbital, involving an excited state $^1\Sigma_u^+$; the Werner band corresponds to an excitation into the π-type molecular orbital, $^1\Pi_u$.[5] Excitations into higher electronic states require photons of shorter wavelengths, which are more likely to ionize hydrogen atoms. In addition to electronic transitions, vibrational and rotational transitions involving both the ground and excited states have to be taken into account in the analysis of the absorption patterns of molecules. All three transition types intimately couple, which fact complicates interpretation and assignments of spectral line patterns.

Parts of the overall transition scheme for a diatomic molecule are depicted in Figure 4.7: The part (a) illustrates transitions between vibrational sublevels (characterized by the vibrational quantum number v) of the electronic ground state and an available electronic excited state. At the low temperatures characteristic of neutral diffuse clouds, around 50–100 K, all of the molecules reside in their vibrational ground state ($v = 0$); only transitions arising from this ground state are therefore indicated.[6] Figure 4.7b, an expansion of the vibrational levels characterized by $v = 0$ and 1, shows the rotational (J) fine structure of these vibrational levels. For a molecule without a permanent dipole moment such as H_2, the Q branch ($\Delta J = 0$) is allowed, along with the branches corresponding to $\Delta J = \pm 2$ (O and S branches; not shown in Figure 4.7). For dipolar molecules, for example, CO and NO, the selection rule is $\Delta J = \pm 1$, that is, the P and R branches are allowed.

---

4) The overall sequence denoted by Eq. (4.47) is known as "Solomon process."

5) Σ and Π indicate the symmetry type of the molecular orbital, that is, the angular momentum along the internuclear axis; the left upper index is the total spin quantum number (1 in this case to indicate a singlet state, i.e., the two electron spins are antiparallel); the upper right index indicates the reflection symmetry along a plane containing the internuclear axis (with the "+" for no change in sign on reflection); and the lower right index the parity (gerade [even] or ungerade [uneven] with respect to the center of symmetry).

6) Figure 5.15b, in Section 5.2.4.6 is an illustrative example for the "vibrational tailing" of the electronic transitions of NO in the atmosphere of Mars.

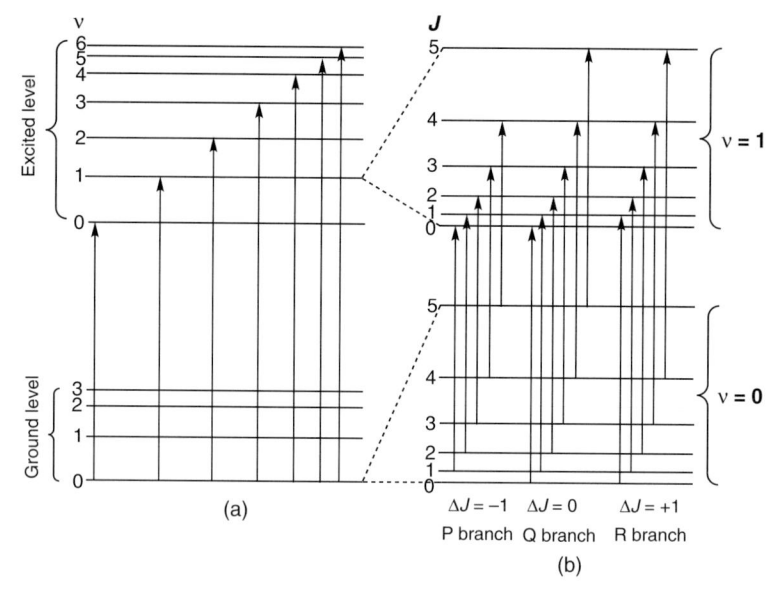

Figure 4.7 Section of the energy level diagram and transitions (vertical arrows) for diatomic molecules. Electronic levels are the ground level and any available excited level; vibrational sublevels are indicated by v, and rotational subsublevels by J. For H_2 (no permanent dipole moment), the rotational Q branch is allowed, for dipolar molecules (e.g., CO), the P and R branches are allowed. See text for additional details.

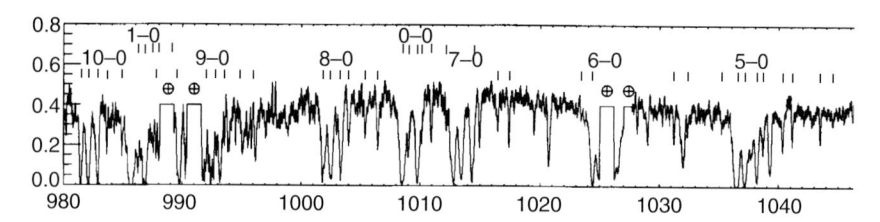

Figure 4.8 Section of the far-UV absorption spectrum of H_2 from an interstellar cloud in the type 1 Seyfert galaxy ESO 141-G55. The wavelengths are given in Å. Upper and lower ticks mark the components of the Lyman and Werner lines, respectively; the notations 1-0, 10-0, etc., indicate transitions between rotational levels J in the vibrational ground state ($v = 0$). The symbol $\oplus$ marks truncated air-glow lines. Reproduced from Ref. [12] by permission of the AAS.

Part of the complex nature of the spectrum of H_2, encompassing a total of ca. 400 absorption lines, is shown in Figure 4.8. The spectrum has been obtained with the Far Ultraviolet Spectroscopic Explorer (FUSE) satellite for a diffuse cloud of 3.3 pc thickness in the galaxy ESO 141-G55 [12].

The simplest stable three-atomic system is the H_3^+ ion. Its dominant formation and destruction processes are represented by Eqs. (4.48)–(4.51). The proton affinity for H_2 (dissociation energy of H_3^+ into $H_2 + H^+$) is 424 kJ mol^{-1}. The cosmic ionization rate for the initiating reaction (4.48) amounts to 2×10^{-16} s^{-1}. H_3^+ acts as a universal proton donor in space, Eq. (4.51), and various examples for this potential have been included in the network charts of the preceding chapter. The equilibrium structure of H_3^+ is an equilateral triangle with three equivalent hydrogen (D_{3h} symmetry) and two paired electrons in the bonding ground state molecular orbital. The only stable electronic excited state is a triplet state (two unpaired electrons) close to the dissociation limit. The lack of a permanent dipole moment excludes detection of rotational transitions in the microwave regime. Detection of H_3^+ therefore relies on the observation of rotational–vibrational transitions in the IR:

$$H_2 + \text{cosmic rays} \rightarrow H_2^+ + e^- \quad (4.48)$$

$$H_2 + H_2^+ \rightarrow H_3^+ + H \quad (4.49)$$

$$H_3^+ + e^- \rightarrow H_2 + H, \text{ or } 3H \quad (4.50)$$

$$H_3^+ + X \rightarrow HX^+ + H_2 \quad (4.51)$$

The interstellar ion H_3^+ is present in the *ortho*-form (o-H_3^+), where all of the nuclear spins are parallel ($I = 3/2$), and the *para*-form (p-H_3^+) with two of the spins paired ($I = 1/2$). p-H_3^+ is more abundant by a factor of ca. 2 than o-H_3^+, contrasting the situation with H_2, where the ratio p-H_2:o-H_2 is 1:3. Spin conversion in H_3^+ is achieved by "scrambling" as shown in Eq. (4.52):

$$H_3^+ + H_2 \rightarrow \{H_5^+\}^* \rightarrow H_2 + H_3^+ \quad (4.52)$$

In Figure 4.9a, the vibrational modes for the ion H_3^+ are illustrated. The v_1 mode of symmetry A_1' is IR-forbidden because there is no change in dipole moment

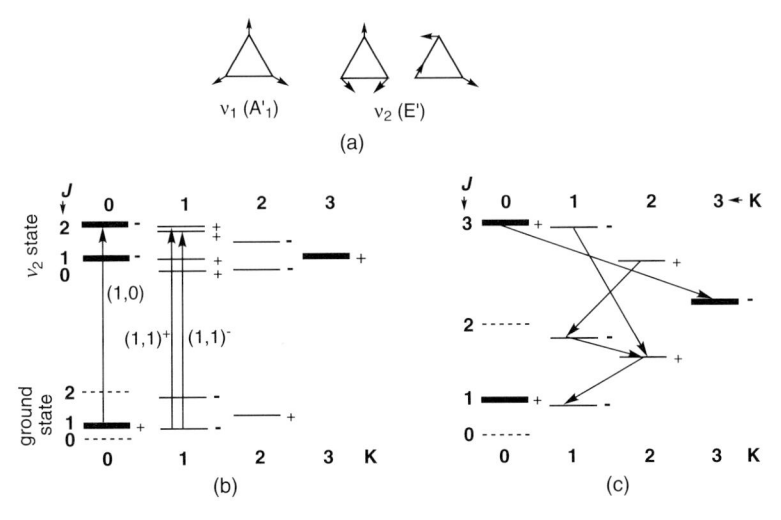

Figure 4.9 Vibrational modes (a), energy level diagram (b), and the rotational levels in the vibrational ground state (c) of the H_3^+ ion, adapted from Refs. [5a, 13]. J is the rotational quantum number, and K its projection on the molecular symmetry axis. The + and − signs indicate parity. Bold lines refer to *ortho*-H_3^+, thin lines to *para*-H_3^+, and dashed lines to forbidden levels. Arrows indicate selected transitions. For the (1,0) and (1,1)$^+$ transitions, see the spectrum in Figure 4.10.

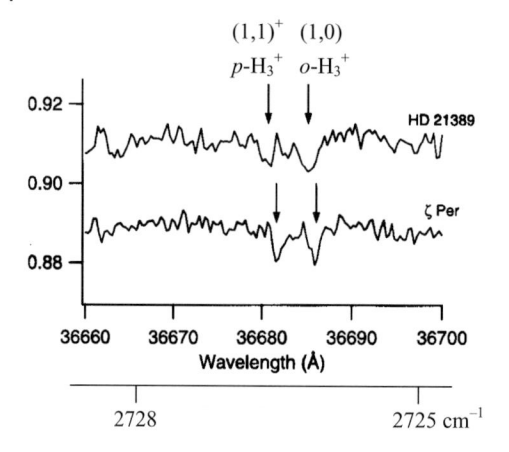

Figure 4.10 The infrared H_3^+ doublet in interstellar diffuse clouds in the sight-lines of the stars HD 21389 and ζ Persei. For assignment of the transitions, see Figure 4.9. Reproduced from Ref. [14] by permission of the AAS.

(this mode is, however, Raman-allowed). The doubly degenerate v_2 of symmetry E' is IR-allowed. In Figure 4.9b, relevant sections of the energy level diagram with the vibrational and rotational (J) substructures are shown, and transitions for both o-H_3^+ and p-H_3^+ are indicated [5, 13]. Figure 4.10 shows the typical doublet for o-H_3^+/p-H_3^+ in interstellar clouds in the line of sight of the stars ζ Persei (a B1 type star) and HD 21389 (an A0 star in the constellation of Camelopardalis).

By evaluating details of the spectroscopic observations of interstellar species, column densities N can be obtained, where "column" relates to the overall length of the line of sight reaching from the observer to the star behind the cloud, the light of which is analyzed in the absorption pattern created by the species present in the cloud. Column densities are commonly reported as the number of particles per square centimeter. More convenient, from a chemical point of view, is the unit mol m^{-2}. For a discussion, see Introduction and Section 5.2.3.2. Selected data for the column densities of atomic and molecular hydrogen, N(HI), N(H$_2$), and N(H_3^+), are collated in Table 4.3 together with the fraction f(H$_2$) of molecular hydrogen as defined by Eq. (4.53):

$$f(H_2) = 2N(H_2)/[N(HI) + 2N(H_2)] \tag{4.53}$$

4.2.3.2 Other Basic Molecules

Diatomic molecules such as the homonuclear dicarbon C_2 (bond order = 2) and dinitrogen N≡N, the heteronuclear molecule carbon monoxide C≡O, and the heteronuclear molecular radicals methylyne ·C–H, cyan ·C≡N, and hydroxyl ·O–H can principally be treated analogously to dihydrogen, with the exception of course, that more electrons are available and – in case of heteronuclear species – the stretching vibration is accompanied by a change in dipole moment and thus is IR-allowed. The radical CH (ground rotational state $^2\Pi_{1/2}$; see Figure 4.11a) was

Table 4.3 Column densities N and molecular hydrogen fractions $f(H_2)$ of selected interstellar clouds.

Star in the sight-line (designation[a])	Location of interstellar cloud	$N(H_2)$ cm^{-2} (mol m^{-2})	$N(HI)$ cm^{-2} (mol m^{-2})	$f(H_2)$	$N(H_3^+)$ cm^{-2} (mol m^{-2})	Reference
SK -67 111 (O7, LMC)	Galactic	2.3×10^{15} (3.8×10^{-5})	6.2×10^{20} (10)	7.4×10^{-6}		[12]
AV 232 (O9, SMC)	Galactic	1.6×10^{16} (2.7×10^{-4})	3.6×10^{20} (6)	8.9×10^{-5}		[12]
AV 232 (O9, SMC)	SMC	2.0×10^{15} (3.3×10^{-5})	–	–		[12]
ESO 141-G55 (Seyfert 1)[b]	Galactic halo	1.9×10^{19} (0.32)	3.5×10^{20} (5.8)	9.3×10^{-3}		[12]
G 191-B2B (hot white dwarf, Capella)	Galactic		$1.4–2.0 \times 10^{18}$[c] $(2.3–3.3 \times 10^{-2})$			[15]
HD 72127B (B2, Vela)	Galactic		1.6×10^{20} (2.7)			[16]
ζ Persei (B1)	Galactic	1.2×10^{21} (20)	1.6×10^{21} (27)	0.6	7×10^{13} (1.2×10^{-6})	[14]
Cyg OB2-12 (B8, Cygnus)	Galactic cloud[d]				3.8×10^{14} (6.3×10^{-6})	

a) The first entry in parentheses refers to the stellar classification (Table 3.1 in Section 3.1.1); the second to its location. LMC = Large Magellan Cloud, SMC = Small Magellan Cloud (the Magellan clouds are companion galaxies of the Milky Way).
b) Seyfert galaxies are particularly luminous light emitters; they are believed to contain supermassive black holes.
c) The D/H ratio in this interstellar material is 1.76×10^{-5}.
d) Dominated by diffuse material, but containing nested cloudlets of higher density material as well (see Ref. [17] in the following chapter).

one of the first molecules to be detected in the interstellar medium by its microwave transition at 9 cm (3.3 GHz) arising from the $J = 1/2$ lambda doublet [18a], and later by the $J = 3/2$ lambda doublet at 4.3 cm (7.3 GHz) [18b] (Figure 4.11b). *Lambda-doubling*, typical for radicals, comes about by the interaction between the molecular rotation and the electronic spin–orbit motion. Commonly, the resulting energy levels are degenerate. For a molecule with an unpaired electron, however, electronic–rotational interaction removes the degeneracy, and the energy level is split into two components.

The respective microwave transition for OH appears at 18 cm (1.7 GHz) for the ground rotational state $^2\Pi_{3/2}$ ($J = 3/2$). This radical can also be detected by its electronic transitions in the UV, two of which, in the near-UV at 307.84 and 308.27 nm, are accessible to ground-based instruments [19]. Spectra for two objects with different $B–V$ color indices are shown in Figure 4.12.

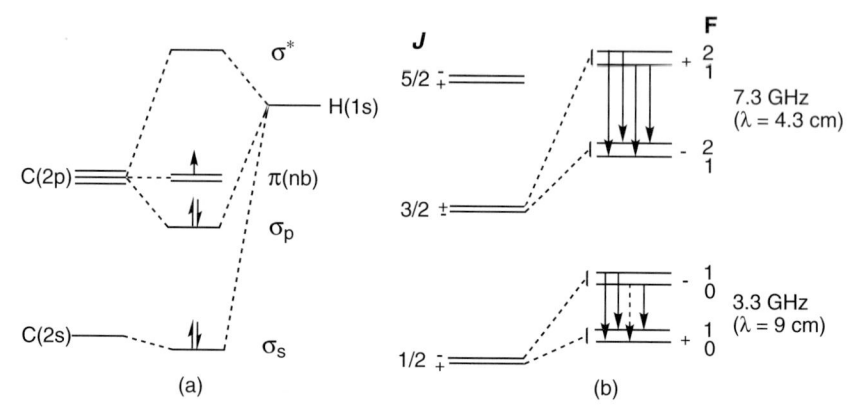

Figure 4.11 (a) Molecular orbital scheme of the CH radical; ground state $^2\Pi_{1/2}$. For OH, there are two additional electrons in the nonbonding (nb) π level. (b) Energy levels (redrawn from Ref. [18b]) of the lower rotational levels J of the $^2\Pi_{1/2}$ state with lambda-doubling and hyperfine structure (total molecular angular momentum quantum number F [including the nuclear spin I]). The signs + and − indicate parities. Transitions are shown by arrows; the (0,0) transition indicated by the broken arrow is forbidden. The energy separations in the diagrams are not to scale.

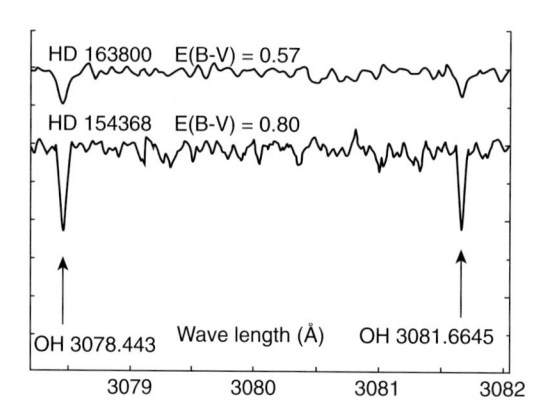

Figure 4.12 Two of the electronic near-UV bands of the hydroxyl radical against two OB stars with differing B–V color indices (the lower B–V index corresponds to the hotter star; Section 3.1.1). Credit: T. Weselak et al., A&A 499 (2009) 783 [19]; reproduced with permission, © ESO.

A molecule of particular interest in the interstellar medium is CO, the most abundant molecule next to H_2, and occasionally employed as a tracer for the abundance of the less easily detected H_2: In diffuse to translucent interstellar clouds, column densities of CO vary between 10^{12} and 10^{16} cm^{-2}, with CO/H_2 ratios between 10^{-7} and 10^{-5}. In dark clouds, this ratio amounts to 10^{-4}. Roughly, the column densities N of these two molecules are related by a power law, Eq. (4.54) [20]:

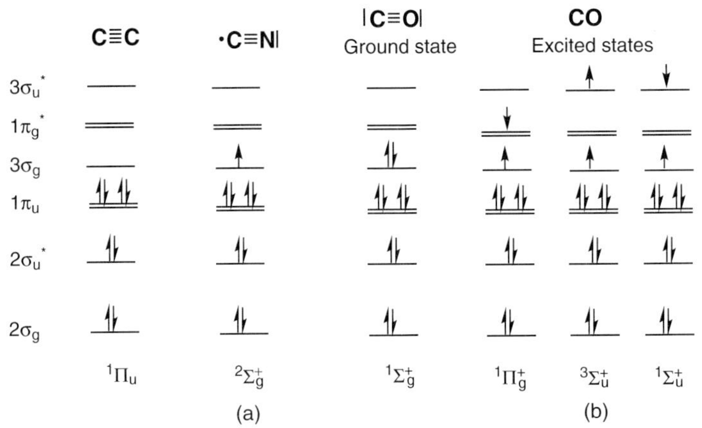

Figure 4.13 (a) Ground state electronic energy levels for C_2, CN, and CO, and (b) three excited states for CO. Energy positions and separations are not to scale. The respective diagram for N_2 compares to that of CO. Note that in C_2 and CN, one of the bonds is antibonding; the bond orders thus are 2 (for C_2) and 2.5 (for the CN radical).

$$N(CO) \propto N(H_2)^\alpha \quad \alpha \approx 2 \tag{4.54}$$

CO is detected by its rotational transitions in the millimeter band (2.6 and 1.3 mm, equivalent to 115 and 231 GHz), its IR characteristics [21], in particular the fundamental CO stretch at 2155 cm^{-1} (4.64 µm) corresponding to the $v = 1 \leftarrow 0$ mode of the R-branch (Figure 4.7), and optical transitions in the UV at $\lambda < 151.0$ nm, reflecting the $^1\Pi - ^1\Sigma^+$ and $^3\Sigma^+ - ^1\Sigma^+$ transitions [20], where $^1\Sigma^+$ is the electronic ground state (in its vibronic ground level, $v = 0$) (cf. Figure 4.13). In Figure 4.14, IR features of CO as obtained in an early [21a] and a more recent [22] investigations are shown. The $v = 1 \leftarrow 0$ band of the vibrational spectrum of ^{12}CO also displays signals of low intensity that are due to ^{13}CO, allowing for the evaluation of the ^{12}C/^{13}C ratio. In the Solar System, this ratio is 89, reflecting the situation 4.6 billion years ago. The mean interstellar ratio is 70 [23], and this lower value is due to supply of ^{13}C to the interstellar medium by red giants, provided by nucleosynthesis in, for example, the Bethe-Weizsäcker cycle (Scheme 3.2 in Section 3.1.3). In circumstellar envelopes of postasymptotic giant stars (and planetary nebulae, in particular), ^{12}C/^{13}C ratios can go down to 30 and even less. There are two main processes by which lower ratios (depletion of ^{12}C) and higher ratios (enrichment of ^{12}C) can be achieved [23]: Lower ratios can be caused by charge exchange, Eq. (4.55), an isotope effect that arises from the fact that the heavier isotopomer has a lower zero point energy; this effect is known as "kinetic isotope effect" in general chemistry. Higher ratios are caused by selective photodissociation; the UV-induced dissociation is more effective for ^{13}CO than for ^{12}CO due to self-shielding by the more abundant ^{12}CO. For the concept of "self-shielding," cf. Section 5.3.3:

$$^{13}C^+ + {}^{12}CO \rightarrow {}^{12}C^+ + {}^{13}CO \tag{4.55}$$

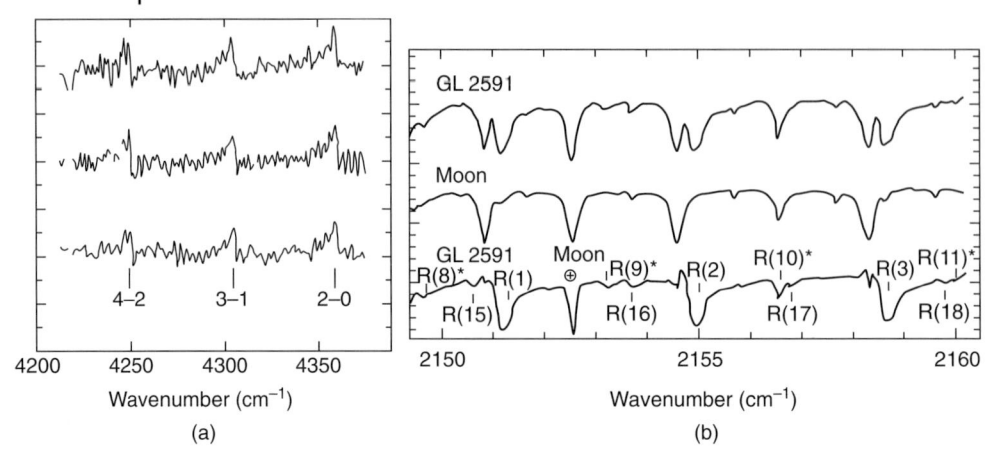

Figure 4.14 Examples of infrared spectra of carbon monoxide. (a) The spectrum obtained for the Becklin-Neugebauer (BN) object, shows three components of the first overtone. The BN object in the Orion nebula is an intensive source of exclusively IR radiation, associated with a dust cloud surrounding newly formed stars. Reproduced from Ref. [21a] by permission of the AAS. (b) The spectrum shows part of the rotational fine structure of the stretching vibration of CO at 2155 cm^{-1} (4.64 μm) toward GL 2591, a massive protostar embedded in a molecular cloud in the Cygnus X region. Spectra are shown without (top) and with correction (bottom) by standard data for Terrestrial features (center). Indicated above and below the bottom trace are the ^{12}CO and ^{13}CO lines, respectively, in the $v = 1 \leftarrow 0$ band. Asterisks mark $v = 2 \leftarrow 1$ lines. Reproduced from Ref. [22] by permission by the AAS.

The simplest multicarbon molecules, C_2 and C_3, can provide information on the pathway of formation of carbon chains and PAHs in interstellar clouds. The electronic ground state of C_2 is a $^1\Pi$ state (Figure 4.13a). Tricarbon, ground state $^1\Sigma_g^+$, is linear, with double bonds between the carbons and a lone pair on each of the two flanking carbon atoms, and thus formally derives from the allene propadiene $H_2C=C=CH_2$ by abstraction of four hydrogen atoms. The tricarbon cation C_3^+ is bent. The radical ·C≡N (ground state $^2\Sigma^+$) can be an important progenitor for the formation of cyanopolyynes. For its interstellar formation, see Scheme 4.4.

Dicarbon C_2 is detected by the vibrational (2,0) and (3,0) bands at 876.5 and 772 nm, respectively, in the electronic $^1\Pi_u - ^1\Sigma_g^+$ transition, where $^1\Pi_u$ is the ground state and $^1\Sigma_g^+$ the first excited electronic state. Part of the rotational fine structure for the (2,0) band is shown in Figure 4.15 [24]. The spectrum has been obtained in the sight-line of the star Cyg OB2-12, a class B8 star in the constellation of Cygnus, embedded in a cloud of nested structure, including cloudlets with a particle density of ca. 7000 cm^{-3} and a temperature of 35 K, conditions that are suitable for the formation of C_2. The corresponding vibrational transitions for CN, near 919 and 790 nm, are the (1,0) and (2,0) bands in the electronic $^2\Pi_u - ^2\Sigma^+$ system. In addition to the $^2\Pi_u - ^2\Sigma^+$ system (where the excited electron is in the $1\pi_g^*$ molecular

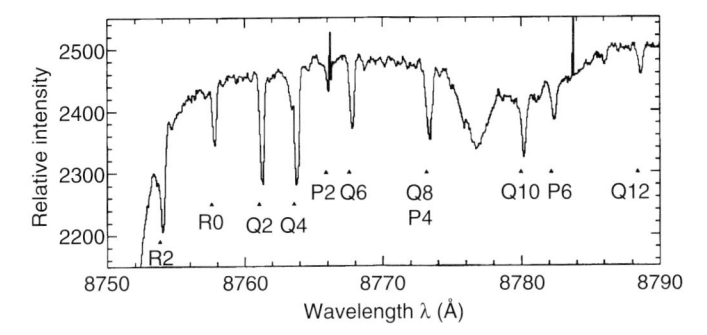

Figure 4.15 Section of the optical absorption spectrum of dicarbon C_2 in the line of sight of Cyg OB2-12, showing the rotational fine structure (the R, P, and Q lines) of the vibrational (2,0) band in the vicinity of 876.5 nm, that is, the electronic $^1\Pi_u - ^1\Sigma_g^+$ system. Cf. Figure 4.7 for details of the energy-level diagram. The strong band at the left-hand margin belongs to the HI Paschen series (Figure 4.5), and the broad line near 877.7 nm to HeI. Reproduced from Ref. [24] with permission from Astronomy & Astrophysics.

orbital; cf. Figure 4.13), the interstellar cyan molecule reveals itself by the (0,0) and (1,0) vibrational bands of the $^2\Sigma^+ - ^2\Sigma^+$ system (excitation into the $3\sigma_g$ level). These two bands are near 387 (0,0) and 358 nm (1,0) [17]. For the band at 387 nm, three rotational bands appear in the spectrum (Figure 4.16).

The smallest oligoatomic carbon chain, tricarbon C_3, has been detected through the rotational lines in the $^1\Pi_u - ^1\Sigma_g^+$ transition near 405 nm in diffuse clouds against stars such as ζ Ophiuchi [25]. A spectrum is shown in Figure 4.17. C_3 is presumed to form by dissociative recombination of C_3H^+, which in turn originates from C^+ insertion into C_2H; Eq. (4.56). A prominent disruption pathway is represented by Eq. (4.57):

$$C_2H + C^+ \rightarrow C_3H^+ + h\nu;\ C_3H^+ + e^- \rightarrow C_3 + H \tag{4.56}$$

$$C_3 + h\nu \rightarrow C_2 + C\ (h\nu < 165.3\ \text{nm}\ (7.5\ \text{eV})) \tag{4.57}$$

As in the case of H_2 and C_2, the homonuclear dinitrogen N_2 has no allowed rotational and vibrational dipole transitions and is thus difficult to detect in the interstellar space. The only viable approach to find interstellar N_2 is by its spectral pattern created by electronic transitions, which lie in the far-UV (shorter than 100 nm) unavailable to Earth-bound telescopes, because short-wave UV is absorbed by the atmosphere (Table 4.2), and also to most space telescopes that cut off at wavelengths well above 100 nm. N_2 was thus the last simple molecule to be traced in interstellar clouds. Detection was finally achieved by locating the J = (0,0) band of the $^1\Sigma_u^+ - ^1\Sigma_g^+$ transition at 95.86 nm [7]. These data were obtained with the FUSE equipment toward the O6 star HD 124314 in the constellation of Centaurus at the satellite's altitude of 750 km, where the telluric density of N_2 does not significantly contribute to the spectral pattern.

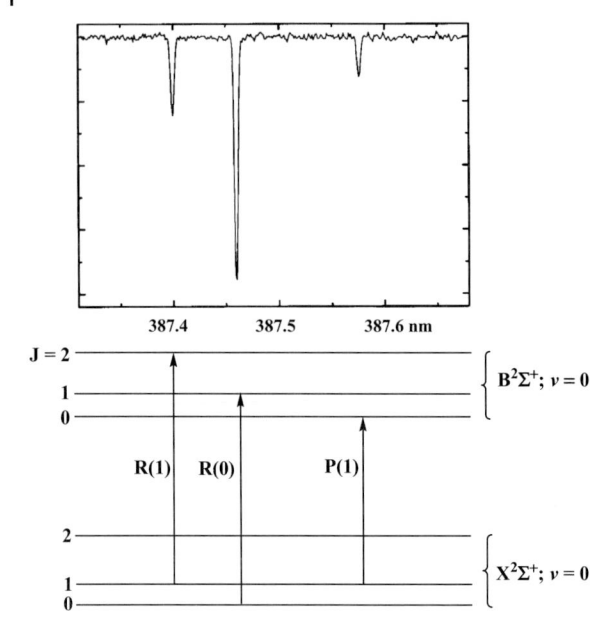

Figure 4.16 Spectrum of the cyan radical in the $v = (0,0)$ band (near 387 nm) of the electronic $^2\Sigma^+ - ^2\Sigma^+$ system, showing the rotational (J) transitions (1,2) and (0,1) of the R-band and (1,0) of the P-band, as indicated in the term diagram (cf. also Figure 4.7). The spectrum has been taken in the sight-line of the star HD 147889 (spectral class B2) behind the outer envelope of the ρ Ophiuchi molecular cloud. Adapted from Ref. [17] Figure 1 with permission, © Mon. Not. Astron. Soc.

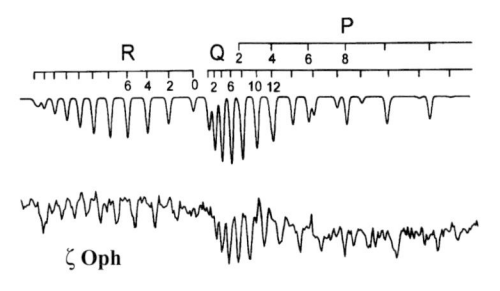

Figure 4.17 Rotational lines of the tricarbon $^1\Pi_u - ^1\Sigma_g^+$ system toward ζ Ophiuchi. The upper spectrum is a simulation. Reproduced from Ref. [25] by permission of the AAS. For the R, Q, and P branches, see Figure 4.7.

4.2.4
Complex Molecules

Several of the more complex molecules have already been briefly addressed in Section 4.2.2: Formaldehyde (H_2CO, Scheme 4.3 and Eq. (4.40)), hydrogen cyanide (HCN, Scheme 4.4 and Eq. (4.41b)), ammonia (NH_3, Scheme 4.4 and Eq. (4.41a)),

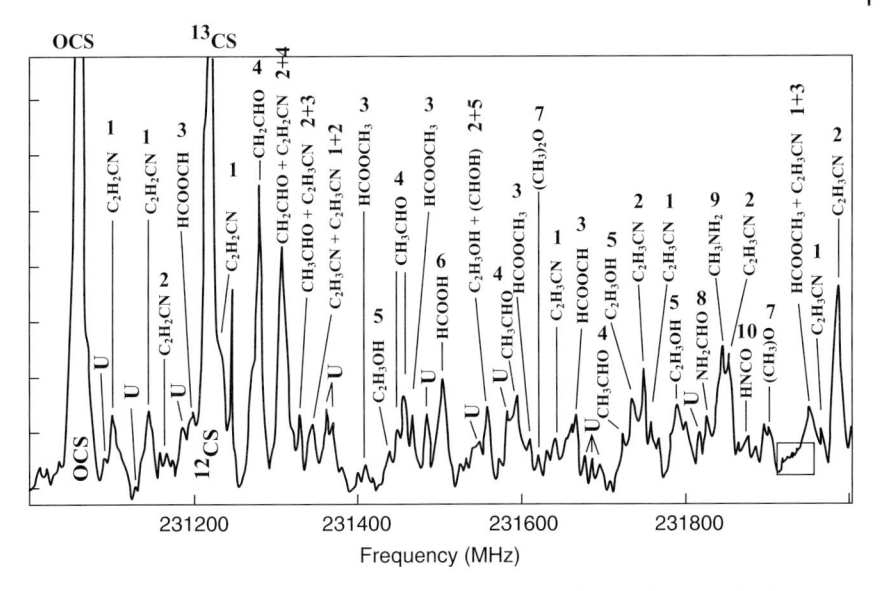

Figure 4.18 Microwave spectrum (1.295 mm band) obtained toward SgrB2(N) in the Large Molecular Heimat. Unidentified features are marked by a lateral "U." Identified molecules are vinylcyanide C_2H_3CN (1), ethylcyanide C_2H_5CN (2), methylformate $HCOOCH_3$ (3), acetaldehyde CH_3CHO (4), ethanol C_2H_5OH (5), formic acid HCO_2H (6), dimethylether $(CH_3)_2O$ (7), formamide NH_2CHO (8), methylamine CH_3NH_2 (9), and isocyanic acid HNCO (10), along with OCS and ^{13}CS. The box on the lower right marks the noise level. Adapted from Ref. [26] with permission by Lucy Ziurys, University of Arizona, Tucson, USA.

methylamine (CH_3NH_2, Scheme 4.4), and acetonitrile (CH_3CN, Scheme 4.4), plus water (H_2O) of course, can be considered basic molecules of life. With the complexity of the molecule, the complexity of the spectra increases, and commonly rotational transitions in the millimeter ($\approx 100\,GHz$) regime are observable for the detection of specific species. Figure 4.18 provides an example.

For the observation of ammonia, its inversion transition can be exploited. Ammonia has trigonal-bipyramidal geometry (C_{3v} symmetry) as shown in Figure 4.19. "Inversion" refers to a flip–flop of the nitrogen in the C_3 axis between equivalent positions above and below the plane spanned by the three hydrogens, hence a reflection at this plane. The nonclassical inversion barrier (allowing for tunneling) is 12.6 kJ mol^{-1}, which goes along with a radio absorption or emission of 1.25 cm (23.7 GHz). In Figure 4.19, the profiles for the $J,K = (1,1)$ transition for two interstellar clouds representing differing conditions for excitations are shown. Column densities of NH_3 are typically in the range of 10^{12}–$10^{13}\,cm^{-2}$ ($\approx 10^{-7}\,mol\,m^{-2}$).

Larger molecules, and in particular those that are saturated or partly saturated (meaning that all or most of the triple and double bonds in the carbon skeleton are annealed by the attachment of hydrogen), cannot efficiently form by cool gas ion–neutral reactions, excluding endothermic reactions and those exothermic

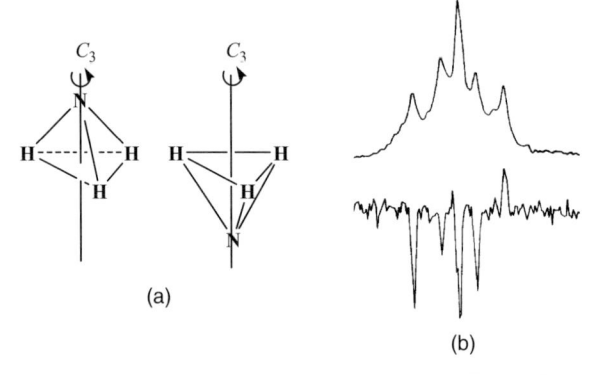

Figure 4.19 (a) Figurative demonstration of the flip-flop (inversion) of ammonia. The C_3 rotation axis is indicated. (b) Two profiles for the $J,K = (1,1)$ inversion transition of ammonia. The top feature is an emission profile from the Orion KL nebula, and the bottom spectrum is an *absorption* profile from the Orion source DR21. © Ref. [50] Figure 5 top and center (Phys. Bl.).

reactions that require activation energy. Examples for complex molecules found in the Sagittarius "Large Molecular Heimat" source,[7] SgrB2(N-LMH) or SgrB2(N) for short, a compact and massive "hot" ($T > 100$ K) molecular core and star forming region at a distance of only ≈100 pc from the Galactic center, are collated in Scheme 4.7. The cloud extends over ca. 1 pc (3 light-years), its "hot core" over ca. 0.1 pc (0.3 light-years), and is hence less extended than the Oort cloud at the outer fringe of our Solar System at approximately 1 light-year.

Most of the organic molecules have been found in the last few years [27–30]. It is likely that many (but not all; vide infra) of the partly saturated molecules shown in Scheme 4.7 are formed from their constituent functional groups on the surface of dust grains present in the hot molecular core (see Section 4.2.5). A prominent representative of these complex molecules is acetic acid [28], which may be considered a precursor compound for the simplest amino acid, glycine $H_2N-CH_2CO_2H$, derived from acetic acid by amination. An even more direct precursor for glycine is the recently discovered amino-acetonitrile H_2NCH_2CN [29] (illustrated on the book cover), which can transform chemically to glycine by hydrolysis, Eq. (4.58), with the water potentially being provided by the icy shells of dust grains. Laboratory simulations have shown that glycine (and alanine and serine) form on UV-photolysis of a mixture of H_2O, CH_3OH, HCN, and NH_3 ices, possibly via amino acetonitrile. Both, vinyl- and ethylcyanide (Scheme 4.7) share the three-carbon backbone of the next to simple amino acid, α-alanine $CH_3CH(NH_2)CO_2H$. Ethylcyanide converts, in principle, to α-alanine by the hydrolysis of the nitrile group (yielding the underlying propionic acid and ammonia) and subsequent or concomitant amination at the Cα (the C atom next to the carboxylic acid

---

7) "Heimat" is German for "home."

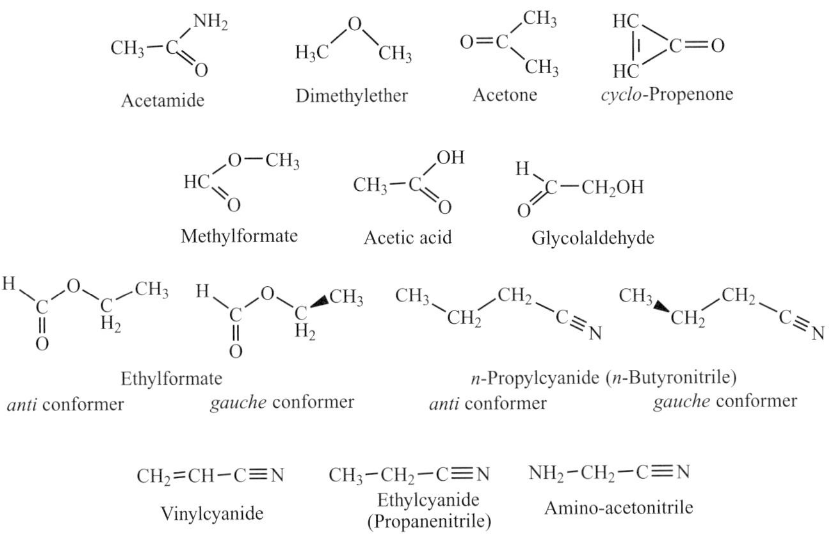

Scheme 4.7 Saturated and partially saturated large interstellar molecules detected in the Sgr B2(N-LNH) "hot" molecular core [27–30] (glycolaldehyde is not restricted to the core area). The *gauche* forms of ethylformate and propylcyanide are slightly higher in energy (by ≈1 kJ mol⁻¹) than the *anti* conformers. Note that methylformate, acetic acid, and glycolaldehyde are isomers (empirical formula $C_2H_4O_2$), with methylformate about 18 time more abundant than acetic acid.

function). No amino acid has so far been detected in the interstellar space; amino acids are present, however, in cometary comas and meteorites (Scheme 5.3 in Section 5.3.2), fragments of asteroids that may have absorbed and thus stabilized these amino acids from the protoplanetary nebula, or picked up and processed amino acid precursor compounds in the course of their evolution and development. Glycolaldehyde, finally, may be considered a "sugar" molecule (liberal use of the term "sugar" provided) since it contains both of the two functional groups that make up an aldose sugar: the aldehyde and the alcoholic functions:

$$H_2N-CH_2-C\equiv N + 2H_2O \rightarrow H_2N-CH_2-CO_2H + NH_3 \quad (4.58)$$

The particularly stable molecule *cyclo*propenone has been proposed to be generated by the reaction between *cyclo*propylidene c-C_3H_2 and oxygen atoms [30], Eq. (4.59), a reaction that requires activation because of the different spin states of the reactants: oxygen is in a triplet (two unpaired electrons) and c-C_3H_2 in a singlet (electron paired) ground state. The unsaturated carbene c-C_3H_2 by itself is very reactive under laboratory conditions, but an abundant species in the interstellar space, where it can form in several successive steps (see the reaction sequence (4.60)): In a first step, c-C_3H^+ reacts with molecular hydrogen, by radiative association, to form linear l-$C_3H_3^+$. This is followed by isomerization of l-$C_3H_3^+$ to c-$C_3H_3^+$. The final step is an electron capture reaction. The starting product in this reaction sequence, C_3H^+, attains a central role, and there are several paths for its generation, represented by, for example, Eq. (4.56) and Eqs. (4.61)–(4.63). For the precursor species C_2H_2 and HC_3N in Eqs. (4.61) and (4.63), respectively, see also Scheme 4.4:

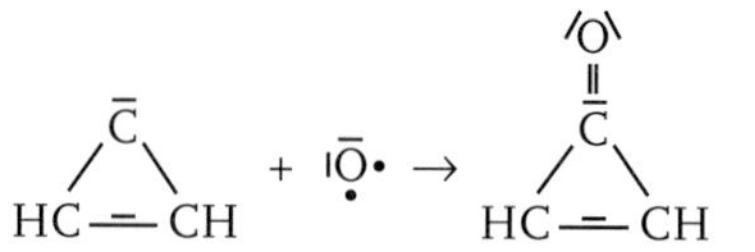

$$\begin{array}{c}\overline{C}\\ /\ \backslash\\ HC = CH\end{array} + |\overline{\underline{O}}\bullet \rightarrow \begin{array}{c}{}^{/O\backslash}\\ \overset{\|}{C}\\ /\ \backslash\\ HC = CH\end{array} \quad (4.59)$$

$$c\text{-}C_3H^+ + H_3 \underset{h\nu}{\overset{}{\rightleftarrows}} l\text{-}C_3H_3^+ \rightleftharpoons c\text{-}C_3H_3^+ \xrightarrow{e^-} c\text{-}C_3H_2 + H \quad (4.60)$$

$$C^+ + C_2H_2 \rightarrow C_3H^+ + H \quad (4.61)$$

$$C_3^+ + H_2 \rightarrow C_3H^+ + H \quad (4.62)$$

$$C^+ + HC_3N \rightarrow C_3H^+ + CN \quad (4.63)$$

Despite of the discovery of the hydride ion H^- (formed from H_2 by heterolytic dissociation through cosmic rays or radiative electron attachment by H), more than 70 years ago, as the main species liable for the optical opacity of the Solar atmosphere, and despite of the prediction, about 30 years ago, that larger negatively charged molecular ions should be around in interstellar clouds, the first of these complex anions, HC_{2n}^- ($n = 2, 3,$ and 4) [31–34] and C_nN^- ($n = 3$ [35] and 5), have only been detected in space recently. These species, the structural and electronic formulae of which are shown in Scheme 4.8 together with selected neutral precursors, are present both in the dark molecular cloud TMC-1 in the constellation of Taurus and in the envelope of a carbon-rich AGB star, IRC+10216, in the constellation of Leo. The fractional abundance ratios of these anions amount to a few percent of their neutral, radical counterparts HC_{2n} and NC_3.

Two mechanisms of formation are discussed: *dissociative* electron attachment, exemplified for NC_3^- in Eqs. (4.64a) and (4.64b), and *radiative* electron attachment, exemplified for the same anion in Eq. (4.65). The precursor molecule in Eq. (4.64a) is HNC_3 (with H attached to N), a metastable isomer of cyanoacetylene HC_3N (with H attached to C), the precursor molecule in Eq. (4.64b) (see Scheme 4.8 for structural formulae). Dissociative attachment requires bond braking, and

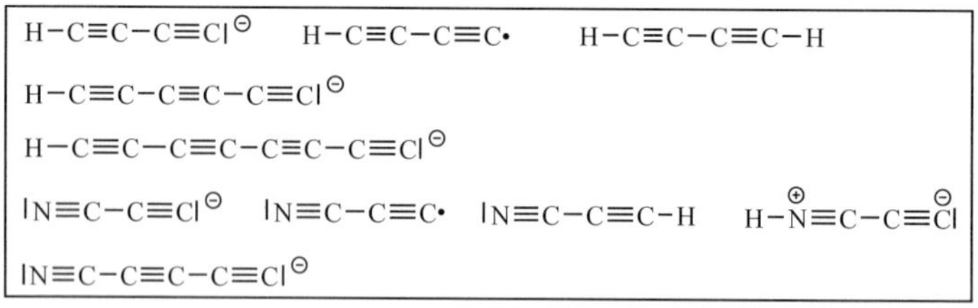

$H-C\equiv C-C\equiv C|^\ominus$ $H-C\equiv C-C\equiv C\bullet$ $H-C\equiv C-C\equiv C-H$

$H-C\equiv C-C\equiv C-C\equiv C|^\ominus$

$H-C\equiv C-C\equiv C-C\equiv C-C\equiv C|^\ominus$

$|N\equiv C-C\equiv C|^\ominus$ $|N\equiv C-C\equiv C\bullet$ $|N\equiv C-C\equiv C-H$ $H-\overset{\oplus}{N}\equiv C-C\equiv\overset{\ominus}{C}|$

$|N\equiv C-C\equiv C-C\equiv C|^\ominus$

Scheme 4.8 Molecular anions and precursor compounds (for HC_3^- and NC_3^-) found in interstellar space.

the energy required to break a C–H bond is usually not balanced by the electron affinity of the molecule, that is, the overall energy balance is such that the reaction is endothermic. In the case of Eq. (4.64b), the overall energy of formation is +126 kJ mol^{-1},[8] rendering formation of C_3N^- along this reaction path unlikely. Hence, Eqs. (4.64a) (which is exothermic) and (4.65) represent the more likely scenario for the generation of NC_3^-. Likewise, the isoelectronic anion HC_4^- is supposed to form by radiative attachment, Eq. (4.66), rather than by dissociative attachment, Eq. (4.67):

$$HNC_3 + e^- \rightarrow NC_3^- + H \qquad (4.64a)$$

$$HC_3N + e^- \rightarrow NC_3^- + H \qquad (4.64b)$$

$$NC_3 + e^- \rightarrow NC_3^- + h\nu \qquad (4.65)$$

$$HC_4 + e^- \rightarrow HC_4^- + h\nu \qquad (4.66)$$

$$H_2C_4 + e^- \rightarrow HC_4^- + H \qquad (4.67)$$

As the molecules grow larger, dissociative attachment becomes energetically feasible. Size confers stability, and the cross section for electron capture increases with size. Further, increasing size is accompanied by an increase of the dipole moment, allowing for higher electron affinities. Thus, the energy of formation for HC_6^- from H_2C_6, Eq. (4.68), is -16 kJ mol^{-1}; the formation energy for HC_8^- from H_2C_8, Eq. (4.69), is -47 kJ mol^{-1}. Equation (4.70) provides a reaction sequence to H_2C_8 and HC_8, the precursor compounds for the formation of the anion HC_8^- by dissociative or radiative electron attachment, respectively [33]. The dipole moments of the precursor molecule H_2C_6 is 5.6 D (8.2 D for HC_6^-). The dipole moment of H_2C_4, 0.9 D, appears to be too small to allow the formation of HC_4^- by radiative dissociation. There is, however, a low-lying excited electronic state (with a sufficiently high dipole moment) available in H_2C_4 that can mix with the ground state. Hence, Eq. (4.67) may become exergonic and thus provides an alternative path toward HC_4^-. The rate coefficients for the formation of the anions HC_{2n}^- by radiative electron attachment have been calculated: At 300 K, $k = 1.1 \times 10^{-8}$ for $n = 2$, and 6.2×10^{-8} cm^3 s^{-1} for $n = 3$ and 4 [31]. Rate constants of this order of magnitude provide an at least qualitative clue for the observed rather high column densities of 10^{11} (in TMC-1) and 4×10^{12} cm^{-2} (in IRC+10216) for HC_6^- [34]:

$$H_2C_6 + e^- \rightarrow HC_6^- + H \qquad (4.68)$$

$$H_2C_8 + e^- \rightarrow HC_8^- + H \qquad (4.69)$$

$$H_2C_6 + HC_2 \underset{H}{\rightarrow} H_2C_8 \underset{H}{\overset{h\nu}{\rightarrow}} HC_8 \overset{e^-}{\rightarrow} HC_8^- + h\nu \qquad (4.70)$$

---

8) By convention, the energy consumed in an endothermic (endergonic) reaction is quoted positive (+), and the energy liberated in an exothermic (exergonic) reaction negative (−). 1 kJ mol^{-1} equals 1.036×10^{-2} eV.

4.2.5
Chemistry on Grains

A substantial part of matter in interstellar clouds consists of submicrometer-sized dust particles, or grains. In molecular clouds, the grain fraction amounts to ca. 1%; the presence of grains is, however, not restricted to dense clouds. Grains are provided by stellar winds and ejecta, with AGB stars as the main factories. But grains are also subject to destruction by blast-waves driven by supernovae. Grain processing through growth, erosion, sputtering, and coagulation, resulting in inhomogeneity and porosity, has also to be taken into account [36]. Lifetimes for grains are in the order of magnitude of 10^8 years [37]; In order to account for the grain density found in interstellar clouds, it is therefore necessary that grains regrow in the interstellar medium. The main fraction of grains has radii <0.3 µm, going down to nanosized grains containing just a few hundred atoms. Grain temperatures typically range from 10 K in molecular clouds with hydrogen densities $n(H_2)$ between 10^4 and $10^6 \, cm^{-3}$, to ca. 100 K in "hot cores."

There is direct and indirect evidence for the presence of grains. A *direct* proof is scattering and extinction of star light by, for example, reflection nebulae and dark clouds. Extinction decreases with increasing wavelength λ, responsible for what is known as "interstellar reddening." For $\lambda \to \infty$, extinction goes to zero. The visual extinction A_V is related to the column density (the length of the line of sight) of hydrogen, $N(H)$, by $A_V = 10^{-21} N(H)$. For a hydrogen density $n(H) = 1 \, cm^{-3}$ and a distance of 1 kpc (3.26×10^3 light-years), the extinction amounts to three magnitudes. Optical luminescence (also termed photoluminescence) and IR emission features directly prove the presence of dust, and additionally provide information on the size distribution and, eventually, the chemical nature of dust particles. These emissions come about by absorption of high-energy starlight photons that raise the grain to an excited state, from which it decays by spontaneous emission of lower energy photons. Polarization effects resulting from the alignment of the charged dust particles with the cosmic magnetic field (typically $\approx 1 \, nT$[9]) is additional intriguing evidence. *Indirect* proof comes from the depletion of elements such as O, C, Si, Fe, Al, and Mg from the interstellar gas. These elements are underrepresented relative to their overall cosmic abundance, indicating that they are integrated in solid carbonaceous and silicate matrices. In addition, deuterium depletion in the gas phase hints toward fixation of ^2H in grain-bound reactions [37]. Further, as noted in Section 4.2.4, the presence of comparatively complex saturated and partially saturated organic molecules in the interstellar gas can, in many cases, only be explained by assuming that these molecules assemble (and are subsequently desorbed from the grain surface) from building blocks generated by surface reactions. Hydrogenation reactions dominate surface reactions at low temperatures because H atoms are by far the most abundant and most mobile species. These surface reactions are often closely related to reactions known for

---

9) $1 \, nT = 10 \, \mu G$. The strength of the terrestrial magnetic field is $60 \, \mu T$ (0.6 G).

heterogeneous catalysis such as the Fischer–Tropsch reactions, that is, the surface activates the reactants after adsorption or absorption[10] and thus reduces the activation barrier for the reaction.

4.2.5.1 The Hydrogen Problem

An indirect proof for the presence of grains is based on the abundance of molecular hydrogen H_2 in the interstellar gas: The fastest *gas-phase* reaction for the formation of H_2 is the reaction between hydrogen atoms and hydride ions, Eq. (4.71), with a rate constant of $k_1 = 10^{-9}\,cm^3\,s^{-1}$:

$$H + H^- \rightarrow H_2 + e^- \quad k_1 = 10^{-9}\,cm^3\,s^{-1} \tag{4.71}$$

The formation rate of H_2 according to Eq. (4.71) is limited by the abundance of hydride, $n(H^-)$, which forms by radiative electron capture, Eq. (4.72), and is destroyed, in diffuse clouds, by photodissociation, Eq. (4.73), with the rate constants k_2 and k_3, respectively. In denser clouds, neutralization with C^+, the most abundant ion, is an alternative route of destruction for H^-, Eq. (4.74):

$$H + e^- \rightarrow H^- + h\nu \quad k_2 = 10^{-16}\,cm^3\,s^{-1} \tag{4.72}$$

$$H^- + h\nu \rightarrow H + e^- \quad k_3 = 10^{-7}\,s^{-1}\ (\text{depends on } T^{1/2}) \tag{4.73}$$

$$H^- + C^+ \rightarrow H + C \quad k_4 = 10^{-7}\,cm^3\,s^{-1} \tag{4.74}$$

Employing the rate constants k_2 from Eq. (4.72) and k_3 from Eq. (4.73), and assuming densities $n(H) = 10^2$ and $n(e^-) = 10^{-4}\,cm^{-3}$, the density of H^-, $n(H^-)$, is

$$n(H^-) = [k_2 n(H) \times n(e^-)]/k_3 = 10^{-11}\,cm^{-3} \tag{4.75}$$

The rate of formation for H_2, $R(H_2)$, is given by

$$R(H_2) = k_1 \times n(H^-) \times n(H) = 10^{-18}\,cm^{-3}\,s^{-1} \tag{4.76}$$

To account for the observed density $n(H_2)$, the requisite minimum rate is, however, $10^{-17}\,cm^{-3}\,s^{-1}$. Hence, a more efficient mechanism for the formation of molecular hydrogen has to be considered, and this mechanism is provided by *surface reactions* on grains, Eq. (4.77). Corresponding experiments have been simulated in the laboratory under ultrahigh vacuum conditions at low temperature on water-ice-coated surfaces of olivine ($FeMgSiO_4$) and carbon:

$$H + H + \text{grain} \rightarrow H_2 + \text{grain} \tag{4.77}$$

Assuming an idealized, viz. a spherical shape of the grains with radius r (and hence a cross section πr^2 and a volume of $4/3 \pi r^3$), the formation rate for H_2 on a grain is given by Eq. (4.78a), in which k_g is the rate of collision between the grain and a hydrogen atom, and $n(g)$ the space density of grains. The collision rate in turn is defined by Eq. (4.78b), where v is the average atomic speed, and P the probability

---

[10] Physical *ad*sorption (physisorption) occurs with a binding energy of ca. $k_B \times 400\,K$ (k_B is the Boltzmann constant), and *ab*sorption (or chemisorptions) with a binding energy of ca. $k_B \times 20{,}000\,K$.

that an H atom colliding with a grain is absorbed and reacts to form a hydrogen molecule, hence the efficiency of the process:

$$R_g(H_2) = k_g \times n(g) \times n(H) \quad (4.78a)$$

$$k_g = 1/2\pi r^2 \times v \times P \text{ cm}^3 \text{ s}^{-1} \quad (4.78b)$$

Substituting k_g in Eq. (4.78a) by the expression provided in Eq. (4.78b), the rate of formation becomes

$$R_g(H_2) = 1/2\pi r^2 n(g) \times v \times P \times n(H) \quad (4.78c)$$

Here, $\pi r^2 n(g) \times v$ is the collision rate. Assuming a fractional grain density of $f(g) = n(g)/n = 3 \times 10^{-27}/4\pi r^3$, $n \approx n(H)$, a collision speed $v = 10^5 \text{ cm s}^{-1}$, and a particle radius of 0.1 μm (10^{-5} cm), a direct relation between the formation rate of H_2, the probability for a successful reaction P, and the number density of hydrogen atoms $n(H)$ is obtained, Eq. (4.78d):

$$R_g(H_2) = 3/8 \times 10^{-17} \times P \times n^2(H) \quad (4.78d)$$

As noted earlier, a minimum $R(H_2)$ of 10^{-17} is required to explain the interstellar H_2 abundance. A density $n(H) \approx 10 \text{ cm}^{-3}$ thus suffices to account for the observed abundance of H_2 even for $P \approx 0.1$. To arrive at sufficiently high P, a high degree of roughness and/or amorphism of the surface is required.

To what extent a collision with concomitant reaction is "successful" not only depends on the collision rate $\pi r^2 n(g) v$ but also on (i) the kinetic temperature of the H atom and on the grain temperature, (ii) the time interval until another H atom is adsorbed and the atoms' mobility on the surface (the energy barrier for site exchange), allowing for an encounter and combination of the two atoms, and (iii) the desorption and ejection of H_2 from the surface prior to destruction or conversion by successive reactions.

4.2.5.2 Grain Structure, Chemical Composition, and Chemical Reactions

The size of interstellar grains ranges from about 5 to 250 nm, with a size population distribution that goes as $r^{-3.5}$. The shape of the grains is irregular, and commonly the grains are porous. Grains in dense clouds are coated with a mantle of ice. This coating is missing in the warmer diffuse and reflection nebulae. To a certain extent, desorption of volatiles from the ice coatings in dense clouds can occur by cosmic ray heating: grains are subjected to bombardment by cosmic rays that, contrary to photons, penetrate deep into the clouds; for additional details, see below. Ice mantles also evaporate in the process of contraction of and star formation within a dense cloud, accompanied by a change in chemistry from unsaturated to more saturated compounds.

Figure 4.20a features a typical built up of a dust grain in a dense interstellar cloud, and Figure 4.20b provides an example of a computer-simulated grain [38]. Depending on their origin, that is, whether they are produced in the envelopes of carbon-rich (C > O) or oxygen-rich stars (C < O), the cores are predominantly composed of carbonaceous materials or silicates. "Silicates" may also be mixtures

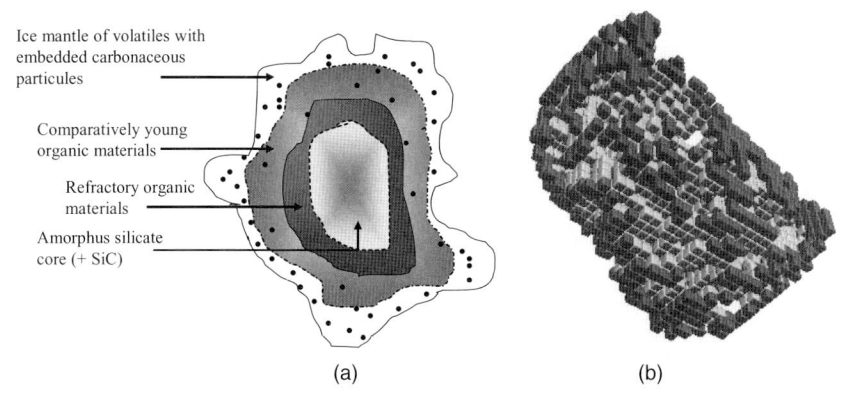

Figure 4.20 (a) Idealized assembly ("four-shell model") of a silicate grain in a molecular cloud. The ice coating on the outer surface is absent in grains from diffuse clouds. Note that most grains are porous. (b) Computed ("Gaussian Random Field") model of a porous, carbon-coated (dark cubes) silicate particle; reproduced from Ref. [38] with permission by Astronomy & Astrophysics.

of silica (SiO_2), SiO, and metal oxides such as MgO, FeO, NiO, TiO_2, and Al_2O_3. Iron is depleted in silicate grains with respect to its overall abundance, suggesting that it should be present in the form of refractory species such as metallic iron, FeO, and FeS in other than silicate-based grains. The core is embedded in a matrix of refractory carbonaceous matter (e.g., polycyclic aromatic compounds, graphite, and diamond), surrounded by a mantle of less unsaturated organic material. Finally, in dense clouds, a shell of ice is formed by volatiles, mainly water, but including substantial amounts of CO_2, CO, NH_3, formaldehyde H_2CO, formic acid HCO_2H, methanol CH_3OH, CH_4, COS, and possibly other, in particular CN-bearing molecules. Embedded into this icy shell are tiny carbonaceous particles.

As far as the chemical composition of the grains is concerned, findings on interplanetary dust particles collated in the Solar System (interplanetary grains, Section 5.3.3), or locked in meteorites, provide signatures that can be extrapolated to interstellar grains – with a grain of salt, since the main part of inter*planetary* grains has undergone some processing. Direct evidence for the chemical composition is derived from spectroscopic features, mainly emission signatures in the IR, summarized in Table 4.4 [37–39]. IR emissions are commonly reported in units of μm in the literature related to phenomena in astronomy. For IR emissions in the common vibrational domain, wave numbers in units of cm^{-1} have also been included in Table 4.4.

Valuable insight into reaction paths and chemical composition in the ice mantle of grains can be obtained by laboratory experiments (for details vide infra). However, any laboratory support, or backup, of a synthetic path in interstellar ice analogs should take into account that there are constraints for the transfer of laboratory results to what is actually occurring in ice coatings of grains. Constraints to be considered are, among others, as follows: (i) the chemical composition and

Table 4.4 Wavelengths and assignments of spectroscopic features from interstellar grains (from Refs. [37–39]).

	Wavelength	Wave number (cm^{-1})	Assignment
X-ray	2.35 nm (528 eV)		O in oxides and silicates
UV	217.5 nm		$\sigma \rightarrow \sigma^*$ of graphene/graphite and large PAH^{n+} [a], carbon onions[b]; surface-OH$^-$ (?)
Vis	~455 nm		Diffuse interstellar band, possibly not associated with dust
Vis to near-IR	650–800 nm		Si nanocrystallites?
IR	2.3 and 4.6 µm	4350 and 2170	ν(CO) (first overtone and normal mode)
IR	2.9 µm	3450	ν(NH$_3$)
IR	3.03 µm	3300	ν(sp-CH) of acetylenes
IR	3.08 µm	3250	ν_s(H$_2$O) of water ice
IR	3.24 µm	3085	Nonaromatic ν(C=CH$_2$)
IR	3.28 µm	3050	ν(sp^2-CH), aromatic CH
IR	3.4 band (3.38, 3.41, and 3.48 µm)	2940	ν(CH$_3$, sp^3-CH$_2$) in paraffins
IR	3.472 µm	2880	ν(sp^3-CH), unsaturated CH
IR	3.55 and 3.67 µm	2820, 2725	ν(CH), aldehydes
IR	4.27 and 15.2 µm	2340 and 660	ν(CO$_2$) and δ(CO$_2$) in ice
IR	4.62 µm	2165	ν(C≡N) of "XCN"[c]
IR	5.8 µm	1725	ν(C=O), aldehydes and ketones
IR	6.25, 7.7, 8.6, 11.3, and 12.7 µm	1600, 1300, 1160, 885, 790	ν(C=C), ν(C–C) and δ(=C–CH) of PAH[a]; cf. Figure 4.21
IR	6.85 and 7.25 µm	1460, 1380	δ(sp^2-CH)
IR	9.7 µm	1030	ν(Si-O) in silicates
IR	10.6 µm	940	SiC, spherically shaped
IR	11.25 µm	890	SiC, irregularly shaped
IR	11.2 µm	890	Crystalline olivine FeMgSiO$_4$[c]
IR	18 µm	555	δ(Si–O–Si) in silicates
Far-IR	21 µm		Disputed (large PAH clusters, nano-TiC, nano-SiC, …?)
Far-IR	23.5 µm		FeS (?)
Far-IR	30 µm		Unidentified

a) PAH = polycyclic aromatic hydrocarbon; cf. Scheme 5.2 in Section 5.3.2 for visualization of selected PAHs, and Scheme 4.9 for coronene.
b) Concentric shells of fullerene molecules.
c) Detection in interstellar grains under debate.

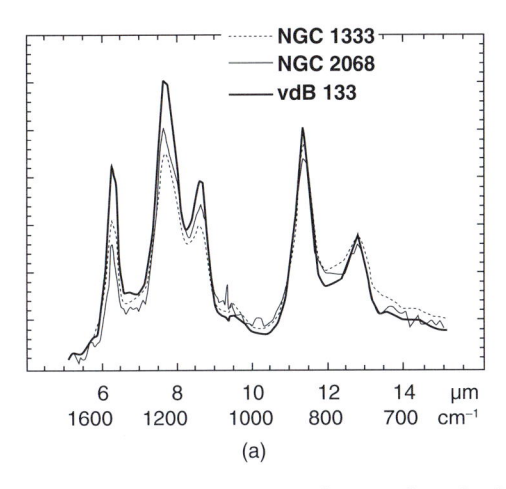

Figure 4.21 (a) IR-spectroscopic features of dust in the 6–13 µm domain, corresponding to stretching (6.25, 7.7, and 8.6 µm) and bending vibrations (11.3 and 12.7 µm) of PAHs (Table 4.4). Sources are the reflection nebula NGC 1333 (Perseus), NGC 2968 (Orion), and vdB 133 (Cygnus). Adapted from Ref. [40] by permission of the AAS. (b) The picture has been taken by the Spitzer space telescope; it shows star formation regions in the nebula NGC 1333, at a distance of 10^3 lightyears, in the constellation of Perseus. Credit: R. A. Gutermuth (Harvard-Smithonian CfA) et al. JPL-Caltech, NASA.

physical state of the ice analog are idealized, (ii) compact surfaces instead of nano- to microsized dust particles are employed, (iii) densities in the surrounding gas are orders of magnitudes above those in the gas in which the dust grains are embedded, (iv) the time factor.

The most intriguing IR emissions are those that are diagnostic for the various types of structure elements occurring in organics, including alkynes, alkenes, and alkanes, tertiary, secondary, and primary carbons, aromatic compounds, and condensed aromatic compounds. Figure 4.21 represents a typical fingerprint in the 6–13 µm range for three reflection nebulae [40]. The spectra show five prominent IR emissions that are indicative of the presence of C—C stretching and C—C—H deformation modes of PAHs (Scheme 5.2 in Section 5.3.2). PAHs were found to cluster in three mass envelopes, of molecular masses 60, 250, and 370 g mol^{-1}. The 3.28 µm C—H stretching vibrations of sp^2-hybridized aromatic carbons of the PAHs require, to produce a strong emission, an excitation to an effective temperature of 700–1000 K, certainly a temperature out of reach for a nano- to microsized dust particle. The problem can, however, be circumvented if it is assumed that the PAHs are subdivided into insulated islets by partial loss of the π-delocalization through partial hydrogenation, that is, by breaking up part of the aromatic structure via formation of bridging elements constituting sp^3-hybridized (aliphatic) carbons [39a]. The resulting islets, which might be symbolized by {PAH-(CH$_2$)$_n$}, have been found in grains associated with carbonaceous chondrites and Antarctic micrometeorites. Laboratory studies have shown that PAHs such as coronene and pyrene are generated in analogs of interstellar ice on UV and proton irradiation. These PAHs are further processed to form functionalized organic molecules (Scheme 4.9). The detection method applied in these experiments, microprobe two-step laser mass spectrometry, or $\mu L^2 MS$ for short, employs a laser beam for ablation, and another one for ionization of the material to be analyzed by mass spectrometry [41], which substantially increases the detection sensitivity.

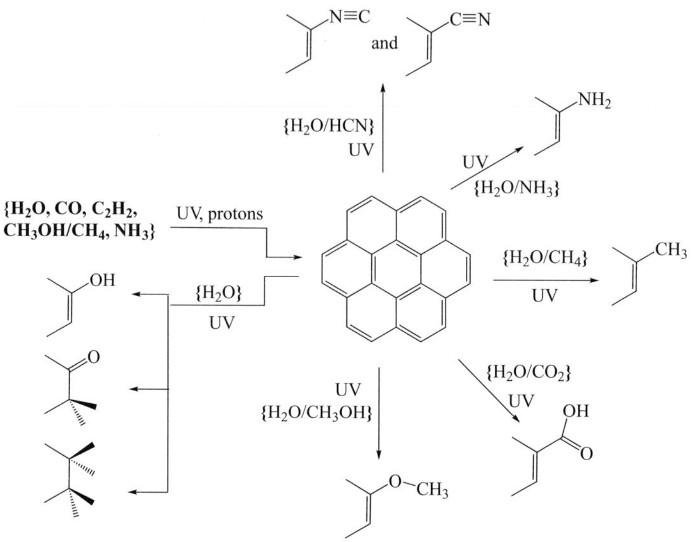

Scheme 4.9 Model reactions of coronene (a representative of PAHs) in interstellar ice analogs of varying composition, subjected to UV irradiation. Based on Ref. [41]. Curly brackets indicate ices.

A 3.4-μm feature essentially representing *aliphatic* hydrocarbons (paraffins) appears at the long-wavelength wing of the water ice band (characterized by a symmetric stretch at 3.08 μm) in dense clouds, whereas, in diffuse sight-lines, it appears without the water band [39b]. Dry ice (carbon dioxide), detected by its stretching and bending modes (4.27 and 15.2 μm, respectively), is surprisingly abundant in the icy layers of grains, where it makes up about one-fifth of the amount of water. In the gas phase, CO_2 is a minor component only, suggesting that by far the main amount of CO_2 is produced in grain surface reactions and remains bonded to the dust rather than being evaporated into the surroundings. Lab experiments have shown that CO_2 is formed from CO and O, if both are confined to water ice, and from CO and OH radicals generated from water if the reaction is supported by UV. Other diagnostic emissions allotted to ice components have been detected, such as the CO stretch at 4.6–4.7 μm and the NH stretch of ammonia ice at 2.9 μm.

Emission features at 9.7 (ν(Si–O)) and ~18 µm (δ(Si–O–Si)) are well-established indicators of the presence of silicates. Good fits to the 9.7 µm band have been obtained for magnesium-rich amorphous olivine, $Mg_{2x}Fe_{2-2x}SiO_4$, where $x \geq 0.9$ [38], hence a composition that comes close to fosterite Mg_2SiO_4. In this respect, interstellar grains resemble inter*planetary* dust particles. Contrasting interplanetary grains, silicates in grains from inter*stellar* clouds are, however, essentially amorphous. Otherwise, the IR signatures of grains from both interplanetary and interstellar origin are more closely related to each other than to terrestrial silicates. The shape and the exact position of the 9.7 band largely depend on the shape of the grains. A long-wavelength shoulder on the 9.7-µm silicate band, at about 11 µm, has successfully been modeled by assuming about 3% admixture of crystalline SiC to the olivine [38]. An additional feature at 21 µm has tentatively been assigned to nanoparticles of SiC, but there are other candidates for this band as well. The assignment of additional broad and unstructured features in the far-IR is disputed. Iron sulfide (FeS) is a candidate for the 23.5 µm band. Factually, sulfur is depleted in the interstellar gas of dense clouds, and thus may be lodged in grains in the form of FeS.

Similar problems for the assignment of broad bands arise from absorptions in the range spanning the near-IR, visible, and UV. A variety of so-called diffuse interstellar bands throughout the visible range of reddened stars, time, and again associated with dust, likely do *not* originate from grains [39b]. On the other hand, a "bump" of varying depth and width in the UV region, centered at 217.5 nm, very likely goes back to carbonaceous matter such as represented by (cationic) PAHs, graphitic carbon, and fullerene-based compounds. Strength and width of this band are linked to the size of the grains: high populations of very small grains increase the intensity and decrease the width.

In principle, X-ray absorption by dust particles is a powerful tool for the identification of elements and their chemical environments in solid matrices, if a sufficiently bright continuous stellar X-ray source is available in the line of sight. The method, known as X-ray absorption spectroscopy, relies on the excitation, by synchrotron radiation, of an electron residing in the K shell ($n = 1$, $l = 0$ (s-electrons)) or L shell ($n = 2$, $l = 0$ (s-electrons) or $l = 1$ (p-electrons)) into an empty unoccupied level and further into the continuum. The corresponding energy defines the K- and L-edges, and the method is usually referred to as X-ray absorption *near* edge structure (XANES).[11] The edge position is very characteristic and hence indicative for a specific element, and also sensitive of the oxidation state (e.g., Fe^{2+} vs. Fe^{3+}) and the nature of the coordinating functions (e.g., "soft" ligands such as S^{2-} vs. "hard" ligands such as O^{2-}). Figure 4.22 provides iron K- and L-edge XANES spectra as obtained for mixtures of graphite and olivine [42]. Broad iron K- and L-edge

11) The region beyond the edge shows characteristic features produced by interference of the outcoming electron wave with the valence electrons of the bonding partners of the atom from which the electron became emitted. These features allow for a determination of bond lengths and, to a certain approximation, coordination numbers (and geometries). The method is known as *extended X-ray absorption fine structure* (EXAFS).

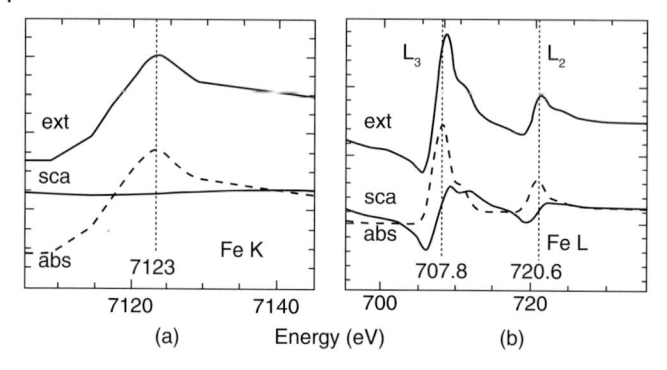

Figure 4.22 Iron K- (a) and L-edge spectra (b) of a model consisting of graphite and the silicate olivine, $Fe^{II}MgSiO_4$; ext = extinction mode, sca = scattering mode. L_2 and L_3 refer to the $2p_{3/2} \to 3d$ and $2p_{1/2} \to 3d$ transitions, respectively (dipole allowed excitations from 2p levels into empty 3d positions). Reproduced from Ref. [42] by permission of the AAS.

features have also been observed in emissions from other Seyfert galaxies, from accreting stellar-mass black holes, and from stars.

Ice mantles on grains can effectively deplete species from the interstellar medium by selectively freezing out less-volatile components from the gas phase, and thus also contribute to isotope fractionation, exemplified here for the fractionation of deuterated H_3^+ [5c]: The ion H_3^+ forms rapidly by the reaction between H_2^+ (generated by ionization of H_2 through cosmic rays) and H_2 ($H_2^+ + H_2 \to H_3^+ + H$). Reaction of H_3^+ with HD, Eq. (4.79a), readily produces H_2D^+, but the reverse reaction, Eq. (4.79b), is too slow at low temperatures to allow for an appreciable reconversion. As a consequence, $H_2D^+/H_3^+ \gg HD/H_2$, as long as there are no efficient destruction paths available. Destruction of H_3^+ and H_2D^+ readily occurs by reaction with the particularly abundant carbon monoxide, Eq. (4.80). If, however, the polar CO molecules are frozen out, the destruction path will not work and H_2D^+ remains enriched:

$$H_3^+ + HD \to H_2D^+ + H_2 \tag{4.79a}$$

$$H_2D^+ + H_2 \to H_3^+ + HD \tag{4.79b}$$

$$H_3^+ + CO \to H_2 + HCO^+ \tag{4.80a}$$

$$H_2D^+ + CO \to (HD \text{ and } H_2) + (HCO^+ \text{ and } DCO^+) \tag{4.80b}$$

Ice mantles on grains have also been recognized as production facility for the formation of many of the more complex molecules found in the interstellar medium. This issue has already been briefly addressed in the context of the generation of CO_2 and the formation and processing of PAHs. Additional examples, backed up by laboratory simulations, will be introduced in the following text. If the molecules are to escape from the grain surface, "heating" is essential. The necessary energy for desorption can be provided in "hot cores," that is, dense clouds in star formation regions containing protostars. But heating, and thus desorption, may also be achieved by cosmic ray bombardment. Cosmic rays have a composition similar to that of the Universe: About 87% are protons, 12% are helium nuclei, and scantly 1% is represented by other nuclei, in particular iron, produced by the α process (Eq. (3.18) in Section 3.1.4). In addition, electrons and γ quanta are associated with cosmic rays. The energy of cosmic rays, when stemming from supernovae, is in the range of 10^8–10^{10} eV, and can go up to 10^{20} eV and even more when originating from, and accelerated by, exploding stars and active galactic cores [43]. Cosmic rays can penetrate deeply into even the densest clouds, and their collision with dust particles heats up the grain, in particular when the colliding nucleus is iron: The energy deposited into a grain is proportional to the square of the nuclear charge z, and this makes iron nuclei ($z = 26$) particularly effective despite of the sufficiently lower flux when compared to protons ($z = 1$).[12] For very small grains, viz. grains measuring a few nanometer only, peak temperatures can go up to 350 K both in silicate and in carbonaceous grains, and to 150 K in grains measuring ca. 50 nm [44].

In "hot cores," a characteristic absorption centered at 4.62 μm is indicative of nitriles of the general composition X–C≡N (where X preferentially is an organic residue), possibly with admixtures of isonitriles C≡N–X and cyanates, such as the cyanate anion NCO⁻ and organic cyanates N≡COR [45]. The formation of "XCN" in ices containing H_2O, CO, CO_2, CH_4, and other organics, together with NH_3 or N_2 as the nitrogen source, exposed to UV photons and/or high-energy protons has been simulated in the laboratory; see also Scheme 4.9 for two examples. Annealing and radiative processing in "hot cores" are supposed to add to the diversification of cyanide species. The 4.62 μm feature is not distinctly present in grains from diffuse or reflection clouds, which are lacking the ice mantle. Interestingly, in some interstellar sources where ices can be assumed to be subjected to intense processing, the abundances of "XCN" and methanol appear to be correlated. For the formation of methanol on grains, see below.

Nitriles are potential precursors for amino acids. Basically, their bulk synthesis can proceed through the cyanhydrine reaction via the hydroxy nitrile, Eq. (4.81). Of particular interest in the context of amino acid precursors is the recent discovery of interstellar amino-acetonitrile [29], the hydrolysis of which, step 2 in Eq. (4.81) for R = H, leads to the formation of glycine; see also Scheme 4.7 and Eq. (4.58) in Section 4.2.4. Glycine, recently detected in particles retrieved from the coma of comet Wild 2 by the spacecraft Stardust (Section 5.4.2), has not (yet) been found in space, but this may well be due to the photolytic lability of amino acids. For any organic residue R in Eq. (4.81), such as R = CH_3 in the case of alanine, the cyanhydrine synthesis delivers a racemate, that is, an equal mixture of the D- and L-form of the amino acid. In principle, preferential destruction of one of the two enantiomers by circularly polarized UV would leave an excess of one of the isomers; for details, see Section 7.4.2. Polarization of UV is caused by the alignment of nonspherical grain particles in the cosmic magnetic field. If the

---

12) The flux ratio Fe/H is 1.6×10^{-4}.

residue R by itself is chiral, for example, for R = $-CH(CH_3)-CH_3$ (*secondary propyl*), asymmetry is already imprinted into the aldehyde serving as the amino acid precursor:

$$R-C\overset{H}{\underset{O}{\diagdown}} \xrightarrow{H-C\equiv N} HO-\underset{R}{\overset{H}{\underset{|}{C}}}-C\equiv N \xrightarrow{NH_3} H_2N-\underset{R}{\overset{H}{\underset{|}{C}}}-C\equiv N + H_2O$$

$$\downarrow {-NH_3}$$

$$\left\{ \begin{array}{cc} CO_2H & CO_2H \\ | & | \\ HC-NH_2 & + \; H_2N-CH \\ | & | \\ R & R \\ \text{R-enantiomer} & \text{L-enantiomer} \end{array} \right\} \quad (4.81)$$

Several alternative synthetic pathways to amino acids have been discussed and supported by laboratory simulations of chemical reactions in interstellar ice analogs and/or by computation [46]. Among these are ion–molecule reactions involving formic acid or acetic acid, initially leading to protonated glycine, Eqs. (4.82) and (4.83), which undergoes dissociative recombination to form glycine, Eq. (4.84). The reaction between the cationic radical $CH_3NH_2^+$ and acetic acid Eq. (4.82), is exothermic (about $113\,kJ\,mol^{-1}$). Glycine may further form from CH_2NH_2, CO_2, and H as depicted in Eq. (4.85), a multistep process.

$$CH_3NH_2^+ + HCO_2H \rightarrow {}^+H_3N\text{-}CH_2\text{-}CO_2H \quad (4.82)$$

$$NH_2^+ + CH_3\text{-}CO_2H + M \rightarrow {}^+H_3N\text{-}CH_2\text{-}CO_2H + M \quad (4.83)$$

$${}^+H_3N\text{-}CH_2\text{-}CO_2H \xrightarrow{e^-} H_2NCH_2CO_2H + H \quad (4.84)$$

$$CH_3NH_2 + CO_2 + H \xrightarrow{\text{(grain)}} \underset{\text{Glycine}}{H_2N\text{-}CH_2\text{-}CO_2H} \quad (4.85)$$

The hydrogenation of dicyan in a reaction cascade comprising four successive steps, Eq. (4.86), may be considered an alternative path (alternative to the cyanhydrine synthesis) for the formation of amino-acetonitrile. So far, dicyan has, however, not been detected in the interstellar medium:

$$N\equiv C-C\equiv N + H \rightarrow N\equiv C-C\overset{H}{\underset{\|N}{\diagdown}}$$
$$\xrightarrow{+H} N\equiv C-C\overset{H}{\underset{NH}{\diagdown}}$$
$$\xrightarrow{+H} N\equiv C-C\overset{H}{\underset{NH_2}{\diagdown}}$$
$$\xrightarrow{+H} N\equiv C-CH_2NH_2$$
$$\text{Amino-acetonitrile} \quad (4.86)$$

Finally, radical reactions starting from ethene (C_2H_4) and cyan radicals $^{\bullet}CN$ may come in. The corresponding reaction sequence, shown in Eq. (4.87), involves the radical intermediates $^{\bullet}CH_2CH_2CN$ and $CH_3^{\bullet}CHCN$. One way by which these radicals can stabilize is by forming vinylcyanide $CH_2{=}CH{-}C{\equiv}N$ through abstraction of H atoms. The overall reaction ($C_2H_4 + CN \rightarrow CH_2{=}CHCN + H$) is exergonic. Vinylcyanide (Scheme 4.7) is a potential precursor for alanine, Eq. (4.88), and has recently been detected in space:

$$^{\bullet}CN + H_2C{-}CH_2 \rightarrow {}^{\bullet}CH_2{-}CH_2{-}CN \tag{4.87}$$

(Vinylcyanide)

$$H_2C{=}CH{-}CN \xrightarrow{+ NH_3, + H_2O} CH_3{-}\underset{NH_2}{\overset{H}{C}}{-}CO_2H + HCN \tag{4.88}$$

Vinylcyanide → α-Alanine

Radicals such as $^{\bullet}CH_2CH_2CN$ (Eq. (4.87)), $CH_3^{\bullet}CHCN$, and $^{\bullet}CH_2CN$ can form by one-electron oxidation from the anions $^{-}CH_2CH_2CN$, $CH_3^{-}CHCN$, and $^{-}CH_2CN$, which in turn are produced in the ion source of a mass spectrometer from OH^{-} and propionitrile or acetonitrile, respectively [46]; for the reactions, see Eqs. (4.89a) and (4.89b). Both propionitrile and acetonitrile molecules are present in interstellar sources. Further reaction with $^{\bullet}NH_2$, a known interstellar radical, can produce amino-nitriles, the hydrolysis of which gives rise to the formation of β-alanine, α-alanine, or glycine, Eq. (4.90):

$$H_2O \xrightarrow{UV} OH^{-} \xrightarrow[-H_2O]{+ CH_3CN} {}^{-}CH_2CN \rightarrow {}^{\bullet}CH_2CN + e^{-} \tag{4.89a}$$

$$OH^{-} + CH_3CH_2CN \underset{H_2O}{\rightleftarrows} \begin{Bmatrix} {}^{-}CH_2CH_2CN \\ CH_3{}^{-}CHCN \end{Bmatrix} \rightarrow \begin{Bmatrix} {}^{\bullet}CH_2CH_2CN \\ CH_3{}^{\bullet}CHCN \end{Bmatrix} + e^{-} \tag{4.89b}$$

$$^{\bullet}NH_2 + \begin{cases} {}^{\bullet}CH_2CN \rightarrow NH_2CH_2CN \\ CH_3{}^{\bullet}CHCN \rightarrow CH_3CH(NH_2)CN \\ {}^{\bullet}CH_2CH_2CN \rightarrow NH_2CH_2CH_2CN \end{cases} \xrightarrow[-HCN]{H_2O} \begin{cases} NH_2CH_2CO_2H \text{ (Glycine)} \\ CH_3CH(NH_2)CO_2H \text{ (α-Alanine)} \\ NH_2CH_2CH_2CO_2H \text{ (β-Alanine)} \end{cases} \tag{4.90}$$

Apart from the synthesis of amino acids by the cyanhydrine (or Strecker) synthesis, Eq. (4.81), and via nitriles formed in radical reactions as exemplified by Eqs. (4.87) and (4.89), an alternative intricate radical mechanism may come in, or even be the more prominent synthetic path to amino acids, in "hot cores" where a photon flux is available: UV photolysis of ices containing H_2O, NH_3, HCN, and CH_3OH have been shown to generate glycine, α-alanine, and serine [47]. As revealed by application of ^{13}C-labeled methanol ($^{13}CH_3OH$) and hydrogen cyanide ($H^{13}CN$), the major products that form are those produced by radical–radical reaction. Scheme 4.10 illustrates the main reaction paths for the formation of glycine and serine. In serine, for example, the central and the carboxylate carbon are delivered by HCN, whereas the side-chain carbon stems from methanol. This radical type of reaction is also supported when using ^{15}N-labeled ammonia ($^{15}NH_3$) and hydrogen cyanide ($HC^{15}N$): Most of the labels in the amino function of the amino acid originate from HCN. In contrast, in the cyanhydrine synthesis, HCN would provide the acid carbon, methanol the central carbon, and ammonia the amine group.

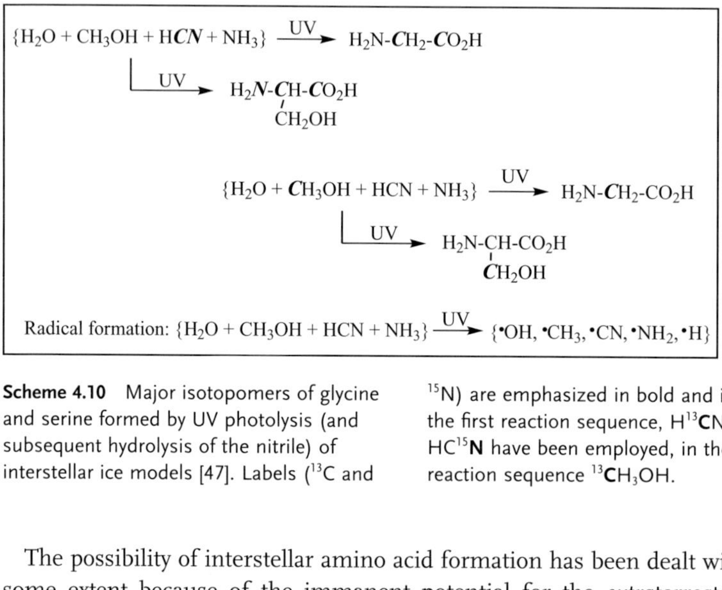

$$\{H_2O + CH_3OH + H\boldsymbol{\mathit{CN}} + NH_3\} \xrightarrow{UV} H_2N\text{-}CH_2\text{-}CO_2H$$
$$\xrightarrow{UV} H_2N\text{-}CH\text{-}CO_2H$$
$$\qquad\qquad\qquad |$$
$$\qquad\qquad\quad CH_2OH$$

$$\{H_2O + \boldsymbol{\mathit{CH_3OH}} + HCN + NH_3\} \xrightarrow{UV} H_2N\text{-}CH_2\text{-}CO_2H$$
$$\xrightarrow{UV} H_2N\text{-}CH\text{-}CO_2H$$
$$\qquad\qquad\qquad |$$
$$\qquad\qquad\quad CH_2OH$$

Radical formation: $\{H_2O + CH_3OH + HCN + NH_3\} \xrightarrow{UV} \{\cdot OH, \cdot CH_3, \cdot CN, \cdot NH_2, \cdot H\}$

Scheme 4.10 Major isotopomers of glycine and serine formed by UV photolysis (and subsequent hydrolysis of the nitrile) of interstellar ice models [47]. Labels (^{13}C and ^{15}N) are emphasized in bold and italics. In the first reaction sequence, $H^{13}CN$ and $HC^{15}N$ have been employed, in the second reaction sequence $^{13}CH_3OH$.

The possibility of interstellar amino acid formation has been dealt with here to some extent because of the immanent potential for the *extra*terrestrial origin of these basic protein building blocks. Nitriles as likely amino acid precursors have been pinned down in interstellar clouds. Several sources with high nitrile abundances also have high methanol contents, suggesting that methanol is also formed in regions where processing of ice is important [45]. Methanol is a ubiquitous interstellar gas-phase species, and condensed methanol has been found as an essential constituent in the ice coatings of grains in cold dark clouds. The fractional abundance of methanol, that is, the abundance relative to hydrogen, is $\approx 10^{-9}$ in cold clouds and $\approx 10^{-6}$ in "hot cores". This abundance cannot be explained by gas-phase reactions alone, such as radiative combination of methyl cations and water, Eq. (4.91a), followed by dissociative recombination; Eq. (4.91b). The gaseous water abundance for the production of $CH_3OH_2^+$ according to reaction (4.91a) is too low to account for the observed methanol abundance, as is the fraction of methanol produced by dissociative recombination, Eq. (4.91b): the main components produced as $CH_3OH_2^+$ recombines with an electron are CH_3, OH, and H:

$$CH_3^+ + H_2O \rightarrow \{CH_3^+ \cdot H_2O\} \rightarrow CH_3OH_2^+ + h\nu \tag{4.91a}$$

$$CH_3OH_2^+ + e^- \rightarrow CH_3OH + H \tag{4.91b}$$

Hence, reactions on surfaces, as well as effective desorption from these surfaces, have to be considered. Among the possible surface-supported reactions, two have been favored, viz. (i) the formation of methanol from fragments generated by proton irradiation of H_2O–CO ice or by electron irradiation of H_2O–CH_4 ice [48], and (ii) the stepwise hydrogenation of CO by H on the grain surface [49]. Let us consider electron radiation first:

Low-energy electrons (ca. 100 eV ≈ 10^4 kJ mol^{-1}) are cosmologically produced in situ as "secondary electrons" via ionization of ice constituents by protons and other nuclei constituting cosmic rays. The irradiation of H_2O/CH_4 10/1 at 10 K under lab conditions yields methanol along with formaldehyde and ethane (plus the dehydrogenation products of ethane: ethene and ethyne) by the reaction sequences provided in Eq. (4.92) for methanol and in Eq. (4.93) for formaldehyde. Methanol forms from the products of homolytic fission of water and methane, a view which is supported by the formation of CD_3OH plus CHD_2OH as H_2O–CD_4 ice is employed instead of H_2O–CH_4. Along with electrons, cosmic ray protons and a secondary UV field can initiate dissociation. A secondary UV field is induced as H^+ and He^+, produced by cosmic ray ionization, recombine with electrons to H and He, plus a photon. Formaldehyde forms by carbon insertion into water and subsequent rearrangement, Eq. (4.93):

$$H_2O \rightarrow H + OH$$

$$CH_4 \rightarrow H + CH_3, \text{ and: } CH_3 \rightarrow H + CH_2 \tag{4.92}$$

$$CH_3 + OH \rightarrow CH_3OH, \text{ and: } CH_2 + H_2O \rightarrow CH_3OH$$

$$C + HOH \rightarrow HCOH \rightarrow H_2CO \tag{4.93}$$

Alternatively or additionally, surface production of methanol occurs by stepwise hydrogenation of CO, Eq. (4.94):

$$CO + H \xrightarrow{\text{grain}} HCO \xrightarrow{+3H,\text{ grain}} CH_3OH \tag{4.94}$$

An approach similar to that used for the grain surface formation of H_2 (Section 4.2.5.1) can then be employed. Adapting Eqs. (4.78b) and (4.78d) from Section 4.2.5.1 to the grain surface formation of methanol, we obtain

$$k_g(CH_3OH) = 1/2\pi r^2 \times v(CO) \times P_{CO} \tag{4.95a}$$

$$\text{and } R_g(CH_3OH) \approx 10^{-17} P_{CO} \times n(CO) \times n(H) \tag{4.95b}$$

where k_g and R_g are the rate constant and formation rate, respectively, for the formation of methanol, πr^2 the cross section of the grain (r = radius), v(CO) the average velocity of the CO molecules hitting the grain, $n(CO)$ and $n(H)$ the densities (in cm^{-3}) of CO and H, respectively, and P_{CO} the probability or efficiency factor which, for the four-step reaction leading to methanol, Eq. (4.94), is ca. 0.003 and hence two to three orders of magnitude smaller than the efficiency factor for H_2 formation, but still sufficiently large to account for the methanol abundance. A higher efficiency is provided if H atoms striking a CO-rich grain surface are considered.

The fractional surface abundance for CO thus obtained, ca. 10^{-6}, compares to the fractional CO abundance in the gas phase of "hot cores." To account for the gas-phase abundances in cold clouds, effective desorption is essential. As delineated above (Ref. [44]), cosmic ray heating can give rise, in sufficiently small grains, to peak temperatures that suffice to initiate desorption. For larger grains, the exothermicity of a reaction can provide the energy necessary for desorption. As an example, the reaction enthalpy for the final hydrogenation step in Eq. (4.94), $CH_2OH + H \rightarrow CH_3OH$, is $-395\,kJ\,mol^{-1}$; the desorption energy for methanol is just $\approx +17\,kJ\,mol^{-1}$.

Summary

The interstellar medium is essentially organized in clouds. These can be diffuse clouds with either neutral hydrogen (HI) and number densities n of 20–50 cm^{-3} (reflection nebulae), or ionized hydrogen (HII) and $n \approx 10^3$ cm^{-3} (emission nebulae), or molecular clouds (dark nebulae) with n up to 10^6 cm^{-3}. Molecular clouds are characterized by rich chemistry, initiated by cosmic rays, and supported by ice-coated dust grains, which make up ca. 1% in these clouds. The overall number of molecules and molecular ions detected in interstellar space is ~170, the largest one is cyanodecapentayne $HC_{10}CN$. In addition, dust grains harbor complex polyaromatic compounds, graphite, and inorganics such as silicates. The precursor molecule for the latter is SiO, formed through a reaction chain starting with $Si^+ + OH \rightarrow SiO^+ + H$.

The formation of interstellar molecules follows complex and interconnected reaction networks. Central species from which chemical reaction cascades can start are C^+, H_2, H, H^+, H_3^+, He, and e^-. Typical reactions in cold (10–20 K) clouds should be exergonic and devoid of an appreciable activation barrier, and allowing for the off-transport of the kinetic energy of the impacting reaction partners. Most ion–neutral reactions fulfill these conditions, as do many reactions where a radical is involved, and electron capture by cationic species. For the generation of organic compounds, the most important initiating reaction is the formation of CH_2^+ by radiative association of C^+ and H_2. The major nitrogen-bearing molecule is N_2, which mainly forms by neutral exchange between N and NH.

The various species are detected by emission and absorption spectroscopy over the complete electromagnetic spectrum, preferentially with satellite-based telescopes. HII regions glow in the red-violet ($\lambda = 656.3$ nm) Balmer Hα light of excited H atoms formed by radiative recombination between H^+ and e^-. Mapping of HI areas is achieved with the help of the 21-cm line corresponding to a spin flip of the electron spin relative to the spin of the proton. Hydride is detected by its absorption pattern in the far-IR and molecular hydrogen H_2 by the UV bands in the range 110.8–91.2 nm (Lyman and Werner bands). H_2 is formed by gas-phase reaction according to $H + H^- \rightarrow H_2 + e^-$ or, more efficiently, by surface reaction (from two H atoms) on grains. Among the more complex molecules found in dense and comparatively hot ($T > 100$ K) clouds such as the Large Molecular Heimat in the constellation of Sagittarius are glycolaldehyde (the simplest sugar), the glycine precursor aminoacetonitrile, and vinyl cyanide, a potential precursor for alanine. Another recent discovery is large anions such as HC_8^- and NC_5^-.

Interstellar dust grains are nano- to micrometer-sized, usually porous particles that, in dense clouds, are coated with ice (H_2O, CO_2, CO, NH_3, CH_4, CH_3OH, and HCN). The cores of the grains consist of (partially hydrogenated) PAHs, graphitic carbon, SiC, and (essentially amorphous) magnesium-rich silicates. Other grain types have cores rich in iron sulfides. The interstellar abundance of a couple of molecules can only be accounted for if grain surface reactions (with subsequent desorption from the surface) are taken into consideration. A prominent example is methanol, which forms by grain-assisted reaction between H and CO, and – in the ice coatings – from $H_2O + CH_2$ or $OH + CH_3$. The generation of the radicals CH_2 and CH_3 is initiated by UV and cosmic rays.

References

1. Hoyle, F. (1957) *The Black Cloud*, Heinemann, London.
2. Ziurys, L.M. (2006) The chemistry in circumstellar envelopes of evolved stars: following the origin of the elements to the origin of life. *Proc. Natl. Acad. Sci. U. S. A.*, **103**, 12274–12279.
3. Klemperer, W. (2006) Interstellar chemistry. *Proc. Natl. Acad. Sci. U. S. A.*, **103**, 12232–12234.
4. Dalgarno, A. (2006) The galactic cosmic ray ionization rate. *Proc. Natl. Acad. Sci. U. S. A.*, **103**, 12269–12273.
5. (a) Oka, T. (2006) Interstellar H_3^+. *Proc. Natl. Acad. Sci. U. S. A.*, **103**, 12235–12242.; (b) Roberts, H., Herbst, E., and Millar, T.J. (2003) Enhanced deuterium fractionation in dense interstellar cores resulting from multiply deuterated H_3^+. *Astrophys. J.*, **591**, L41–L44.; (c) Roberts, H., and Millar, T.J. (2006) Deuterated H_3^+ as a probe of isotope fractionation in star-forming regions. *Philos. Trans. R. Soc. A*, **364**, 3063–3080.
6. Herbst, E. (2005) Molecular ions in interstellar reaction networks. *J. Phys. Conf. Ser.*, **4**, 17–25.
7. Knauth, D.C., Andersson, B.-G., McCandliss, S.R., and Moos, H.W. (2004) The interstellar N_2 abundance towards HD 124314 from far-ultraviolet observations. *Nature*, **429**, 636–638.
8. Dalgarno, A., and Black, J.H. (1976) Molecule formation in the interstellar gas. *Rep. Prog. Phys.*, **39**, 573–612.
9. Guélin, M., Muller, S., Cernicharo, J., McCarthy, M.C., and Thaddeus, P. (2004) Detection of the SiNC radical in IRC+10216. *Astron. Astrophys.*, **426**, L49–L52.
10. Richardson, N.A., Yamaguchi, Y., and Schaefer, H.F. III (2003) Isomerization of the interstellar molecule silicon cyanide to silicon isocyanide through two transition states. *J. Chem. Phys.*, **119**, 12946–12955.
11. Lucas, R., and Liszt, H.S. (2002) Comparative chemistry of diffuse clouds.

III Sulfur-bearing molecules. *Atronom. Astrophys.*, **384**, 1054–1061.
12 Shull, J.M., Tumlinson, J., Jenkins, E.B., Moos, H.W., Rachford, B.L., Savage, B.D., Sembach, K.R., Snow, T.P., Sonneborn, G., York, D.G., Blair, W.P., Green, J.C., Friedmann, S.D., and Sahnow, D.J. (2000) Far ultraviolet spectroscopic explorer observations of diffuse interstellar molecular hydrogen. *Astrophys. J.*, **538**, L73–L76.
13 McCall, B.J., Geballe, T.R., Hinkle, K.H., and Oka, T. (1999) Observation of H_3^+ in dense molecular clouds. *Astrophys. J.*, **522**, 338–348.
14 Indriolo, N., Geballe, T.R., Oka, T., and McCall, B.J. (2007) H_3^+ in diffuse interstellar clouds: a tracer for the cosmic-ray ionization rate. *Astrophys. J.*, **671**, 1736–14747.
15 Vidal-Madjar, A., and Ferlet, R. (2002) Hydrogen column density evaluation towards Capella: consequences on the interstellar deuterium abundance. *Astrophys. J.*, **571**, L169–L172.
16 Welty, D.E., Simon, T., and Hobbs, L.M. (2008) Spatial and temporal variations in interstellar absorption towards HD 72127AB. *Mon. Not. R. Astron. Soc.*, **388**, 323–334.
17 Słyk, K., Bondar, A.V., Galazutdinov, G.A., and Krełoeski, J. (2008) CN column densities and excitation temperatures. *Mon. Not. R. Astron. Soc.*, **390**, 1733–1750.
18 (a) Rydbeck, O.E.H., Elldér, J., and Irvine, W.M. (1973) Radio detection of interstellar CH. *Nature*, **246**, 466–468.; (b) McCarthy, M.C., Mohamed, S., Brown, J.M., and Thaddeus, P. (2006) Detection of low frequency lambda-doublet transitions of the free ^{12}CH and ^{13}CH radicals. *Proc. Natl. Acad. Sci. U. S. A.*, **103**, 12263–12268.
19 Weselak, T., Galazutdinov, G., Beletsky, Y., and Krełowski, J. (2009) The relation between interstellar OH and other simple molecules. *Astron. Astrophys.*, **499**, 783–787.
20 Burgh, E.B., France, K., and McCandliss, S.R. (2007) Direct measurement of the ratio of carbon monoxide to molecular hydrogen in the diffuse interstellar medium. *Astrophys. J.*, **658**, 446–454.
21 (a) Scoville, N.Z., Hall, D.N.B., Kleinmann, S.G., and Ridgeway, S.T. (1979) Detection of CO band emission in the Becklin-Neugebauer object. *Astrophys. J.*, **232**, L121–L124.; (b) Geballe, T.R., and Wade, R. (1985) Infrared spectroscopy of carbon monoxide in GL 2591 and OMC-1:IRc2. *Astrophys. J.*, **291**, L55–L58.
22 Van der Tak, F.F.S., van Dishoeck, E.F., Evans, II, N.J., Bakker, E.C., and Blake, G.A. (1999) The impact of the massive young star GL 2591 on its circumstellar material: Temperature, density, and velocity structure. *Astrophys. J.*, **522**, 991–1010.
23 Sheffer, Y., Rogers, M., Federman, S.R., Lambert, D.L., and Gredel, R. (2007) Hubble Space Telescope survey of interstellar ^{12}CO/^{13}CO in the solar neighborhood. *Astrophys. J.*, **667**, 1002–1016.
24 Gredel, R., Black, J.H., and Yan, M. (2001) Interstellar C_2 and CN towards the Cyg OB2 association: a case study of X-ray induced chemistry. *Astron. Astrophys.*, **375**, 553–565.
25 Maier, J.P., Lakin, N.M., Walker, G.A.H., and Bohlender, D.A. (2001) Detection of C_3 in diffuse interstellar clouds. *Astrophys. J.*, **553**, 267–273.
26 Ziurys, L.M. (Oct: 2006) Initial observations with an ALMA band 6 mixer-preamp: exciting prospects for the future. *NRAO Newsletter*.
27 Snyder, L.E. (2006) Interferometric observations of large biologically interesting interstellar and cometary molecules. *Proc. Natl. Acad. Sci. U. S. A.*, **103**, 12243–12248.
28 Belloche, A., Garrod, R.T., Müller, H.S.P., Menten, K.M., Comito, C., and Schilke, P. (2009) Increased complexity in interstellar chemistry: detection and chemical modelling of ethyl formate and n-propyl cyanide in Sagittarius B2(N). *Astron. Astrophys.*, **499**, 215–232.
29 Belloche, A., Menten, K.M., Comito, C., Müller, H.S.P., Schilke, P., Ott, J., Thorwirth, S., and Hieret, C. (2008) Detection of amino acetonitrile in Sgr

B2(N). *Astron. Astrophys.*, **482**, 179–196.

30 Hollis, J.M., Remijan, A.T., Jewell, P.R., and Lovas, F.J. (2006) Cyclopropenonen (c-H_2C_3O): a new interstellar ring molecule. *Astrophys. J.*, **642**, 933–939.

31 Herbst, E., and Osamura, Y. (2008) Calculations on the formation rates and mechanisms for C_nH anions in interstellar and circumstellar media. *Astrophys. J.*, **679**, 1670–1679.

32 Cordiner, M.A., Millar, T.J., Walsh, C., Herbst, E., Lis, D.C., Bell, T.A., and Roueff, E. (2008) Organic molecular anions in interstellar and circumstellar environments, in *Organic Matter in Space. Proc. IAU Symp*, vol. **251** (eds S. Kwok and S. Sandford), pp. 157–160.

33 (a) Ramijan, A.J., Hollis, J.M., Lovas, F.J., Cordiner, M.A., Millar, T.J., Markwick-Kemper, A.J., and Jewell, P.R. (2007) Detection of C_8H^- and comparison with C_8H towards IRC+10216. *Astrophys. J.*, **664**, L47–L50.; (b) Brünken, S., Gupta, H., Gottlieb, C.A., McCarthy, M.C., and Thaddeus, P. (2007) Detection of the carbon chain ion C_8H^- in TMC-1. *Astrophys. J.*, **664**, L43–L46.

34 McCarthy, M.C., Gottlieb, C.A., Gupta, H., and Thaddeus, P. (2006) Laboratory and astronomical identification of the negative molecular ion C_6H^-. *Astrophys. J.*, **652**, L141–L144.

35 Thaddeus, P., Gottlieb, C.A., Gupta, H., Brünken, S., and McCarthy, M.C. (2008) Laboratory and astronomical identification of the negative molecular ion C_3N^-. *Astrophys. J.*, **677**, 1132–1139.

36 Voshchinnikov, N.V., Il'in, V.B., and Henning, T. (2005) Modelling the optical properties of composite and porous interstellar grains. *Astron. Astrophys.*, **429**, 371–381.

37 Draine, B.T. (2003) Interstellar dust, in *Carnagie Observatories Astrophysics Ser.*, vol. **4** (eds A. McWilliam and M. Rauch), Cambridge University Press, Cambridge, pp. 1–18.

38 Min, M., Waters, L.B.F.M., de Koter, A., Hovenier, J.W., Keller, L.P., and Markwick-Kemper, F. (2008) The shape and composition of interstellar silicate grains. *Astron. Astrophys.*, **462**, 667–676, and Erratum, *ibid.* 486, 779–780.

39 (a) Duley, W.W., and Grishko, V.I. (2003) Some' solid facts' on interstellar dust. *Astrophys Space Sci.*, **285**, 699–708.; (b) Snow, T.P. (2004) Absorption spectroscopy of interstellar dust, in *Astrophysics of Dust. ASP Conf. Ser.*, vol. **309** (eds A.N. Witt, G.C. Clayton, and B.T. Draine), Astronomical Society of the Pacific, pp. 93–114.

40 Uchida, K.I., Selgren, K., Werner, M.W., and Houdashelt, M.L. (2000) Infrared space observatory mid-infrared spectra of reflection nebulae. *Astrophys. J.*, **530**, 817–833.

41 Spencer, M.K., Hammond, M.R., and Zare, R.N. (2008) Laser mass spectrometric detection of extraterrestrial aromatic molecules: mini-review and examination of pulsed heating effects. *Proc. Natl. Acad. Sci. U. S. A.*, **105**, 18096–18101.

42 Draine, B.T. (2003) Scattering by interstellar dust grains: II. X-rays. *Astrophys. J.*, **598**, 1026–1037.

43 Bauleo, P.M., and Martino, J.R. (2009) The dawn of the particle astronomy era in ultra-high-energy cosmic rays. *Nature*, **458**, 847–851.

44 Herbst, E., and Cuppen, H.M. (2006) Monte Carlo studies of surface chemistry and nonthermal desorption involving interstellar grains. *Proc. Natl. Acad. Sci. U. S. A.*, **103**, 12257–12262.

45 Whittet, D.C.B., Gibb, E.L., and Nummelin, A. (2001) Interstellar ices as a source of CN-bearing molecules in protoplanetary disks. *Orig. Life Evol. Biosph.*, **31**, 157–165.

46 Andreazza, H.J., Fitzgerald, M., and Bowie, J.H. (2006) The formation of the stable radicals •CH_2CN, $CH_2^•CHCN$ and •CH_2CH_2CN from the anions $^-CH_2CN$, CH_{3b}^-CHCN and $^-CH_2CH_2CN$ in the gas phase. A joint experimental and theoretical study. *Org. Biomol. Chem.*, **4**, 2466–2472.

47 Elsila, J.E., Dworkin, J.P., Bernstein, M.P., Martin, M.P., and Sandford, S.A. (2007) Mechanism of amino acid formation in interstellar ice analogs. *Astrophys. J.*, **660**, 911–918.

48 Wada, A., Mochizuki, N., and Hiraoka, K. (2006) Methanol formation from electron-irradiated mixed H_2O/CH_4 ice at 10 K. *Astrophys. J.*, **644**, 300–306.

49 Garrod, R., Park, I.H., Caselli, P., and Herbst, E. (2006) Are gas-phase models of interstellar chemistry tenable? The case of methanol. *Faraday Discuss.*, **133**, 51–62.

50 Winnewisser, M., and Winnewisser, G. (1983) Zur Physik und Chemie der Interstellaren Materie. *Phys. Bl.*, **39**, 9–15.

5
The Solar System

5.1
Overview

According to the nebula theory, our Solar System came into existence about 4.57 billion years ago, when a sluggardly revolving interstellar cloud of gas (essentially hydrogen) and dust became gravitationally instable and collapsed to form a cluster of protostars, among them our protosun, surrounded by a collapsing envelope of gas and dust and a circum-stellar disk in gravitational contact with the protosun, providing material to be accreted onto the growing Sun [1]. To preserve the angular momentum, the contracting matter increasingly accelerated its rotation, and flattened to finally form the proto-planetary disk, today's Solar ecliptic. The envelope gradually became dispersed, and accretion onto the Sun came to a halt, leaving remainders of gas and submicrometer-sized dust grains (about 1% of the overall mass of the presolar system) in an increasingly flattened disk. Coagulation of the grains to larger particles finally led to planetary embryos (planetesimals) and the protoplanets.

The contraction of the central Solar body was accompanied by a steady increase in the core temperature via conversion of the kinetic energy of the atoms and ions into heat. At this time of the evolution, the Sun probably resembled a *T* Tauri star (Section 3.1.1). The temperature increase to $>5 \times 10^6$ K finally initiated nuclear fusion: the Sun became a hydrogen burner, producing energy, and thus sustaining its temperature by fusion of four protons to form helium (=α particles) plus two positrons plus two neutrinos; see Eq. (5.1) for the net reaction; the energy stemming from the conversion of the mass defect accompanying this fusion to energy. The very high temperature caused a fractionation of the elements in the surrounding disk: the less volatile "metals"[1] condensed in areas comparatively close to the Sun, where they later formed the terrestrial planets, mainly constituting iron, nickel, silicates, and alumosilicates, while hydrogen and helium (but also methane, ammonia, water, and carbon oxides) were blown outward by the Solar wind, where they formed the gas giant planets (Jupiter and Saturn) and the ice giants (Uranus and Neptune). The formation of plannets from the circumsolar

---

1) In astronomy, the term *metal* subsumes all elements beyond helium.

Chemistry in Space: From Interstellar Matter to the Origin of Life. Dieter Rehder
© 2010 WILEY-VCH Verlag GmbH & Co. KGaA, Weinheim
ISBN: 978-3-527-32689-1

cloud took place in several successive steps: accretion of dust particles, by gentle collision and electrostatic forces, to mm- and cm-sized grains, grains to rocks and km-sized clumps, from which the planetisimals (>100 km) and ultimately the planets formed:

$$4\,{}^{1}_{1}H \rightarrow {}^{4}_{2}He + 2\beta^{+} + 2\nu\,(+2.7\times 10^{9}\,\text{kJ mol}^{-1}) \tag{5.1}$$

This overall development lasted a few million years only, plus additional several 100 million years until the complete system achieved today's relative stability. Intermittently, that is, in the time span between ca. 4.6 and 3.9 billion years (Ga) ago, reorganization of the planetary system took place: following the generally accepted Nice model (named for the French Mediterranean metropolis Nice, at the University of which this model was developed around 2005), the giant (or Jovian) planets moved from an initially more compact configuration closer to the Sun into their present positions. This reorganization by the dynamic evolution of the orbits of the giant planets also initiating the grouping of small bodies in the outer Solar System (such as the Trojans and the Kuiper belt objects; see below), the translocation of Kuiper belt objects and their insertion into the outer asteroid belt, and the "late heavy bombardment," 3.9 Ga ago, of the inner (or terrestrial) planets and our Moon by planetisimals gravitationally scattered by the giant planets.

The objects that constitute our Solar System (cf. Figure 5.1) can be categorized in the following way:

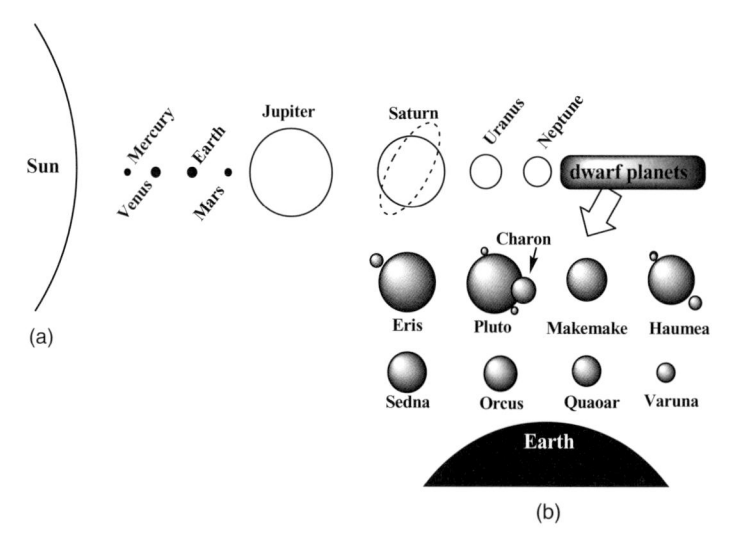

Figure 5.1 The size of planets is shown in relation to the Sun (a), and the size of the dwarf planets (without Ixion, whose size is still uncertain) in relation to Earth (b). For the dwarf planets, the moons are also shown.

- *The Sun* as the central star, accounting for the main mass of the solar system, viz. 332 946 Earths, or 99.86% of the overall mass. As other stars in our Universe, the Sun mainly consists of hydrogen (78.5% by mass; H atoms in the very outer sphere, H^+, i.e., naked protons, in the inner zones) and helium (19.7%; essentially $^4He^{2+}$ nuclei). The main minor constituents are oxygen (0.86%), carbon (0.4%), and iron (0.14%). Our Sun is classified as a G2 main sequence population I star (see Section 3.1.1), with a surface (photospheric) temperature of ca. 5800 K and a core temperature of 15.6×10^6 K. The energy produced in the core (which extends to ca. 0.25 Solar radii, $r_\odot$) is transported by thermal radiation up to 0.7 $r_\odot$, and further by thermal convection to the photosphere. The density changes from 160 g cm^{-3} in the core to 0.2×10^{-6} g cm^{-3} in the photosphere. The coolest region, with temperatures of ca. 4100 K, lies just beyond the photosphere, which is followed by the hot corona and the heliosphere. The heliosphere, defined by the Solar wind (essentially protons), extends to its outer limit, the heliopause in the Kuiper belt at ~100 astronomical unit (AU). Here, the protons bump into particles of the local interstellar cloud.

- *The planets* and their moons. Most of us are used to nine planets (Mercury, Venus, Earth (Terra), Mars, Jupiter, Saturn, Uranus, Neptune, and Pluto, going outward from the Sun). Planets have originally been defined as sizable bodies circling the Sun (or any sun, as extra-solar systems are included). The term *planet* has been redefined by the International Astronomical Union (IAU) in 2006, and although this definition is not without controversy, it will be considered here: (1) a planet is a body orbiting the Sun; (2) a planet has sufficient mass to attain thermodynamic equilibrium, that is, it assumes an about globular shape; (3) a planet dominates its orbit (has cleared off its orbit from other objects). With this definition, Pluto, which fulfils conditions (1) and (2) only, is excluded. According to size and composition, the remaining eight planets fall within two subcategories: (a) the small inner, terrestrial, or earthean planets Mercury, Venus, Earth, and Mars, all "rocky" and with an iron–nickel core, and densities between 3.9 and 5.5 g cm^{-3}; and (b) the huge outer planets (Jupiter, Saturn, Uranus, and Neptune), sometimes subdivided into the gas giants Jupiter and Saturn (with a composition not unlikely that of the Sun) and the ice planets (Uranus and Neptune), which have densities between 0.7 and 1.6 g cm^{-3}. For an overview of some of the characteristics of the planets, see Table 5.1.

- *Moons* are objects which circle the planets, and some of the larger moons of the outer planets are of specific interest because they feature criteria which resemble those of Earth and thus might enable primitive life. There are several mechanisms by which moons came to exist, among these coformation with the planet, capture of an extraplanetary object, and impact debris. Except of Mercury and Venus, all planets (and some of the dwarf planets) do have moons. Gas planets, in particular Saturn, are additionally circumvented by belts of small objects. The rings of Saturn, for example, consist of dust and particulate matter (including water ice), and further accommodate *moonlets*.

Table 5.1a Body and orbital characteristics of the eight planets of our Solar System.

	Diameter at equator[a),b)]	Mass[a),c)]	Mean density ($g\,cm^{-3}$)	Orbital radius[a),d)] (AU)	Orbital period[a),e)] (years)	Rotational period[a)] (days)	Axial tilt (°)
Mercury	0.382	0.055	5.427	0.39	0.241	58.64	0.01
Venus	0.949	0.815	5.204	0.72	0.615	243.02[f)]	177.3
Earth	1	1	5.515	1	1	1	23.44
Mars	0.532	0.107	3.934	1.52	1.881	1.03	25.19
Jupiter	11.21	318	1.326	5.20	11.86	0.41	3.13
Saturn	9.45	95	0.687	9.54	29.46	0.43	26.73
Uranus	4.01	14.6	1.27	19.22	84.0	0.72[f)]	99.97
Neptune	3.88	17.2	1.638	30.06	164.8	0.67	28.32

a) Relative to Earth = 1.
b) The equatorial diameter of Earth is 12756 km.
c) Earth's mass, $m_\oplus$, is 5.97×10^{24} kg.
d) The eccentricity varies from 0.007 (Venus) to 0.206 (Mercury), Earth: 0.017; the inclination with respect to the Sun's equator is between 3.38° (Mercury) and 7.25° (Earth).
e) Indicated is the sidereal orbital period. The sidereal orbital period refers to one full orbit of the planet around the Sun, while the synodic orbital period is the time it takes the planet to reappear at the same point in the sky (relative to the Sun), as observed from Earth.
f) Retrograde rotation, i.e., the planet spins in the opposite direction as it orbits the Sun.

Table 5.1b Additional characteristics of the eight planets of our Solar System.

	Symbol[a)]	Main atmospheric constituents	Atmospheric surface pressure[b)] (MPa)	Magnetic field strength (μT)	Surface temperature (mean) (K)[c)]	Spherical albedo[d)]
Mercury	☿	None[e)]	–	0.5×10^{-3}	90–700 (443)	0.12
Venus	♀	CO_2, N_2	9.3	–	710–770 (735)	0.75
Earth	⊕	N_2, O_2, H_2O, Ar	0.101	30–60	184–331 (287)	0.305
Mars	♂	CO_2, N_2, Ar	0.8×10^{-3}	–	186–268 (227)	0.26
Jupiter	♃	H_2, He, CH_4	$20–200 \times 10^{-3}$	$0.42–1.4 \times 10^3$	(165)	0.343
Saturn	♄	H_2, He, CH_4	n.d.	20	(134)	0.342
Uranus	⛢	H_2, He, CH_4	n.d.	10–110	(76)	0.3
Neptune	♆	H_2, He, CH_4	n.d.	1.4	(72)	0.29

a) Variants and alternatives of these symbols exist.
b) n.d. = not defined.
c) Add 273 to arrive at °C.
d) Albedo ("whiteness"), a surface property of the planets, is the ratio of diffusely reflected to incident light.
e) For Mercury's exosphere see Section 5.2.2.

- *The asteroids.* Most of the asteroids assemble in the asteroid belt between Mars and Jupiter, but some cross Earth's orbit (Alten-type asteroids), the orbits of Earth, Mars, and Venus (Apollo-type asteroids), or approach Earth's orbit from beyond without crossing it (Armor asteroids). Further asteroids have been found with highly eccentric orbits, crossing between Saturn and Uranus and beyond (*centaurs*, e.g., Chiron, also classified as a comet). *Trojans* are asteroids gravitationally locked into synchronization with, and thus sharing, the orbits of Mars (Mars trojans), Jupiter (Jupiter trojans), or Neptune (Neptune trojans). Asteroids encompass numerous rocky (achondrite), carbonaceous, and metallic (iron–nickel) bodies of varying size and shape. Its largest representative is *Ceres* in the asteroid belt, now classified, together with Pluto and other Kuiper belt objects, as dwarf planet. The asteroid belt probably represents remainders from the period of planet formation in a region where accretion to planets had been prevented by gravitational forces exerted by Jupiter, plus objects dislocated from the Kuiper belt. The asteroid objects are therefore valuable messengers, carrying information from the very beginning of our Solar System.

- *The interplanetary medium* includes grains, dust, and gas, the latter mainly in its ionized form (plasma). Its mean density in the Earth's orbit amounts to five particles per cm^{-3} (which compares to extremely diffuse interstellar clouds), its overall mass to that of a small asteroid. Despite of its extremely low volume density, this medium fulfils a couple of important tasks, for example, the stabilization of the Solar magnetic field. Grains hitting Earth's atmosphere are called *meteoroids*; the light streak they produce on burning *meteors*. Remnants from larger interplanetary chunks, which have (partially) survived the impact with our planet, are termed *meteorites*. These again are intriguing messengers from outer worlds. The age of most meteorites has been dated to 4.6 billion years, that is, the age of the Solar System itself.

- *The Kuiper belt* is the trans-Neptunian area in which Pluto (density ca. $2\,g\,cm^{-3}$) circles the Sun. Pluto is not the only celestial body in the Kuiper belt: among the recently discovered objects of comparable size (Figure 5.1) are Eris (originally named Xena), Makemake [pronounced 'maːkiˈmaːki], Haumea, Varuna, Orcus, and Quaoar. All these objects are now classified as *dwarf planets* (for a comparison of characteristics cf. Table 5.2), together with other objects not belonging to the Kuiper belt, such as the asteroid Ceres, or not necessarily being a Kuiper belt object, such as Sedna (see below). The Kuiper belt extends into the *Scattered Disk*, a distant region sparsely populated by icy bodies, extending to about 100 AU and overlapping with the *Extended Scattered Disk*, which reaches out to 1000 AU.

- *The* (hypothetical) *Oort cloud*, an outer halo of our Solar System, is at an average distance of 1–2 light-years (60–120×10^3 AU; cf. Figure 5.2, and Table 5.3 for an overview of units for distances employed in astronomy). The Oort cloud is thus subjected not only to gravitational forces exerted by the Sun, but also from

Table 5.2 Characteristics of dwarf planets in the Solar System[a].

	Mass[b] $10^{-3} m_\oplus$	Orbital radius[c] (AU)	Eccentricity	Inclination to ecliptic (°)	Orbital period[b] (years)	Rotational period[b] (days)
Ceres	0.16	2.5–3.0	0.080	10.59	4.60	0.38
Pluto[d]	2.2	29.7–49.3	0.249	17.14	248.1	6.39[e]
Ixion	(0.1)	30.1–49.4	0.243	19.58	250	Unknown
Orcus	0.13	30.3–48.1	0.226	20.59	245.3	0.57
Haumea	0.7	35.2–51.5	0.189	28.19	285.4	0.16
Makemake	0.7	38.5–53.1	0.16	29	310	Unknown
Eris	2.8	37.8–97.6	0.442	44.2	557	ca. 0.3
Varuna	(0.06)	40.9–45.3	0.051	17.2	283.2	ca. 0.2
Quaoar	(0.4)	42.0–45.3	0.038	7.98	288	Uncertain
Sedna	(0.5)	76.2–975.6	0.855	11.93	12.1×10^3	0.42

a) Ceres belongs to the asteroid belt, Pluto (formerly the ninth planet), Orcus, Haumea, Makemake, Eris, Varuna, and Quaoar are Kuiper belt objects, and Sedna might be classified as an object of the inner Oort cloud or Extended Scattered Disk.
b) Relative to Earth = 1. Indications provided in parentheses are preliminary data.
c) Perihelion and aphelion are indicated.
d) With an albedo of 0.49–0.66, Pluto is almost as brilliant as Venus.
e) Retrograde rotation.

our Sun's nearest stellar neighbor, Proxima Centauri (at a distance of 4.24 ly; a companion of α-Centauri, a binary system at 4.37 ly). These and other gravitational disturbances give rise, from time to time, to disconnections of icy dust bodies ("dirty snowballs") constituting the cloud, then moving toward the Sun and achieving the status of a *comet* (however, not all the comets originate from the Oort cloud).

- *Sedna*, resembling the dwarf planets in the Kuiper belt to some extent, has an extremely eccentric (i.e., elliptical) orbit, which reaches far beyond the Kuiper belt and thus into the space of the Scattered Disk/inner Oort cloud. Sedna's aphelion (the furthest point to the Sun) amounts to 976 AU, its perihelion (the closest point to the Sun) to 76 AU.

The names of the planets and dwarf planets are connected to mythology. As indicated by the symbols of the planets (Table 5.1b), they have also been related to (al)chemical elements (and vice versa): "*Sol gold is, and Luna silver we declare, Mars yron, Mercurie is quyksilver, Saturnus leed, and Jupiter is tyn, and Venus coper, by my fathers kyn*" (Chaucer 1386). The so-called planetary metals were thus dominated, or ruled, by one of the seven planets (Mercury, Venus, Earth, Mars, Jupiter, Saturn, and Uranus) familiar to our ancestors:

- *Mercury*. Messenger of the Roman Gods, known for *speed* and ability, which also stands for the chemical element mercury (*quick*silver), the only metallic element which is liquid at room temperature. The astronomical symbol for

$$\boxed{\tan 1'' = d(\text{AU})/d(\text{pc})}$$

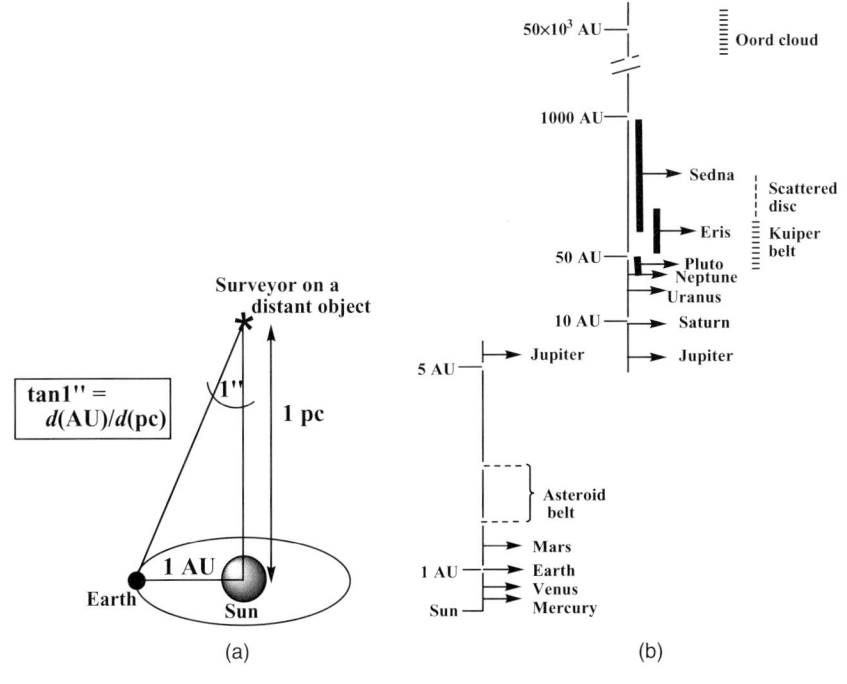

Figure 5.2 (a) Definition of the parsec (parallax of arcsecond, pc) and the AU (the mean distance between the Earth and the Sun). For distance units employed in astronomy and their conversion see Table 5.3. (b) Distances of objects in our Solar System; logarithmic scale for the upper right part (Jupiter to Oort cloud); for distances, cf. also the orbital radii provided in Table 5.1.

Table 5.3 Units for distances employed in astronomy.

	Meter (m)	Astronomical unit (AU)	Parsec (pc)	Light-year[a] (ly)
AU	149.6×10^9	–	4.85×10^{-6}	1.58×10^{-5}
pc	3.086×10^{16}	206.3×10^3	–	3.26
ly	9.46×10^{15}	63.2×10^3	0.307	–

a) 1 ly is the distance the light covers within a year in vacuum. Speed of light c (*celeritas*) = 299.8×10^6 m s^{-1}.

mercury is derived from the Messenger's herald's staff (*caduceus*), a winged stick entwined by two serpents forming left-handed helices.

- *Venus* is (apart from the Moon) the brightest object in the sky. Its brightness is related to its particularly high albedo (and this again to its exceptionally dense CO_2 atmosphere). Apparently based on this brightness, the planet has been named after the Roman Goddess of beauty and love. In its dawn appearance ("Morning Star"), Venus has also been identified with Lucifer (the light-bringer,

Latin translation of the Greek Phosphorus), an angel kicked out of, and tumbled from heaven, after having fallen out of favor with God. Nowadays, Lucifer is rather a synonym for Satan. The symbol, the stylized hand mirror of the Goddess Venus, also represents femininity. It has also been used for the element copper: mirrors had been manufactured from polished copper.

- *Earth*. The symbol used for the planet Earth goes back to historical times and appears to be multicultural. It possibly symbolizes a wheel. Alchemists employed the symbol, along with the one for Venus, for the element copper.

- *Mars*, the red (due to ferric oxide) planet, carries the name of the Roman God of Wars. The symbol, composed of a shield and a spear (or an arrow), representing masculinity, is associated with the element iron.

- *Ceres*, the largest object of the asteroid belt, is named after the Roman Goddess of growing plants, harvest, and motherly love. As for Saturnus, the God of agriculture, a sickle (⚳, upside down with respect to the symbol for Saturn) symbolizes this dwarf planet.

- *Jupiter*, the biggest of the planets, was already named by the Romans after their highest God Jove, ruler and protector of the home and the family life. Jupiter alludes to "the generous good nature of tin, Jove's metal" (Primo Levi, *Il Sistema Periodico*, Einaudi Publ. 1975). Jupiter thus also attains the role of the preserver; tin bins have been – and are still being – used to preserve food. The origin of the symbol may go back to the letter *Z* for Zeus, the Greek equivalent of Jupiter. Alchemists used the Jupiter symbol for zinc, and for tin and tin oxide.

- *Saturn* is named after the Roman God Saturnus (who became the namesake of Saturday). In the Greek mythology, the planet was made sacred to Kronos; Saturnus, the God of agriculture, gardening, and harvest, is the Roman pendant to Kronos. The symbol may stand for a sickle or scythe; it attributes to a gardener. But Saturn, holding an hourglass, is also associated with death, as is lead, the Saturnian element, the hazardous properties of which had well been recognized already in olden times.

- *Uranus* (Latinized for Ouranus, the Greek God of the Sky) is the farther of Kronos (Saturn) and the grand farther of Zeus (Jupiter). Its symbol is a hybrid of those for the Sun (☉) and Mars, the combined power of which dominates the Sky. The symbol is associated with platinum, which the alchemists believed to be an amalgam of silver and gold (☉).

- *Neptune*, at its time of discovery (1846) the planet "deepest in the eternal sea" was named after the Roman God of the Sea. It is symbolized by this God's stylized trident, a symbol which, along with that of the planet Mercury, is conjointly used for the "watery" element quicksilver.

- The Roman God of the Underworld was the inspiration for the name of the planet *Pluto*. Pluto's astronomical symbol (♇) is also supposed to be evocative of the astronomer Parcival Lowell, who became famous for fueling the conception that there were canals on Mars.

- For the remaining dwarf planets in the Kuiper belt, also mythological names were chosen: *Ixion* (in the Greek mythology, King Ixion became famous for assassinating his father-in-law), *Orcus* (Roman God of the Underworld), *Haumea* (the patron Goddess of the island of Hawai'i), *Makemake* (the creator of humanity and God of Fertility in the myths of the native people of Easter Island), *Eris* (the Greek Goddess of Discord, responsible for the Trojan War; proposed symbol: Eris' hand ⚳), *Varuna* (the ancient Indian deity of the waters and the heavens, and guardian of immortality), *Quaoar* (the Creator God of the Tongva, the native people around the area of Los Angeles).
- *Sedna* is named after the Inuit Goddess of the Sea, believed to live in the cold depths of the Arctic Ocean.

5.2
Earth's Moon and the Terrestrial Planets: Mercury, Venus, and Mars

As a consequence of the genesis of our Solar System, the planets Mercury, Venus and Mars, together with our home planet, disclose pretty similar features as far as their core and mantle composition is concerned; they differ, however, as it comes to surface structure and atmosphere. The similarities in mantle composition extend to the moons (Earth's Moon, and Mars' Phobos and Deimos) and the greater part of the objects of the asteroid belt. Phobos ("fear") and Deimos ("dread") are, in Greek mythology, the sons of Ares (Mars for the Romans), who is the God of War. They are tiny irregularly formed rocks with a largest elongation of 27 (Deimos) and 10 km (Phobos). It is believed that these objects have been captured by Mars from the asteroid belt.

5.2.1
The Moon

Contrasting the Martian moons, Earth's Moon (Luna; astronomical symbol ☾) is quite sizable: its mean diameter amounts to 3476 km (about ¼ of Earth), its mass to 0.013 $m_\oplus$, which is 5.6 times the mass of the dwarf planet Pluto. Lacking an extensive iron–nickel core, the Moon's density is just 3.3 g cm^{-3}, as compared to 5.52 g cm^{-3} of its mother planet. Luna's age has been determined to be 4.527 ± 0.01 billion years [2], that is, about the age of the Solar System as a whole, but it still formed a little bit later (about 30–50 million years) than Mars and the asteroids. These findings challenge, among other evidence, the popular theory according to which our Moon separated from early Earth, leaving apart the Pacific basin, and support the giant impact/ejected ring theory, proposing that the Moon formed from the material of a disk orbiting proto-Earth, material which became ejected on occasion of a collision with an about Mars-sized protoplanet in the early stage of planet formation.

The dating of the Moon's age or, rather, the time when the Lunar magma ocean crystallized is based on the ratio of the isotopes tungsten-182 and tungsten-184 in silicate and metal constituents of Lunar rock samples brought back to Earth by the Apollo mission. ^{182}W is formed by β^- decay of ^{182}Hf (half-life $t_{1/2} = 8.9 \times 10^6$ years; via ^{182}Ta, $t_{1/2} = 114.4$ days), Eq. (5.2a), but also results from cosmic neutron capture by ^{181}Ta; Eq. (5.2b). In order to exclude errors caused by ^{182}W formed through neutron capture, the ^{182}Hf–^{182}Ta chronometry requires resorting to material which does not contain tantalum-181.

Hafnium is a lithophile (silicate loving) element, while tungsten is siderophile (iron loving), that is, during fractionation in the very early period of the formation of the minerals constituting the Lunar crust (and the Lunar magma ocean in particular), hafnium became enriched in silicates and depleted in the refractory metals. The fractionation of ^{182}Hf originally present in the stony and metallic constituents of various rock types furnished, within the first about 50 million years (the effective lifetime of ^{182}Hf), variations of the ^{182}W/^{184}W ratio, viz. a relative excess abundance of ^{182}W in the silicates, and a relative deficiency of ^{182}W in the sidereal (metal-rich) fraction. This apportionment can be exploited for dating celestial first hour bodies [3a, b], a method that has also widely been employed to determine the age of the meteorites and of Mars [3c]:

$$^{182}_{72}\text{Hf}(\beta^-,\gamma)^{182}_{73}\text{Ta}(\beta^-,\gamma)^{182}_{74}\text{W} \tag{5.2a}$$

$$^{181}_{73}\text{Ta}(n,\beta^-)^{182}_{74}\text{W} \tag{5.2b}$$

The most prominent extended geological formations on our Moon are (i) the highlands, constituting a crust (between 35 and 65 km thick) that emerged in the course of the solidification within about first 100 million years, and (ii) the basalt maria[2] formed in the aftermath of volcanic eruptions, clearly distinct not only by their origin and, to a certain extent, chemical composition, but also by their albedo: while the highlands have a high albedo, the maria are comparatively dark. The crust is dominated by olivine and pyroxene, which crystallized first, anorthositic (i.e., calcium-rich) plagioclase feldspar, and KREEP-rich magma, where KREEP is short for *k*alium (=potassium), *r*are *e*arth *e*lements (the lanthanides), and *p*hosphorus (in the form of phosphate). The basaltic eruptions, covering about 17% of the Lunar surface, but making up for only 1% of the overall crust, exhibit a higher iron content and more elevated titanium levels (in the form of ilmenite $FeTiO_3$) than basalts from Earth. Pyroclastic glasses, that is, glasses originating from volcanic eruptions, which have developed as secondary products, can have titanium contents between 1% (appearing greenish) and about 14% (reddish appearance). The chemical compositions of the minerals mentioned above are provided in Table 5.4. Included is a mineral, armalcolite, which occurs on the Moon along with ilmenite. Armalcolite, named after the three Apollo astronauts (*Arm*strong, *Al*drin, and *Col*ins), was originally believed to be a mineral typical of the Moon. Meanwhile, this mineral has also been traced on Earth.

The dark maria so characteristic of the Moon's near side are almost completely missing on the far side. Other asymmetries are the predominance of KREEP and a higher concentration of iron and titanium on the facing side, and a mightier

---

[2] Singular: mare, Latin for "sea."

Table 5.4 Chemical composition (approximate) of some minerals in the lunar crust.[a]

Plagioclase feldspar[b]	Solid solution of $NaAlSi_3O_8$ and $CaAl_2Si_2O_8$ (anorthite)
Anorthosite	Plagioclase feldspar rich in Ca
Olivine	$(Mg,Fe)_2SiO_4$
Pyroxene	$M,M'(Si,Al)_2O_6$; M,M' mainly Mg^{2+}, $Fe^{2+/3+}$, Co^{2+}, Mn^{2+}, Na^+
Pyroxferroite	$CaFe_6Si_7O_{21}$
Ilmenite	$FeTiO_3$
Armalcolite	$Mg,FeTi_2O_5$
Christobalite, tridymite	SiO_2
Basalt	Mixture of Fe- and Mg-silicates, plus olivine, pyroxene, and plagioclase

a) Except of armalcolite, only main minerals (>1%) are collated here.
b) When the magma oceans cooled, feldspars crystallized first and rose to the top, forming a floating crust, today's highlands.

crust on the far side. The reasons for these asymmetries are not quite clear. A "lop-side" convection as the magma oceans started to cool down and solidify from the top has been proposed to have caused this imbalance, lop-side because the facing side of the Moon, subjected to the additional faint irradiation received from the sunlight reflected from Earth, was slightly warmer than the opposite side. This explanation presupposes that the Moon was in synchronous rotation (sidereal rotation period = orbital period = 27.3 days) and thus locked in its orientation to Earth already at an early stage.

Secondary features on the Moon's surface are due to impact cratering and volcanism, and weathering as a consequence of permanent bombardment by cosmic particles and the Solar wind (impact erosion), generating a finely grained, pulverized surface material termed *regolith*, which also covers Mercury's surface, and gives rise to a comparatively low albedo. Based on the number of impact craters on the far side of the Moon, evaluated by the SELENE mission, volcanic activity continued until 2.5 billion years ago. SELENE stands for *Selenological and Engineering Explorer*; this Japanese Lunar orbiter spacecraft was launched in 2007. Selene is also synonym for the Greek Goddess of the Moon, and the namesake for the chemical element selenium, referring to its chemical similarity to previously discovered tellurium (*Tellus* = Earth). Volcanic activity included fire fountain eruptions, which are well known from active volcanic areas on Earth. Contrasting Earth, were these eruptions are due to extrusion of CO_2 (plus some SO_2 and eventually H_2O) originally trapped in olivine-rich basalts, Lunar samples point to a CO-rich gas phase and thus to a less oxidizing potential of Lunar basalts. CO forms by oxidation of minor graphite inclusions in depths of about 8 km. Oxidizing agent is mainly ferrous oxide FeO, which is converted to metallic iron.

Iron-rich, μm-sized FeNiCo metal spherules have in fact been found in picritic glasses of Lunar magmas [3d]. "Picritic" refers to enrichment of high-Mg olivine.

Recent investigations of rock samples brought back to Earth in the frame of the Apollo mission 1969 revealed the presence of several ppm of water in volcanic glass and tiny crystals of apatite. On its flyby of the Moon in 1999, the Cassini spacecraft detected IR absorptions at 2.8 and 3 μm, typical of the fundamental absorption of the OH group, and hinting toward absorbed (on mineral surfaces) water in case of the 3-μm band. To explain the presence of water on the Moon, essentially three possibilities became discussed: Solar wind; the steady precipitation of micrometeorites; impacts of cometary meteorites in particular in the early phase, about 4 billion years ago. The main component of the Solar wind is protons, which can interact with oxides in the Lunar surface to form hydroxyl and water, supported by $Fe^{2+/3+}$. While this source of water, as well as water originating from micrometeorites, is likely to contribute, the main amount of water found on the Moon appears to originate from impacting comets. Comets are "dirty snowballs" (Section 5.4) mainly consisting of water-ice, dust, and mineral debris. In October 2009, the *Lunar Crater Observation and Sensing Satellite* (LCROSS) crashed two impactors into the crater Cabeus, releasing a plume of dust and vapor, which was analyzed by instruments on the satellite. Cabeus, a 90-km diameter crater near the South Pole, is permanently shaded, and thus a cold trap (of an estimated 40 K) for volatiles. The detection of considerable amounts of water in the plumes, both by absorptions in the IR and emissions in the UV, along with trace amounts of methane, ethanol, ammonia, and carbon dioxide, clearly hints toward a cometary origin of the assumed ice sheets at the bottom of Cabeus, water-ice deposits which likely cover permanently shaded floors of other craters as well. Permanently shadowed craters at the Moon's North Pole can be as cold as 26 K, but several tens of cm of Lunar soil are enough to preserve deeper water ice even outside permanently shadowed craters.

5.2.2
Mercury

Our Moon does have several features in common with Mercury, in particular as it comes to surface structures, and this is a consequence of both celestial bodies being devoid of an atmosphere comparable to that of the other terrestrial planets, smoothing off pregnant features due to volcanism and impacts by meteorites. Contrasting the Moon, Mercury has been resurfaced by volcanic flows and experienced tectonic crunching throughout its history [4a]. The geology and the chemical composition of Mercury's surface features, although generally comparable to that of the Moon, thus also show distinct peculiarities. Contrasts between bright and dark areas are much less pronounced on Mercury than on the Moon.

With a mean diameter of 4879 km, Mercury is considerably smaller than Earth (12 735 km), but somewhat larger than Luna (3476 km). Its density, 5.43 g cm^{-3}, is almost as high as that of Earth (5.52 g cm^{-3}), indicating that a particularly high proportion of Mercury, about 65% of its overall mass (ca. 40% of its overall volume), is represented by an iron-rich core. This is about twice as much as for any other terrestrial planet. The disparity (with respect to Earth, Mars, and Venus) between metallic core and rocky mantle has been explained by a giant impact event early in Mercury's history, an impact which has partly stripped off the rocky mantle. Despite its comparatively low mass (0.055 that of Earth), Mercury's surface gravity, 0.38 g,[3] is rather high (Moon: 0.16 g).

Mercury, as Earth, has a permanent, centered dipolar magnetic field, though just about 1% of the strength of Earth's magnetic field. The presence of a permanent magnetic field indicates that the outer layer of the iron core is liquid, producing a dynamo effect[4] as the planet rotates. The magnetic field is, however, considerably weaker than expected, and this has been traced back to a stagnant layer at the top of the molten outer core, partly suppressing the dynamo effect. While there is a massive iron core, possibly also containing some nickel and sulfur, the average abundance of iron in Mercury's surface material, and by inference its crust and mantle, is less than 6% (by weight) [4b], and hence about, or beyond, the percentage of iron in the crust of the other inner planets. Mercurian olivine and pyroxene contain very little iron, whereas iron constitutes up to 20% of these minerals on Earth. The iron contents in Mercury's surface materials were estimated by the Gamma Ray and Neutron Spectrometer (GCNS) of Messenger: highly energetic protons emitted by the Sun and coming from the inter-space (cosmic rays) dislodge highly energetic neutrons on impacting the elements of the surface materials. These neutrons are partly slowed down by interaction with other atoms, loosing much of their energy, and hence becoming thermal neutrons, that is, their energy of motion compares to that of gas molecules at ambient temperature. Thermal neutrons can be captured by ^{56}Fe. The resulting excited $\{^{57}$Fe$\}$* is a γ emitter, Eq. (5.3), and the characteristic 7.6 MeV γ rays were detected by the GCNS:

$$^{56}_{26}\text{Fe} + ^{1}_{0}\text{n} \rightarrow \{^{57}_{26}\text{Fe}\}^* \rightarrow ^{57}_{26}\text{Fe} + \gamma \tag{5.3}$$

Orbital features of Mercury are markedly distinct from those of the other planets. Mercury's orbit is rather elliptic: the distance from the Sun in its aphelion is 0.467 AU, in its perihelion 0.307 AU. Its orbital period amounts to 88 terrestrial days, its rotational period to 58.6 days, hence a 3:2 orbit–spin resonance (rotating three times for every two revolutions around the Sun). A day on Mercury lasts 176 Earth days. The slow precession of the perihelion of Mercury's orbit (i.e., the change of the orientation of the perihelion within the orbit's plane) is a valuable empirical foundation for the general theory of relativity introduced by Einstein in 1915. Accordingly, this shift is a consequence of the curvature of spacetime brought about by the Sun's gravity.

Spectra ranging from the ultraviolet to infrared, taken by Messenger's Atmospheric and Surface Composition Spectrometer, reveal some general similarities in

---

3) 1 g = 9.80665 m/s² is the mean surface gravity on Earth.

4) A rotating, electrically conducting fluid induces and maintains a magnetic field, if there is also convection in the fluid. This effect, termed *dynamo effect*, is responsible for Earth's and Mercury's magnetic fields. Convection within the liquid outer core occurs between the solid inner core and the solid mantle. The conducting fluid can be metallic (iron or iron/nickel) as in the case of Mercury and Earth, or ionic as in the case of the Sun. For Venus, the rotation is too slow to produce a magnetic field. With the Sun, the dynamo effect comes about by differing rotation periods for equatorial vs. higher latitude areas (a body rotating like a solid ball does not induce a magnetic field).

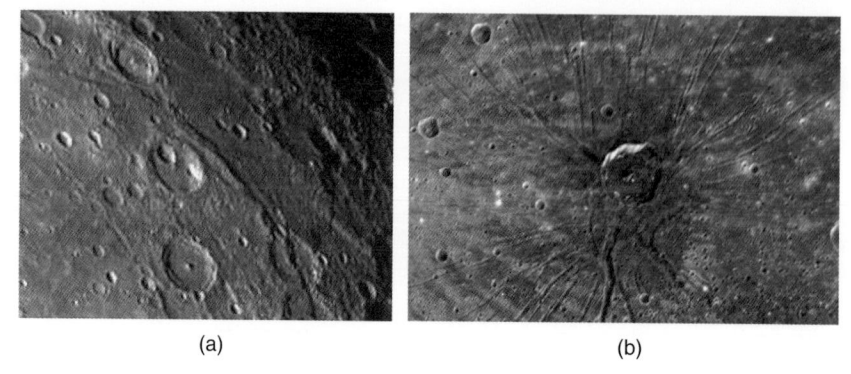

Figure 5.3 Details of Mercury's surface as photographed during the Messenger flyby in 2008. Shown are impact craters and a curved cliff (lobate scarp), (a) about 3 km high, extending for more than 300 km (possibly formed as the planet shrank when cooling), and a spider-like network (b) of uncertain origin. The spider arms, narrow troughs (the Pantheon Fossae) spanning out from a central crater about 40 km wide, named Apollodorus, is located in the planet's Caloris basin, which is 1550 km in diameter. Apollodorus is a reputed architect of the Pantheon, a temple in Rome; *fossa* is Latin for trench. Credit: *NASA/John Hopkins University Applied Physics Laboratory/Carnegie Institution of Washington.*

mineral composition of the surface of Earth, Moon, and Mercury but also distinct differences that have yet to be analyzed. "Messenger" stands not only for the fleet-footed God who became the namesake of this planet, but is also an acronym for *MErcury Surface, Space ENvironment, GEochemistry, and Ranging* mission. The spacecraft had its flybys in January and October 2008, and September 2009, and will eventually enter Mercury's orbit in 2011. (In 1974/1975, Mercury was encountered in the frame of the Mariner-10 mission). Distinct differences between the surface *structures* of the Moon and Mercury are (i) shallower craters, (ii) troughs radiating from around impact craters (Figure 5.3b) in the huge lava-filled Caloris[5] basin, formed by a major impact about 3.8 billion years ago (and later modified by volcanic activity), and (iii) cliffs crisscrossing Mercury's surface (Figure 5.3a). These features are believed to indicate a period of shrinking as the planet's iron core progressively cooled.

Since Mercury's axis of rotation is almost perpendicular to the plane of its orbit around the Sun, very much in contrast to our home planet and Mars with an inclination of the axis relative to the ecliptic of 23.4° and 25.2°, respectively, the shadowed floors of its craters in the polar regions never are touched by the warmth of direct radiation of the Sun, and it is possible that there is water ice at the permanently shadowed crater bottoms. At least this is suggested by specific radar signatures of these spots. Temperatures otherwise range from highs of 450 to lows of −180 °C. The water may have been introduced by comets and/or formed by

---

5) Every second time Mercury passes through its perihelion, the Sun is overhead and thus heating up (heat = *calor* in Latin) this basin.

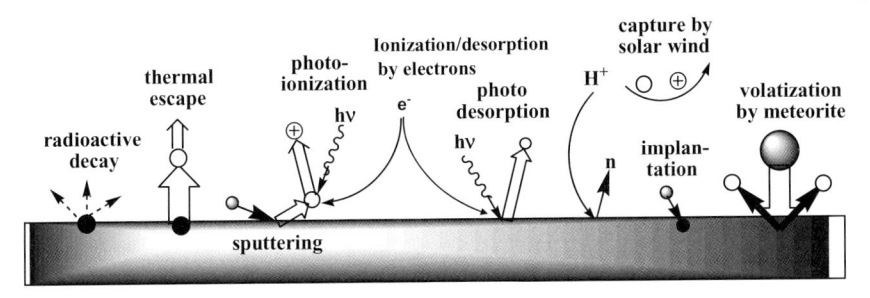

Figure 5.4 Sources and sinks for material in the exosphere of Mercury (modified from Ref. [5a]). H^+ = protons, hv = photons, n = neutrons, e^- = electrons.

chemical sputtering; cf. Eq. (5.6) below. Its overall amount is about 10^{15} kg, comparing to 10^{16} kg in the southern polar cap of Mars, and 10^{18} kg of ice represented by the Earth's Antarctic ice shield.

Mercury does not have a significant permanent "atmosphere". The pressure at ground level is 10^{-15} bar (10 nPa), corresponding to the pressure in Earth's exosphere, and Mercury's "atmosphere" is, therefore, usually referred to as an exosphere. The mean density is about 10^4 particles (mainly ions; see below) per cubic centimeter. The mean free path is approximately 10^8 m, allowing for an essentially free motion of the particles. For comparison, the particle density in Earth's atmosphere at ground level (mainly molecules) is ca. 10^{19} cm^{-3}, the mean free path is just about 10^{-7} m!

Several processes are responsible for providing matter to Mercury's exosphere [5a], and these are summarized in Figure 5.4: hydrogen (H^+) and helium (doubly charged He^{2+}) are mainly supplied by the Solar wind, but some helium, originating from radioactive α-decay, is also provided by thermal desorption, and present as singly charged He^+. Water-related ions like O^+, OH^-, H_2O^+, and H_3O^+ are believed to have been blasted off the surface by Solar winds, and on occasion of impacts with icy meteorites, and/or by secondary reaction cascades (see also below). The abundance of these ions amounts to about one-third of that of sodium. Also noteworthy is the presence of C^+. Among the heavier ions, the metals Na^+, K^+, Mg^+, Ca^+, and Al^+, and the nonmetals S^+ (and H_2S^+), Si^+, O_2^+, and Cl^+ have been detected (Figure 5.5) [5b]. Neutral Na (as well as K and Ca) has been observed remotely from Earth. Sodium makes up about 35% of the exospheric material. The presence of these metallic and nonmetallic species suggests that they originate from minerals in the surface of Mercury, such as pyroxene-like alumosilicates (cf. Table 5.4 in Section 5.2.1; for the generation of Na, K, Ca, Al, and Si, but also O), carbonates (C, O, Na, K, Ca), graphite and carbides (C), oxides (O and metal ions), sulfide deposits, or sulfur of volcanic origin (S). Chlorine may also stem from volcanic activity, but minerals containing (small) amounts of chlorine also come in.

Processes by which these originally nonvolatile exospheric materials are delivered from the regolith and other exposed surface materials encompass sputtering and vaporization by the Solar wind, cosmic rays and micrometeorite impact,

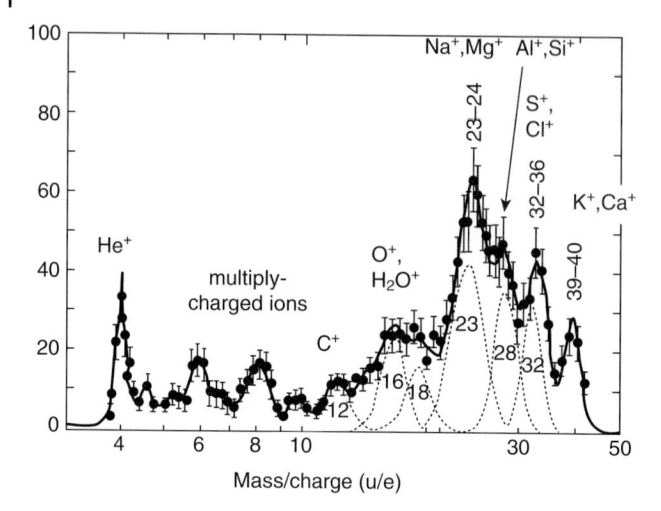

Figure 5.5 Ions detected in Mercury's exosphere on occasion of the Messenger flyby in January 2008. Counts are plotted against the ratio mass (m, in atomic mass units, u) to charge (q, in units of e). The peak for m/q = 32–36 u/e, assigned to S^+ and Cl^+, may also represent H_2S^+. The thin, dashed curves are Gaussian fits to C^+ (m = 12), O^+ (16), H_2O^+ (18), Na^+ (23), Si^+ (28), and S^+ (32 u/e). Multiply-charged ions are likely due to ionization by plasma sheet electrons. Credit: NASA/JHU Appl. Physics/CIW (Mercury Flyby 2, Oct. 6, 2008), modified.

photon (UV impact) stimulated desorption, and meteoritic volatilization (Figure 5.4). Sputtering can be a physical or chemical process. Physical sputtering is the release of material by particle impact, while chemical sputtering presupposes a chemical reaction between the surface-bound mineral and the energetic particle, followed by the (thermal) desorption of the reaction product. Sodium and water may have originated from chemical sputtering initiated by protons, as shown symbolically for a sodium alumosilicates (feldspar) in Eq. (5.4); Si^+ and O_2^+ by dissociative ionization of silica, Eq. (5.5a). A possible reaction path leading to the formation of negatively charged hydroxide OH^- is depicted in Eq. (5.5b). Negatively charged ions, formed by electron capture or electron transfer, have also been detected in interstellar clouds (Sections 4.2.1 and 4.2.4). For a possible reaction sequence leading to the formation of H_3O^+ and further to H_2O, cf. Eq. (5.6), a reaction sequence which, however, relies on the presence (or intermittent formation) of molecular hydrogen:

$$NaSi_3AlO_8 + H^+ \rightarrow Na^+ + \text{"}Si_3AlO_7\text{"} + OH; \quad OH + H^+ \rightarrow H_2O^+ + h\nu \tag{5.4}$$

$$SiO_2 + h\nu \rightarrow Si^+ + O_2^+ + 2e^- \tag{5.5a}$$

$$H_2O + e^- \rightarrow OH^- + H \tag{5.5b}$$

$$O \underset{h\nu}{\overset{H^+}{\rightleftarrows}} O^+ \underset{H}{\overset{H_2}{\rightleftarrows}} OH^+ \underset{H}{\overset{H_2}{\rightleftarrows}} H_2O \underset{H}{\overset{H_2}{\rightleftarrows}} H_3O^+ \overset{e^-}{\rightarrow} H_2O + H \tag{5.6}$$

Since a sizable part of the exospheric material is present in the cationic form, hence forming a plasma, Mercury's dipolar magnetic field helps, at least to a certain extent, to sustain the exosphere in a complex interaction between the plasma and the magnetosphere.[6] Nonetheless, depletion of Mercury's exosphere occurs by thermal escape, mainly of the light elements hydrogen and helium, by entrainment in the Solar wind, by Solar radiation pressure, and by interaction with interplanetary material. The Solar wind is only partly hold at bay by Mercury's weak magnetic field: the magnetopause, at 1700 km above the planet's surface, occasionally opens when interacting with the interplanetary magnetic field, giving rise to a flux transfer and thus allowing for interaction of the interplanetary medium with Mercury's surface [6]. Neutral sodium, calcium, and magnesium, which are present in the magnetotail and in the antisunward exosphere, are sputtered off by high-energy electrons provided by, inter alia, these flux events. Some of the materials also get reimplanted into the surface through absorption or chemical reaction.

The high abundance of sodium in the exosphere has allowed for a "mapping" of the sodium occurrence as a function of distance and angular distribution through the ease of the observability of the typical emissions of excited sodium atoms at 589.6 nm (D1 line) and 589.0 (D2 line), corresponding to transitions from the 3p(1/2) and 3p(3/2) levels to the 3s level.

5.2.3
Venus

5.2.3.1 General, and Geological and Orbit Features

"The air was like wine, a little high in oxygen content, but tinglingly sweet and fresh and warm ..." notes Kilgour, after having stepped out of his spaceship and set foot on Venus; in van Vogt's "A Can of Paint" [7a]; a situation which does not quite coincide, in particular as far as the oxygen contents are concerned, with the atmospheric conditions on our neighbor planet. Venus has so far been visited in the frame of about 30 space missions, starting with the first flyby of *Mariner* 2 in 1962, the first landing (*Venera* 4 in 1967) and transmission of pictures showing surface details (*Venera* 9, 1975), followed by a comprehensive radar mapping carried out by the *Magellan* probe during 1989–1994. In 2006, the spacecraft *Venus Express* collected detailed information of, inter alia, atmospheric double-eyed vortex structures in the planet's Southern polar area. The spacecraft *Messenger* also passed by Venus for a gravity assist en route to Mercury, collecting data from Venus' atmosphere.

Due to its extremely dense atmosphere (Figure 5.6a; for further discussion vide infra), surface details taken on occasion of flybys and orbiting events can only be made visible by radar techniques, revealing a structuring of the surface not so much different from that of Earth (Figure 5.6b), with the exception that there is

---

6) The magnetosphere of a planet is usually defined as the volume (surrounding the planet) that is controlled by the planet's magnetic field.

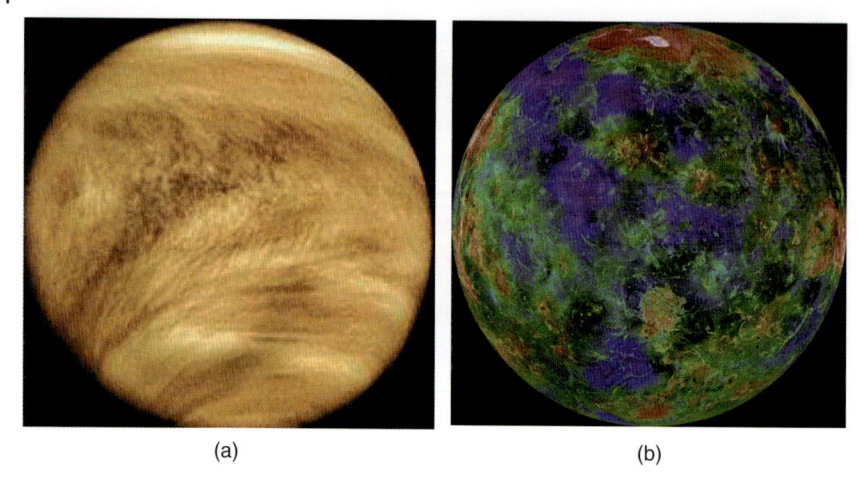

Figure 5.6 (a) Cloud structure of Venus. Credit NASA/USGS (public domain). (b) False color radar topography of Venus at 3 km resolution, taken by Magellan. Red represents mountainous areas, while blue typifies valleys. The large reddish area in the north is Ishtar Terra with the highest mountain on Venus, Maxwell Montes, 10.5 km above average altitude. The large highland regions are analogous to continents on Earth. Credit: Magellan Spacecraft, Arecibo Radio Telescope, NASA/USGS (public domain).

no liquid water on Venus–very much to the distress of those who considered Venus a sister planet of Earth, habitable though plunged with swamps and oceans, and subjected to incessant rain [7b, c]. Venus constantly looses, and has lost, water by cosmic ray dissociation and ionization, forming H^+ and O^+ which permanently escape into space.

One of the intriguing peculiarities of Venus is its retrograde rotation (clockwise; opposite direction with respect to its motion in orbit), which clearly distinguishes it from other planets, save Uranus. As a consequence, the Sun *rises* in the West! A possible reason for the retrograde rotation is a major impact with a very large celestial body early in Venus' history. Another unusual feature is Venus' very slow rotation period of 243 Earth days, only around 9% more than its orbital period of 224.7 days. Every 5 years, Venus comes closest to Earth. Since Venus rotates almost exactly five times between one closest approach to Earth and the next, Venus then always presents nearly the same face to Earth. The length of a Venus day, from noon to noon, is 116.8 Earth days. The slow rotation may have been imposed, during the planet's development within the past about 4.5 billion years, by tidal effects going back to its dense atmosphere, gradually slowing down the motion. The particularly long rotation period of Venus has two important consequences: (i) a negligibly small oblateness, that is, Venus is practically ideally spherical, and (ii) the lack of a permanent magnetic field as known for Mercury and Earth. Further, Venus' orbit is only faintly eccentric, and the inclination of its axis of rotation towards its orbit is just 2.6°; Venus thus does not experience pronounced seasons as Earth or Mars.

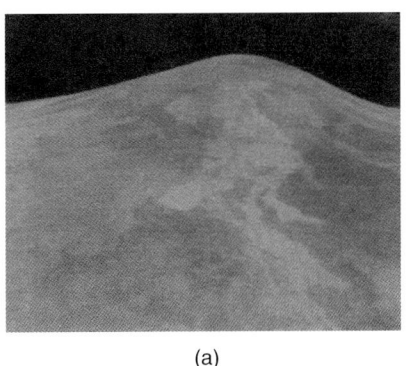

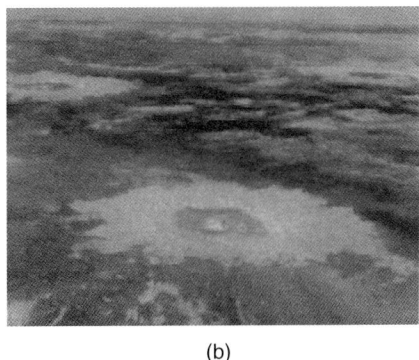

Figure 5.7 Geological details of Venus. (a) Image of the summit of Sif Mons, a shield volcano, formed by lava flows of low viscosity. A series of bright and dark lava flows is visible in the foreground. The brightest flows are associated with the most recent volcanism in the region. They overlay older lava flows which appear darker. (b) Perspective view of the "crater farm" on Venus; the crater in the foreground has a diameter of 37.3 km. The image was created by superimposing Magellan images on topography data; and on data returned from Venera 14. Credit: NASA (public domain).

Venus is devoid of a permanent moon. Currently, however, a quasi-satellite, the asteroid 2002 VE68 is a co-orbital companion of Venus [8]. Its orbit has a high eccentricity of ca. 0.4, that is, this asteroid is also dipping into Mercury's and Earth's sphere of influence. The asteroid has been captured by Venus ca. 7000 years ago, and will eventually escape.

Venus' surface, rolling plains with modest highlands and lowlands, is essentially smooth. The main geological features on Venus are two continent-sized highlands (see Figure 5.6b), Ishta Terra and Aphrodite Terra.[7] There is no evidence for plate tectonic activity as on Earth. Ishta Terra shelters the highest elevation on Venus, the Maxwell Montes massive, which rises up to about 10.5 km above mean altitude. The volcano Sif Mons (Figure 5.7a) reaches a height of ca. 2 km. The features observed at the slopes of Sif Mons clearly indicate volcanism, while the "crater farm," consisting of three craters of 37–63 km diameter, Figure 5.7b, represents impact craters, produced by three major fragments of a meteorite. Other geological features, marked by concentric rings and troughs originating from rising currents in the mantle, are unique to Venus. Also typical to Venus is the limitation of the volcanic ejecta blankets to the immediate vicinity of the craters: the dense atmosphere suppresses ballistic expulsion and distribution of ejecta.

The density and the depth of the atmosphere also provide a protective shield for high-speed objects entering from space. Whether or not a meteoroid surviving

---

7) *Ištar* is an Akkadian Goddess, *Aphrodite* the Goddess of beauty, love and lust in Greek mythology. The Maxwell mountain has been named after James Clerk *Maxwell*, "creator" of the Maxwell equations in electromagnetic theory. *Sif* is the Nordic Goddess of vegetation.

ram pressure,[8] burning and/or evaporation, is torn apart after entering a planetary atmosphere is a matter of its size, its composition, the angle by which it enters and, predominantly, the aerodynamic stress, viz. the dynamic pressure it is subjected to in the atmosphere. The dynamic pressure p (in Pa) is defined by

$$p = 1/2\,\rho v^2 \qquad (5.7a)$$

where ρ is the density (in g cm^{-3}) of the atmosphere, and v the speed (in m s^{-1}) of the object entering the atmosphere. The density ρ is obtained from the molar mass M (in g mol^{-1}) of the gas or gas mix and the temperature T (in K) by

$$\rho = (M \cdot 298/22.26 \cdot T) \times 10^{-3} \text{ g cm}^{-3} \qquad (5.7b)$$

at the 1 bar (10^5 Pa) level, that is, at a height in Venus' atmosphere of about 50 km. The factor 22.26 l mol^{-1} (for an ideal gas at 273 K and 101.3 kPa: 22.41 l mol^{-1}) reflects the volume (in L) occupied by 1 mole (6.022×10^{23} molecules; the Avogadro constant N_A) of CO_2. Assuming a temperature of 330 K at the 1 bar level, and taking into account that Venus' atmosphere almost exclusively consists of CO_2 ($M = 44$ g mol^{-1}), the density is $\rho \approx 1.8 \times 10^{-3}$ g cm^{-3}. Assuming further a speed of entry for a meteoroid of ca. 20 km s^{-1}, we arrive at a dynamic pressure $p \approx 3.6 \times 10^8$ bar (3.6×10^4 GPa), enough to dismember even sizable and massive chunks, producing fragments such as those which have created the "crater farm" shown in Figure 5.7b.

5.2.3.2 Venus' Atmosphere

At ground level, the atmospheric pressure is 92 kbar (9.2 GPa). This corresponds to a ground level pressure exerted by a ca. 900 m column of sea water on Earth. Temperatures can go up to 750 K (480 °C), and this is due to the particularly effective greenhouse effect[9] on Venus, mainly affected by the 96.5% of carbon dioxide constituting Venus' atmosphere. In part, the green house effect is counter-acted by sunlight reflected away from Venus' surface by sulfuric acid clouds. This cooling effect may have been more efficient in earlier times of the evolution of Venus' atmosphere, where a more pronounced volcanic activity provided more SO_2 (and water) for the production of H_2SO_4 (see below). The particularly high amount of CO_2 (which is in the supercritical state[10]) may have been provided by thermal decomposition of carbonate minerals such as siderite $FeCO_3$, magnesite

---

8) The shock wave generated by the rapid compression of air in front of the meteoroid, the so-called ram pressure, heats the air which in turn heats the meteoroid.

9) Sunlight penetrating the atmosphere is absorbed by the soil and re-emitted at longer wavelengths, mainly in the IR domain which we perceive as warmth. This IR radiation is reabsorbed by gases such as CO_2, CH_4, and N_2O (the main "greenhouse gases"), and by clouds. From there, part of the radiation escapes into space, and part is irradiated back into the lower atmosphere. On Venus, the latter proportion is particularly high due to the extremely dense atmosphere.

10) In the supercritical state, i.e., beyond a critical temperature T and a critical pressure p, gaseous and liquid states are no longer distinct. For carbon dioxide, the critical data are $T = 304$ K, $p = 7.38$ MPa (73.8 bar), $\rho = 0.47$ g cm^{-3}.

MgCO$_3$ and dolomite CaMg(CO$_3$)$_2$. These minerals decompose to form metal oxides and CO$_2$ at temperatures around 480 °C (decomposition of siderite can already occur at sufficiently lower temperatures), and/or mobilization from carbonates by sulfuric acid and acid gases present in Venus' clouds and atmosphere. Acid-induced CO$_2$ mobilization does also proceed with calcite CaCO$_3$. Calcite is the only thermodynamically stable modification of calcium carbonate at Venus temperatures. If all of the carbon dioxide locked in Earth's carbonates were released thermally and/or by acidity, our atmosphere would be thicker than that of Venus.

In atmospheric chemistry, the abundance (or, more correctly, the *amount*) of a gas is often referred to in terms of the "(volume) mixing ratio" c_X. The quantity c_X is related to the *mole fraction* f_X used in general chemistry. The mole fraction indicates the number of moles of a specific gaseous component X per mole of a gas mixture such as present in the atmospheric gas mix. For a single component gas X, $f_X = 1$. The quantities c_X and f_X are dimensionless; for convenience, however, mixing ratios/mole fractions are often quoted in terms of % (i.e., $f_X \times 100$), ppm ($f_X \times 10^6$), or ppb ($f_X \times 10^9$), etc. In addition, the quantity *number density* n_X is in use, defined as the number of molecules of component X per unit volume of the gas mix; n_X thus is the number of molecules per cubic centimeter (unit: cm^{-3}) or, employing the unit mol, $n'_X = n_X/N_A$ mol cm^{-3} = $10^3 n_X/N_A$ mol l^{-1}. The two quantities n_X and c_X are related to each other by

$$n_X = c_X \cdot n_i \tag{5.8}$$

with n_i the number of all of the molecules of the gas mix per cubic centimeter; cf. also Table 1.1 in Chapter 1.

More conveniently, the amount of a specific gas X is quantified by its column amount (or column abundance, or column density). The column amount of a gas is the amount of the gas (expressed as number of molecules or moles) in a vertical column of unit cross section, extending from the planet's surface to the top of its atmosphere. From a chemistry point of view and employing SI units, column amounts should be quoted in units of moles per square meter, mol m^{-2} [9]. Column amounts N quoted in cm^{-2} can be converted to mol m^{-2} by multiplication with the factor $10^4/N_A$. The ratio of column amount for a specific gaseous species to the total gas column again is the mole fraction f_X (mixing ratio c_X). In Table 5.5, the main gases constituting Venus' atmosphere are collated, and their amounts are given in % (or ppm) and in column amounts. For conversion between column amount N and mole fraction f_X, the following relations, Eqs. (5.9a) and (5.9b), apply:

$$\text{column amount } N \text{ of a gas X}: N = m/a \times 10^3 \, f_X/M \text{ mol m}^{-2} \tag{5.9a}$$

where

- m is the mass of the planet's atmosphere in kg,
- a the surface area of the planet in m^2,
- f_X the fraction of the gas ($\times 10^{-2}$ when quoted in %, $\times 10^{-6}$ (ppm), $\times 10^{-9}$ (ppb)),
- M the molecular mass in g mol^{-1}

Table 5.5 Amounts of gases in the Venusian atmosphere.

	CO_2	N_2	SO_2	Ar	H_2O	CO	He	Ne	HCl	$(HF)_n$
c_X[a]	96.5%	3.5%	150 ppm	70 ppm	20 ppm	17 ppm	12 ppm	7 ppm	0.4(2) ppm	0.03(2) ppm
N[b]	23×10^6	1.3×10^6	2.4×10^3	1.8×10^3	1.1×10^3	0.6×10^3	3×10^3	0.35×10^3	~10	~1.5[c]

Trace gases: OCS, NO, O_2, H_2S, H_2SO_4, SO_3, H_2, Kr, O_3, SO

a) Mixing ratio. The mole fractions f_X are obtained from the c_X by multiplication with 10^{-2} (c_X in %) or 10^{-6} (c_X in ppm).
b) Column amount in mol m^{-2} (see text and Eq. (5.9a)).
c) Calculated for $n = 1$.

For Venus, $m/a = 4.8 \times 10^{20} \text{kg}/4.6 \times 10^{14} \text{m}^2 = 1.04 \times 10^6 \text{kg m}^{-2}$

$$\text{Mole fraction } f_X = \text{column amount } N \text{ (in mol m}^{-2}) \times (m/a)^{-1} \times 10^{-3} M \tag{5.9b}$$

As noted, CO_2 is by far the most abundant gas on Venus. Of the remaining gases in Venus' atmosphere, which goes up to ca. 245 km, nitrogen accounts for almost all the remaining 3.5%, roughly four times the absolute amount on Earth with an atmospheric abundance of 78.1% N_2; or, in terms of column amounts, 2.8×10^5 (Earth) versus 12.5×10^5 mol m^{-2} (Venus). All the remaining gases, SO_2, Ar, H_2O, CO, He, and Ne, and trace amounts of other gases constitute scantly 300 ppm (0.03%); cf. Table 5.5. The low percentage of argon (70 ppm) in the Venusian atmosphere might, on first sight, be considered unreckoned. On Earth, 0.93% (by volume in dry air) of the atmosphere is represented by argon. Relating the abundance of Ar to the overall mass of the atmosphere, 4.8×10^{20} kg for Venus and 5.2×10^{18} kg for Earth, reveals an about equal overall amount of the gas of ca. 4×10^{16} kg in both atmospheres. Column amounts are 1.75×10^3 (Venus) and 2.3×10^3 mol m^{-2} (Earth). Argon has accumulated in Earth's and Venus' atmosphere by decay of ^{40}K ($t_{1/2} = 1.25 \times 10^9$ y) present in potassium-rich minerals (micas, feldspars, hornblendes) during the bygone 4.5 billion years; Eq. (5.10):

$$^{40}_{19}\text{K} \begin{cases} \longrightarrow {}^{40}_{20}\text{Ca} + \beta^- & 89.1\% \\ \xrightarrow{\text{electron capture}} {}^{40}_{18}\text{Ar} \\ \longrightarrow {}^{40}_{18}\text{Ar} + \beta^+ \end{cases} 10.9\% \tag{5.10}$$

An intriguing observation in Venus' atmosphere is the depletion of hydrogen ^{1}H with respect to its heavier isotope deuterium ^{2}H, or D. The D/H ratio on Venus is 0.025, more than 100 times that on Earth (0.000 16). The depletion of ^{1}H is a consequence of Venus lacking the protective permanent magnetic field characteristic of Earth and Mercury, allowing the Solar wind to erode the planet's atmosphere, in particular of lighter particles.

The "cosiest" area in Venus' atmosphere is at the 1 bar level at an altitude of ca. 50 km above ground, with canny temperatures around 300 K (27 °C). The thick clouds (Figure 5.6a) extend between 47 and 70 km altitude, and cover the Venusian sky completely and permanently, driven and restructured by heavy winds with a speed of up to 350 km h^{-1}. The clouds are composed of sulfur dioxide, and a sulfuric acid aerosol, $H_2SO_4 \cdot H_2O$ in the upper, and H_2SO_4 in the lower boundary layer.

Sources of energy to power chemical reactions are the high surface and near-surface temperature, photons ("light energy") above the dense clouds, and cosmic rays (mainly protons) within the clouds. Active species are supplied by thermolysis, photolysis/photoionization or cosmic ray ionization and can be delivered into various atmospheric levels to promote chemistry there. Information on the composition of Venus' atmosphere is mainly based on near IR and microwave spectroscopy, exploiting CO_2 transparency windows at 1.1, 1.18, 1.27, 1.74, and 2.3 µm (9090, 8470, 7870, 5750, and 4350 cm^{-1}); and on the rotational microwave excitation of water in the 1.35 cm range. For the chemistry in planetary atmospheres, as in interstellar clouds, a vast variety of (coupled) model chemical reactions can be formulated. Employing reaction rate constants for the chemical reactions, and observed abundances of planetary chemical species, one arrives at production rates, which can be counter-checked against the actual abundances to test the viability of the original model assumptions. The three main cycles on Venus involve carbon, sulfur, and chlorine, and all three cycles are coupled in a complex manner [10].

5.2.3.3 Chemical Reactions

One way of producing sulfuric acid in the upper atmosphere relies on sulfur dioxide and carbon dioxide, as shown in the reaction sequence (Eq. (5.11)), where M stands for a third collision partner absorbing the kinetic energy of the interacting atoms, often a matrix (hence M) such as provided by a particle surface. The initiating step is the photodissociation of CO_2 by far UV light ($\lambda <$ 169 nm), followed by oxidation of SO_2 to SO_3, hydration of SO_3 to H_2SO_4 and atmospheric downward transport of H_2SO_4. An additional pathway to arrive at sulfuric acid includes NO_x. Nitrogen oxides can catalyze the oxidative (by O_2) conversion of SO_2 to SO_3, very much as in the production of sulfuric acid in the lead chamber process of olden times. As on Earth, nitrogen oxide on Venus can form from N_2 and O_2 in the lower atmosphere by lightening. In deeper parts of the clouds, essentially shielding from UV irradiation, sulfur trioxide can directly form from sulfur dioxide and molecular oxygen at the surface of dust particles containing catalytic amounts of, for example, vanadium oxides; Eq. (5.12). Vanadium oxide V_2O_4/V_2O_5 is the catalyst employed in the contact processes of modern sulfuric acid production:

$$CO_2 + h\nu \rightarrow CO + O \qquad (5.11a)$$

$$O + SO_2 + M \rightarrow SO_3 + M \qquad (5.11b)$$

$$SO_3 + H_2O \rightarrow H_2SO_4 \qquad (5.11c)$$

$$2SO_2 + O_2 + M \xrightarrow{\{VO\}} 2SO_3 + M \qquad (5.12)$$

As mentioned, nitrogen oxides NO_x (N_2O, NO, NO_2) can be generated in the dense regions of Venus' atmosphere by lightening, that is, via electric discharge mediated splitting of nitrogen and oxygen containing precursor molecules, followed by recombination of N and O. A second source for nitric oxide NO is its formation from N and O in Venus' upper atmosphere beyond the clouds. Nitrogen and oxygen atoms can be produced on the dayside by photodissociation of N_2, CO_2, CO, and O_2. N and O are then transported to the nightside through atmospheric circulations, where they recombine to form NO via excited NO*, accompanied by radiation visible as nightglow on the nightside. The same process has been observed on Mars and will be treated in more detail in the respective chapter (Fig. 5.15 in Section 5.2.4.6).

Photodissociation of sulfur-containing compounds, in particular OCS, with a comparatively low dissociation energy of the S=C bond of 301 kJ mol^{-1}, can lead to the formation of elemental sulfur S_n, where n may attain the values 1–8. S_8, S_7, and S_6 are the predominant allotropes at an altitude of 47 km, and S_2 dominates near the surface [11]. The formation of S_n can also proceed according to the net reaction presented by Eq. (5.13a), or via photodissociation of OCS, Eq. (5.13b) and, in the middle and upper layers of the atmosphere, photochemically from SO_2. The dissociation energy of, for example, S_3 is 261 kJ mol^{-1}, corresponding to a limiting wavelength of 461 nm for photolysis. Reaction (5.14) accounts for the fact that fluxes of CO and SO_3 are comparable at ca. 58 km above ground level [11]. Elemental sulfur is comparatively abundant in the lower 20 km of the atmosphere, where it is involved in a plethora of subsequent reactions, two of which, producing reactive Cl and OH radicals, are displayed by Eqs. (5.15a) and (5.15b):

$$H_2SO_4 + 2OCS \rightarrow H_2O + CO_2 + SO_2 + CO + S_2 \qquad (5.13a)$$

$$COS + h\nu \rightarrow CO + 1/n\, S_n \qquad (5.13b)$$

$$CO_2 + SO_2 \rightarrow CO + SO_3 \qquad (5.14)$$

$$S + HCl \rightarrow SH + Cl \qquad (5.15a)$$

$$SH + CO_2 \rightarrow OCS + OH \qquad (5.15b)$$

Along with the photochemically driven reaction (5.11a) in the upper atmosphere, leading to the formation of CO and finally SO_3, Eq. (5.14), and S_n, Eq. (5.13b), thermally driven reactions in the lowest atmospheric layers, such as those represented by Eq. (5.16a), produce carbonyl sulfide OCS. Carbonyl sulfide, a key compound in Venus' atmospheric sulfur cycle, may be lost in the middle atmosphere, Eq. (5.13). Photo- and thermochemical processes balance the OCS abundance within the 30–40 km band as shown in Eq. (5.16b) [11]; for an additional reaction resulting in depletion of OCS see Eq. (5.16b):

$$CO + 1/2\, S_2 \rightarrow OCS \qquad (5.16a)$$

$$2COS + SO_3 \rightarrow CO_2 + CO + SO_2 + S_2 \qquad (5.16b)$$

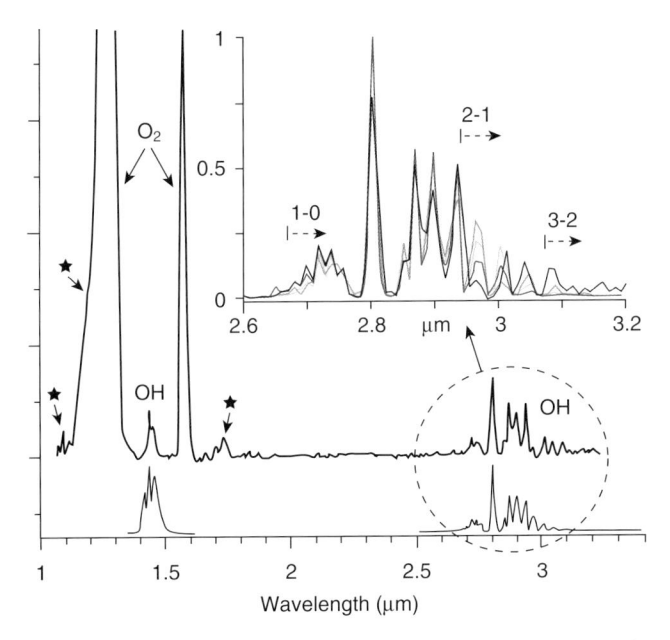

Figure 5.8 Infrared spectrum for O_2 and OH as observed in the air glow at Venus' night side by the *Venus Express* spacecraft in 2007. In the inset, the OH emissions around 2.9 μm are superimposed with synthetic spectra (grey lines) for rotational temperatures T_r of 200, 250, and 300 K. The bottom spectrum is the complete synthetic spectrum for $T_r = 250$ K. Reproduced from Ref. [13] with permission by Astronomy & Astrophysics.

These exclusively atmospheric processes can be coupled to the interaction with surface-bound minerals such as pyrite FeS_2, anhydrite $CaSO_4$, calcite $CaCO_3$, and diopside $CaMgSi_2O_6$; cf. Eqs. (5.17)–(5.19). Equation (5.18) represents a possibly important process for the removal of SO_2 from the atmosphere:

$$FeS_2 + CO_2 + CO \rightarrow FeO + 2OCS \tag{5.17}$$

$$SO_2 + CaCO_3 \rightarrow CaSO_4 + CO \tag{5.18}$$

$$CaMgSi_2O_6 + SO_3 \rightarrow CaSO_4 + MgSiO_3 + SiO_2 \tag{5.19}$$

Oxygen atoms formed according to Eq. (5.11a) can recombine to form O_2, and this has been verified by the intense oxygen airglow[11] in Venus' atmosphere (see also Figure 5.8): recombination initially produces singlet-O_2 ($^1O_2^*$), followed by transition to ground-state triplet-O_2 ($^3O_2$). The reaction is displayed in Eq. (5.20). Overall, the oxygen and CO abundances are low, and in addition to the oxygen depletion

---

11) More generally, the term *airglow* (or nightglow) refers to weak light emission from a planet's atmosphere, due to various light-emitting processes such as recombination of ions, luminescence, and chemiluminescence.

by formation of SO_3 (Eq. (5.11b)), pathways resulting in the reoxidation of CO participate. Likely processes, which have been backed up by laboratory evidence, are the photocatalytic oxidation of CO on aerosols containing TiO_2, and the reoxidation of CO, initiated by thermolysis ($T > 350\,°C$) of an intermediately formed peroxychloroformyl radical ClC(O)OO; Eq. (5.21) [12]. The formation of ClC(O)OO affords chlorine atoms, involving comparatively abundant HCl as the possible source; see Eq. (5.15a). Chlorine atoms are also formed, along with H_2, in the reaction between HCl and H and, above the cloud top, by photodissociation of HCl:

$$2O + M \rightarrow {}^1O_2^* + M;\ {}^1O_2^* \rightarrow 3O_2 + h\nu \tag{5.20}$$

$$Cl + CO + M \rightarrow ClCO + M;\ ClCO + O_2 + M \rightarrow ClC(O)OO + M \tag{5.21a}$$

$$ClC(O)OO \rightarrow ClO + CO_2 \tag{5.21b}$$

Reactions (5.16b) and (5.21b) are examples for the recycling of CO_2. Additional CO_2 can form, in particular on the planet's night side, by reversing reaction (5.11a), or directly from CO and O_2, contributing to the depletion of O_2. Another reaction sequence concomitantly replenishing CO_2 and diminishing O_2 is provided in Eq. (5.22). A further sink for oxygen is its surface interaction with ferrous (Fe^{2+}) to form ferric (Fe^{3+}) minerals:

$$Cl + O_2 + M \rightarrow ClOO + M \tag{5.22a}$$

$$ClOO + CO \rightarrow ClO + CO_2 \tag{5.22b}$$

The particularly reactive OH radical forms as displayed in Eq. (5.15b) and other reactions, a selection of which is provided by Eqs. (5.23)–(5.25). The OH radical has recently been detected in Venus' atmosphere by its air glow emissions at 1.44 and 2.80 µm, along with emissions at 1.27 and 1.58 µm for oxygen, Figure 5.8. The high degree of correlation (in altitude) with the O_2 bands suggests formation of OH according to Eqs. (5.25) and/or (5.26) [13]. Loss of OH at about 90 km can occur by reaction with O (to form H and O_2), and by reaction with CO, Eq. (5.27), likely a major pathway for the production of CO_2 on Mars:

$$S + H_2O \rightarrow OH + SH \tag{5.23}$$

$$Cl + H_2O \rightarrow OH + HCl \tag{5.24}$$

$$H + O_3 \rightarrow OH + O_2 \tag{5.25}$$

$$O + HO_2 \rightarrow OH + O_2 \tag{5.26}$$

$$OH + CO \rightarrow H + CO_2 \tag{5.27}$$

The "coexistence" of H_2S and SO_2 in Venus' atmosphere, that is, of two gases which readily react to form sulfur in the presence of catalytic amounts of water or other catalysts (such as bauxite; cf. the Claus process) has fueled, along with other "evidence," speculations that there may be microbial life in the Venusian atmosphere, continually producing these gases. Liquid water as a habitat for living

Table 5.6 Normative compositions of Venusian minerals;[a] modified from Ref. [15].

Mineral	Composition	Percentage
Orthopyroxene	75% enstatite $MgSiO_3$ + 25% ferrosilite $FeSiO_3$	18.2–25.4
Clinopyroxene	48% wollastonite $CaSiO_3$ + 36% enstatite + 16% ferrosilite	2.5
Diopside[b]	$CaMgSi_2O_6$	9.9–10.2
Olivine	75% forsterite Mg_2SiO_4 + 25% fayalite Fe_2SiO_4	9.1–26.6
Anorthite	$CaAl_2Si_2O_8$	24.2–38.6
Albite	$NaAlSi_3O_8$	3.0–20.7
Orthoclase	$KAlSi_3O_8$	0.5–25.0
Nepheline	$NaAlSiO_4$	8.0
Ilmenite	$FeTiO_3$	0.5–3.0

a) The data are based on findings by *Venera* 13 and 14, and *Vega* 2, all of which landed on lava flows and volcanic constructs.
b) See reaction (5.19) for the interaction of this mineral with atmospheric SO_3.

organisms is not available on Venus today; it may have been present, though, in previous states of Venus' planetary development, billions of years ago when the Sun was less luminous than today. To confidently answer this question, we will have to wait for more detailed results than presently available on analyses of rocks and minerals from Venus' surface. Hydrous minerals, such as tremolite, a water-containing calcium–magnesium silicate belonging to the amphibole family, of net composition $Ca_2Mg_5Si_8O_{22}(OH)_2$, which can have formed during a wetter past, could still exist at current conditions on Venus' surface [14]. Amphiboles are formed when lava and magma interact with water and, although thermodynamically unstable, could have resisted decomposition for eons. The possibility and limitations of primordial life on planets other than Earth, including the eventuality of life having been shuttled to Earth, or from Earth to other planets, will be addressed in more detail in Chapter 7.

So far, our knowledge of Venus' mineralogy is restricted to data collated by the Venus landers *Venera* 13, *Venera* 14 and *Vega* 2, and based on X-ray fluorescence analyses (XRF).[12] The surface, as it appears today, has been reshaped (almost) completely a few hundred million years ago, that is, we are restricted to information reflecting a pretty young Venus. From the *Venera* and *Vega* data, abundances of elements heavier than sodium are available, and based on these abundances, normative minerals and mineral compositions can be calculated. These normative minerals, assembled in Table 5.6, include the three iron minerals fayalite (ferrous

12) Irradiating a material with high-energy X-rays (or with γ rays) leads to the ejection of electrons from inner electronic levels as a result of absorption of the in-coming energy. The empty space is refilled by electrons falling from outer levels, and the energy difference between outer and inner level is emitted. The wavelength of this emitted light is very characteristic of the chemical nature of the atom. The phenomenon is referred to as fluorescence because the energy of the radiation emitted by the exited atom is lower than that of the absorbed energy.

Figure 5.9 (a) Global view of the planet Mars, showing the Northern polar icecap (CO_2 and H_2O), "scars," for example, the Valles Marineris, just South of the equator, and dark areas (which can fade on occasion of sand storms), the plains Acidalia Planitia (upper right, ca. 1500 km diameter), and Isidis Planitia (center right). The dark area just to the West of this plain is Syrtis Major, a volcanic sheet at the boundary between the northern lowlands and southern highland. Olympus Mons (center left, circled), 26 km above the mean Northern surface level (21 km above mean zero), is the highest elevation in the Solar System.[13] (b) The Twin Peaks *(19°N, 34°W)*, and a rock strewn surface, primarily basalt, as photographed by Pathfinder in 1997. Credit: NASA (public domain).

orthosilicate Fe_2SiO_4), ferrosilite (ferrous metasilicate $FeSiO_3$), and ilmenite $FeTiO_3$. Olivine $(Fe,Mg)_2SiO_4$ is the most abundant mineral in Earth's mantle, and has also been found on the Moon, on Mars, and in meteorites.

Calculations and laboratory evidence also point toward magnetite Fe_3O_4, hematite Fe_2O_3, and iron–sulfur phases, varying between the compositions of pyrite FeS_2 and pyrrhotite Fe_7S_8; cf. also Eq. (5.17) for the relevance of pyrite for atmospheric chemistry. Prospective Venus landers equipped with Mössbauer spectroscopy devices should shed more light on the nature of iron minerals in Venus' surface [15].

5.2.4
Mars

5.2.4.1 General

The "red planet" Mars (Figure 5.9), our next neighbor when we move away from the Sun, has fired scientists' (and fiction writers' [16]) imagination more than any

---

13) *Valles Marinieris* = Mariner valley (discovered by the Mariner 9 orbiter 1971–72). *Acidalia Planitia* = Acidalia plain; Acidalia is a mythological fountain in Orchomenus (Greece) where Venus used to bathe with the Graces. *Isidis Planitia* = Isidis' plain; Isidis, or Isis, is the Egyptian pendant of Venus and Aphrodite. *Syrtis major* is the Roman name for the Gulf of Sidra on the Libyan coast. *Olympus Mons*, the Olympic mountain, is the antique residence of the Greek Gods.

other planet in the Solar System. This particular interest in Mars roots, at least in part, in the "discovery" (in 1877) of a network of subtle lines on Mars' surface by the Italian astronomer Giovanni Schiaparelli, which he termed *canali*, channels connecting presumed bodies of water. While some of these structures correspond to canyons and terracing, most are visual illusion, caused by albedo and contrast effects. About one-and-a-half decades later, the American amateur astronomer Percival Lowell, an enthusiastic observer of Mars, took up the case. He translated *canali* as "canals," implying that these are artificial structures, a system of irrigation canals constructed by a Martian civilization to combat the increasing desiccation of their planet. At the latest from this perception, Mars was spotlighted as a dying planet, the civilization of which was about to conquer Earth [17]. We now know that there is water on Mars, hazes of water ice in Mars' clouds and polar caps, and subsurface ice, as recently (2008) confirmed by the Phoenix mission. And there have probably been considerable amounts of liquid water in former times, flows of water which have carved many of the ravines on Mars' surface. Naturally, the confirmation of the presence of water on Mars, as other recent findings such as the discovery of seasonal methane plumes (see below), has refreshed discussions on the possibility of (former) primitive life on the red planet.

5.2.4.2 Orbital Features, and the Martian Moons and Trojans

The rotational period of Mars was already determined in the 17th century by the Dutch astronomer, physicist, and mathematician Christiaan Huygens. In 1659, Huygens observed the movement of the dark, about triangular feature now known as Syrtis Major (Figure 5.9). His observation, 24.5 h, is very close to the exact rotational period of $24^h37^m23^s$ (24.62 h). A day on Mars (a *sol*) thus is only slightly longer than a day on Earth. A Mars year amounts to 1.881 Earth years. The present axial tilt, 25.19°, is again comparable to that of Earth (23.44°). Contrasting Earth, Mars' orbit is rather eccentric. Tycho Brahe's close observations of the projections of the Martian motion to the nightsky enabled Johannes Keppler to derive, based on "irregularities" caused by the high eccentricity, the laws of planet motion 400 years ago (see Chapter 1).

The present eccentricity, 0.093 (as compared to 0.017 for Earth), leads to pronounced "eccentricity seasons," overlaid by the seasons which are due to the axial tilt, that is, the seasons we are familiar with on our home planet. Due to the substantial eccentricity, the intensity of sunlight at perihelion exceeds that at aphelion by a factor of 1.455, and this is the main factor responsible for the strong annual climate cycle. As Earth, Mars experiences long-term periodical variations of orbital characteristics such as its orbit eccentricity, the tilt of its rotational axis with respect to the plane of the orbit (obliquity, and the obliquity cycle), and variations of the orientation of its axis in space (precession), Table 5.7. These periodic variations, also known as the three Milanković cycles, cause long-term variations in climate.

Mars is accompanied by two tiny, irregularly formed moons, Phobos and Deimos (Figure 5.10), and a couple of Trojan asteroid, the best characterized of which is

Table 5.7 Orbital characteristics of Mars in comparison to Earth.

		Mars	Earth
Present characteristics	Orbital period (a)	1.881	1
	Eccentricity	0.093	0.017
	Rotational period (d)	1.026	0.9973
	Axial tilt (°)	25.19	23.44
Cycles	Changes in eccentricity; period (a)	0.0–0.14; 96 000[a]	0.0005–0.0607; 100 000
	Precession (a)	175 000	26 000[b]
	Changes in obliquity (°); period (a)	15–35[c]; 120 000	22.1–24.5; 41 000

a) Superimposed with a period of 2 400 000 years.
b) Overlaid by a nutation period ("wobbling" of the orientation of the axis by 9.2") of 18.6 a.
c) Under debate.

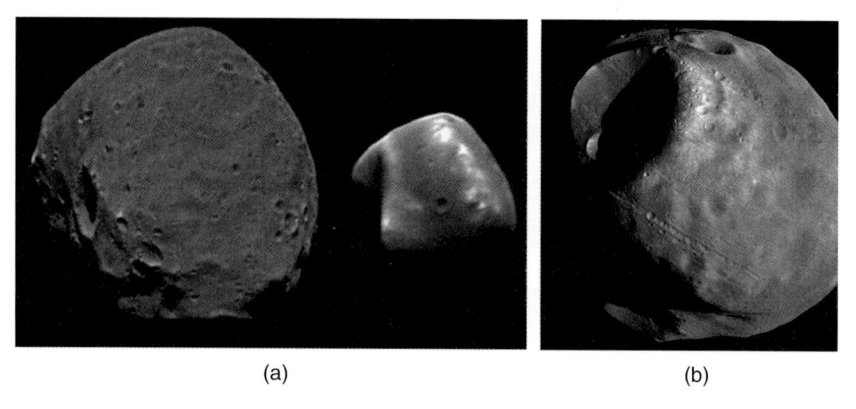

Figure 5.10 (a) Mars' two moons, Phobos (left, 26.8 × 22.4 × 18.4 km) and Deimos (15 × 12.2 × 10.4 km). Credit: NASA (public domain). (b) Another view of Phobos, showing the Stickney crater (Stickney is the maiden name of the wife of the moons' discoverer, Asaph Hall) and grooves, possibly produced by material ejected from Mars on occasion of impact events on Mars. Credit: NASA/Viking 1 Orbiter.

the asteroid 5261 Eureka ("I have found it" [the principle of buoyancy]; ascribed to Archimedes of Syracuse), about 2 km across, and in a 1 : 1 resonance (i.e., same orbital period) with Mars. Classified as an achondrite A-class asteroid (rich in olivine; cf. Section 5.3), it trails Mars in an eccentric orbit (eccentricity 0.065) with an inclination toward Mars' ecliptic of 20.3°. Phobos and Deimos resemble

C/D-type asteroids, that is, they are composed of carbon-rich silicates and have a low albedo. As our moon, they are covered with regolith, which effectively smoothes the surface mainly of Deimos, while the larger Phobos remains heavily cratered. According to composition, the two moons may be captured asteroids, although this assumption has been challenged recently. Their low mean density, 1.89 (Phobos) and 1.47 g cm^{-3} (Deimos), could be due to inclusions of water ice, or may reflect their "rubble pile" nature, providing significant porosity. This latter notion may point towards accretion in their present position. Alternately, primordial asteroids captured by Mars may have undergone transformation. Both moons orbit Mars in its equatorial plane, and their orbital and rotational motions are synchronized. The semimajor axis of Phobos' orbit is 9377 km, that is, this moon is pretty close to Mars, orbiting just 5989 km above the Martian surface, and rising about twice per Mars day in the West (it moves ca. three times faster in its orbit than Mars rotates). The orbit is not stable: Phobos, subjected to tidal forces, slowly approaches Mars and will eventually reach the Roche limit in about 10 million years. The Roche limit is the boundary where a moon (or other celestial body) is torn apart "to pebbles in the sky" by the gravitational forces of the planet. With 23 460 km, Deimos is much farther away; it rises in the East.

5.2.4.3 Geological[14] Features, Surface Chemistry, and Mars Meteorites

Mars is substantially smaller than Earth. Its equatorial diameter is 6787 km (Earth: 12 756 km); the volume is hence just 0.14 that of Earth; the surface gravity is 0.38 g (cf. footnote 3). Mars' topography (see also Figure 5.9) is dominated by volcanic plateaus and impact basins, the surface primarily being composed of basaltic materials very much of the same nature as basalts of Earth. The vesicular nature of the basaltic rocks evidences the pervasive presence of significant amounts of volatile agents such as water, carbon dioxide, and chlorine. Vast areas of Mars are covered with finely powdered material containing ferric oxides, responsible for the rust-red color of Mars, clearly visible even with the naked eye, and responsible for the martial name of this planet. Mars has a liquid iron/nickel-sulfur core, with a substantially higher sulfur amount than Earth's core, mainly accounting for the lower mean density of Mars, viz. just 62% that of Earth. This core is surrounded by a silicate mantle plus crust, the thickness of which increases from the Northern lowlands (40 km) to the South Pole (70 km). Parts of the crust became magnetized at a time where Mars still kept a permanent magnetic field. The magnetic field faded away 4 billion years ago. Apparently, Mars has been lacking, within the last 4 billion years, a solid *inner* core and thus a liquid interlayer zone between inner core and mantle, allowing for convection and, concomitantly, the dynamo effect responsible for the generation and maintenance of a magnetic field (cf. Footnote 4). In contrast, Earth (which is sufficiently larger than Mars) and Mercury (with its more extended core structure and proximity to the Sun) do have a core

---

14) *Areology* should be a more appropriate term for describing Mars' "geolocical" features: *Geo* (from the Greek *Ge* for Earth) refers to our home planet; *are*, derived from *Ares*, the Greek name for the red planet, relates to Mars.

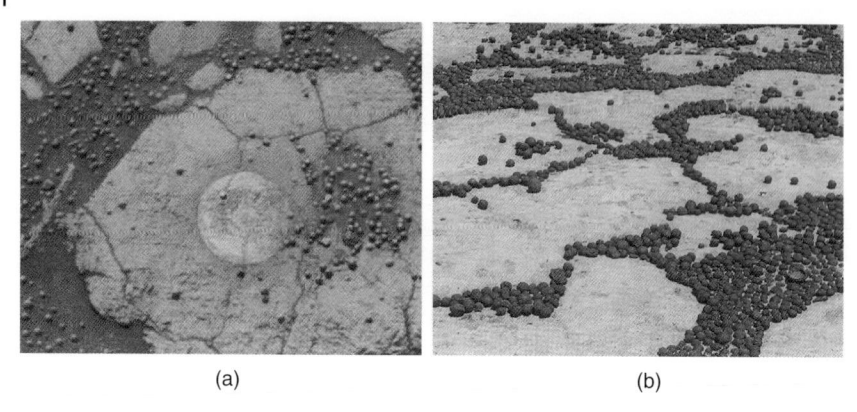

Figure 5.11 Small spherules, rich in (rust-colored) hematite Fe_2O_3 on basalt rock. The spheres are believed to have formed by slow deposition from a bath of water saturated with minerals, followed by water and wind erosion. (a) The picture was taken by the micro imager of the Mars exploration rover *Opportunity* at Eagle Crater in 2004; the granules are up to 5 mm in diameter. Credit: Mars Exploration Rover Mission, JPL, NASA. (b) Similar hematite concretions, 1 mm to 5 cm in diameter, can be observed in the Grand Staircase-Escalante National Monument in Southern Utah, USA. Permission by Brenda Beitler Bowen, Purdue University, Indiana (USA).

differentiated into a solid inner and a liquid outer zone, and thus are accompanied by a permanent magnetic field.

Noticeable is Mars' hemispheric dichotomy: while the Northern Hemisphere is dominated by low-lying planes covered with sand and dust, highlands of geologically more ancient origin strewn with impact craters are characteristic for the Southern Hemisphere. A mega-impact of a gigantic asteroid about 4 billion years ago satisfactorily models this dichotomy. Martian craters exhibit a wider range of morphologies than Lunar or Mercurian craters. Many are surrounded by splash marks, sinuous channels, and lobate structures. Particularly impressive are the Valles Marineris (cf. Figure 5.9), an extended rift and canyon system which stretches over a distance of 4000 km, ramifying toward the West into a dendritic system of chasms and gorges, which might go back, as other characteristic canons, to periods where there were heavy flows of liquid water on Mars. Further indicators for the former presence of liquid water are minerals such as millimeter-sized hematite (reddish-brown Fe_2O_3, Figure 5.11), jarosite, a basic hydrous sulfate of potassium and iron with the chemical formula $KFe_3^{III}(OH)_6(SO_4)_2$, and goethite. Goethite α-FeO(OH) and limonite (a mixture of goethite and hematite) form by weathering of iron minerals such as hematite, Eq. (5.28a), and oxidative weathering of pyrite FeS_2, Eq. (5.28b) in the presence of water, but can also be a primary mineral developing under hydrothermal conditions. The presence of goethite does also have implications for the presences of its isomorphous aluminum analog diaspore, α-AlO(OH): diaspore can form at low temperature and pressure in the presence of goethite:

$$Fe_2O_3 + H_2O \rightarrow 2FeO(OH) \tag{5.28a}$$

$$2FeS_2 + 1\frac{1}{2}O_2 + H_2O \rightarrow 2FeO(OH) + 4S \tag{5.28b}$$

The transformation of minerals such as olivine and pyroxene, containing *ferrous* iron, into hematite, goethite, and limonite, that is, minerals containing *ferric* iron, affords oxidation, most likely by oxygen. This oxygen can be provided by photolysis of CO_2 ($CO_2 + h\nu \rightarrow CO + O$), or by peroxides (including atmospheric H_2O_2), and by the perchlorates recently proposed to be present on Mars, a conjecture which is based on data collected by the Phoenix lander in 2008 in the Northern planes not too far off the North Pole. Lab experiments have shown that perchloric acid $HClO_4$ can be produced by photochemical gas phase oxidation of chlorine volatiles [18]. Peroxides may also be responsible for the intriguing lack of *any* organic material in the Martian soil, including organics of nonbiogenic origin. Perchlorates may take over a similar role due to their high oxidation power, although on Earth, certain extremophiles (bacteria prospering under extreme conditions) can use perchlorate as a primary electron source in respiration. Saturated water solutions of perchlorates can remain liquid down to temperatures as low as $-70\,°C$; surface temperatures on Mars range between -87 and $-5\,°C$. It is thus not too much of a surprise that these Phoenix findings (awaiting confirmation) have again ignited the discussion of, even contemporary, primitive life on our neighbor planet; see also Section 7.5.

Considerable amounts of subsoil water ice, estimated to encompass about two-thirds of the ice crust of Greenland, appear to be around almost all over Mars, particularly widespread in the Southern polar region. Extended salt deposits have been found in the Southern highlands, formed 3.5–3.9 billion years ago by ground water ascending to and evaporating at the surface, and/or water delivered to extended shallow basins by "rivers," as suggested by channel-like structures leading to the salt deposits. Finally, the recent discovery of carbonate minerals not only suggests that there was abundant liquid water on Mars, but also that these water reservoirs must have been slightly alkaline, or neutral at least, very much as the water of Earth's oceans ($pH \approx 7.5$). A more detailed account on water, carbonates, and sulfates on Mars will be provided in Section 5.2.4.5.

An unreckoned feature of Mars' surface composition, when compared to Earth and the Moon, is the low contents of Al and Ca, and high contents of Fe and S.

In contrast to other planets, we do have available *direct* information of Mars' surface composition through Martian meteorites found in several places on Earth, in particular in Antarctica and in the Sahara desert, that is, in locations where weathering is suppressed. These meteorites originate from rocks ejected from Mars by collisions with, for example, asteroids or big meteorites. Their Martian origin arises out of the element compositions of the minerals, the composition of gas inclusions, and the isotopic ratios [19]. Iron oxide contents are higher than in terrestrial silicate rocks, and some chalcophiles (Ni, Co, Cu, and In) are less abundant; siderophiles (Cr, W, Mn) in turn are more abundant in Martian than in terrestrial rock. More than 50 Mars meteorites have so far been discovered; all are achondritic (i.e., stony) meteorites of volcanic origin. There are three subfamilies:

Table 5.8 Composition of Mars meteorites (SNC meteorites).

Group	Main constituents	General composition
Shergottites	Pyroxenes: augite; pigeonite	$(Ca,Mg,Fe,Ti,Al)(Si,Al)_2O_6$; $(Mg,Fe^{II},Ca)Si_2O_6$
Nakhlites	Clinopyroxenes: augite; Fe-rich olivine; titanomagnetite	$(Ca,Mg,Fe,Ti,Al)(Si,Al)_2O_6$; $(Fe,Mg)_2SiO4$; $Fe^{II}(Fe^{III},Ti)_2O_4$
Chassignites	Dunite: >90% olivine; pyroxene (see above), chromite, plagioclase	$(Fe,Mg)_2SiO_4$; $(Fe,Mg)Cr_2O_4$; $NaAlSi_3O_8/CaAl_2Si_2O_8$
ALH 84001	Orthopyroxene:[a] enstatite; ferrosilite	$Mg_2Si_2O_6$, $Fe_2Si_2O_6$

a) See also Table 5.4.

1) Shergottites, by far the largest group, igneous or basaltic rocks similar to terrestrial basalts consisting mostly of pyroxenes, but also containing carbonates and sulfates, suggesting that they have been exposed to water on ancient Mars;

2) Nakhlites, igneous rocks containing augite and olivine crystals;

3) Chassignites similar to terrestrial dunite, a rock containing mostly olivine.

For a more detailed account on the composition, see Table 5.8. The names of these three groups are derived from the first locality where they were found: Sherghati in India; El-Nakhla in Egypt; and Chassigny in France; Mars meteorites are thus also referred to as *SNC* meteorites.

Different to some extent from these meteorites is ALH84001 (ALH stands for Alan Hills, the ice field in Victoria Land, Antarctica, where the meteorite was found in 1984), mainly composed of orthopyroxene: main components are enstatite and ferrosilite, minor components chromite and maskelynite, a gassy plagioclase phase; Figure 5.12. The rock formed originally from molten lava about 4.2 billion years ago (igneous crystallization age), and accordingly possibly stems from an ancient Martian volcano. It was subjected to a shock event 16 million years ago, responsible for its lofting off Mars. After straying in the inner Solar System for 16 million years, it finally was captured by Earth and fell down in Antarctica 13 000 years ago. About 3.6 billion years back from now, rounded globules of carbonate minerals formed within the rock, suggesting infiltration of liquid water. In these carbonate globules, tiny structures were detected (Figure 5.12b) which have originally been thought to represent fossilized bacteria, a notion which received support by the discovery of polyaromatic hydrocarbons (PAHs) and magnetite (Fe_3O_4) crystals in the globules. PAHs can originate from alterations of organic remains, and magnetite grains can be produced by magnetobacteria. It is often argued that, since Mars gradually lost its magnetic field 4 billion years ago, development of magnetotactic bacteria producing magnetite or greigite (Fe_3S_4) for

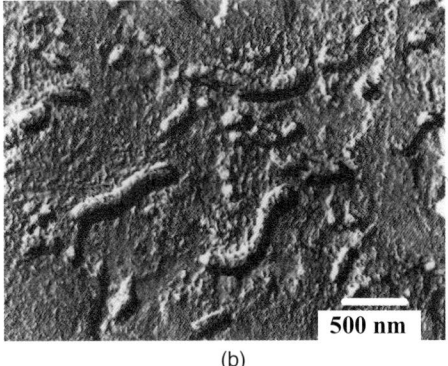

Figure 5.12 (a) The Mars meteorite Alan Hills 84001, found 1984 in Antarctica, weighs 1.94 kg. (b) Transmission electron micrograph, showing a detail from carbonate granules embedded in the meteorite, structures which are reminiscent of fossilized nanobacteria. Credit: JSC/NASA (public domain).

orientation must have developed before this time, which is hard to perceive. However, once a material gets magnetized, the magnetism only slowly fades off. Consequently, magnetized material from Mars' early days may well have survived a sufficiently long time span to "justify" the evolutionary development of magnetosomes in bacteria. Additional arguments have been brought forward in disfavor of magnetite crystallites as biomarkers. PAHs and magnetite have, for example, also been found in non-Martian meteorites. Further, the structures resembling fossilized bacteria are severely smaller (in diameter) – by a factor of about 100 – than any bacterial life forms on Earth, and by a factor of about 10 for the lowest limit of the simplest conceivable self-contained living organisms. The biological nature of supposed "nanobacteria" found on Earth, and advertised as an indication for life forms beyond the generally accepted lower limit for an organizational unit in accordance with the definition of life (as defined in Section 7.1), appears to have been ruled out for good [20a]. All these facts dampen (but do not exclude [20b]; see Section 7.5) the notion that we are dealing with a fossilized extraterrestrial life form.

5.2.4.4 Methane

The detection of methane on Mars is based on its IR absorption centered at 3018 cm^{-1} (3.31 µm), corresponding to the trebly degenerate stretching vibration, the v_3 fundamental of F_2 symmetry; Figure 5.13. The global average of methane in Mars' atmosphere in terms of mole fraction (volume mixing ratio) is 10 ppb, with spatial and time variations between 0 and 30 ppb. The global average corresponds to a column amount of $\approx 6 \times 10^{19}$ m^{-2} ($\approx 10^{-4}$ mol m^{-2}). With a photochemical lifetime of ca. 2×10^{10} s (a maximum of 600 a), a methane flux of ca. 4 g s^{-1} is required to account for this average [21]. Olivine-rich rocks on Mars, as those found in the nakhlite and chassignite meteorites, are likely a major contributor

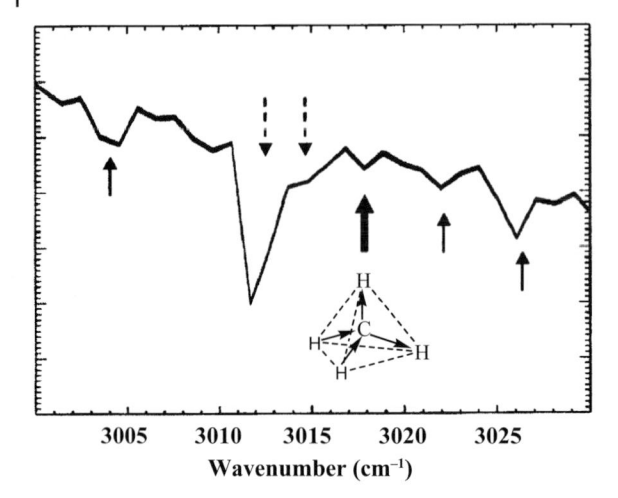

Figure 5.13 IR spectrum (obtained by the Planetary Fourier Spectrometer of the Mars Express spacecraft). The bold arrow (3018 cm^{-1}) and inset indicate the v_3 methane fundamental. The remaining bands correspond to water (3003, 3022, and 3026 cm^{-1}, thin arrows) and Solar lines (3012 and 3014 cm^{-1}, broken arrows). Adapted and modified from © Ref. [21], Figure 1A Reprinted with permission from AAAS.

to the presence of methane in the Martian atmosphere, and its spatial and time variability. But the discovery of methane on Mars and, in particular, the perennial and localized appearance of "outbursts" of methane on Mars, has also inspired people to relate Martian methane to a biotic origin, that is, to the methane production by methanogenic bacteria. Both of these potential sources of methane, the abiotic and the biotic one, will be outlined in this section. An additional source for methane may originate from the conversion of organic matter provided by impacting carbonaceous chondritic micrometeorites.

The overall process for the *abiotic* methanogenesis proceeds via the hydrothermal conversion of olivine $(Fe,Mg)_2SiO_4$ into ferric silicates/hydroxides, magnesium silicates/hydroxides (serpentines such as chrysotile $Mg_3(OH)_4Si_2O_5$), and hydrogen, a process called *serpentinisation*. Concomitantly, or in a successive reaction in the presence of CO_2, methane is formed [22]. The formation of hydrogen involves the reduction of protons, stemming from water; the corresponding reduction equivalents are provided by the oxidation of ferrous iron (from olivine, or fayalite Fe_2SiO_4) to ferric iron in the resulting magnetite, formed along with chrysotile and silica, Eq. (5.29a). An eight-electron reduction is afforded to convert CO_2 to CH_4; the reduction equivalents are again provided by ferrous ions, Eq. (5.29b). Equations (5.29a) and (5.29b) represent thermodynamically and kinetically feasible reactions at subsoil temperatures below 600 K and elevated pressure. Model reactions with rock of shergottite composition showed that these weathering processes can actually occur at temperatures between 0 and 150 °C and the partial pressures of CO_2 (ca. 6 mbar) and O_2 (10 µbar) of the current Martian atmosphere. The direct

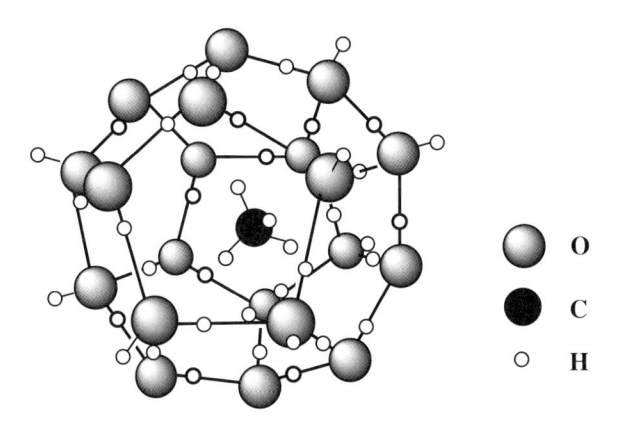

Figure 5.14 Section from the unit cell of methane hydrate (overall approximate composition $(CH_4)(H_2O)_6$), a host–guest compound where methane (black carbon in the center) is captured in a cage of 20 water molecules forming a pentagon dodecahedron. These clathrates, abundant in many deep-sea areas on Earth (and possibly also on the icy moons of the outer planets), may be present in Mars' subsoil and responsible for perennial methane fumes.

reaction between hydrogen and carbon dioxide to form methane, Eq. (5.30), has a high activation barrier; here, chromium spinel (Cr_3O_4), iron–nickel sulfides and magnetite can act as effective catalysts. As noted previously, sulfur is particularly abundant on Mars, and sulfates can be reduced, competitively to carbonates or CO_2, by hydrogen formed via serpentinisation, Eq. (5.31):

$$3(Fe, Mg)_2SiO_4 + 3H_2O \rightarrow Fe_3O_4 + Mg_3Si_2O_5(OH)_4 + SiO_2 + H_2 \quad (5.29a)$$

$$6Fe_2SiO_4 + CO_2 + 2H_2O \rightarrow 4Fe_3O_4 + 6SiO_2 + CH_4 \quad (5.29b)$$

$$4H_2 + CO_2 \rightarrow 2H_2O + CH_4 \quad (5.30)$$

$$H_2 + CaSO_4 \rightarrow CaO + H_2S + 1\frac{1}{2}O_2 \quad (5.31)$$

CH_4 is a sufficiently more effective greenhouse gas than CO_2, and the methane once formed on primordial Mars by hydrothermal processes may well have contributed to a warmer period on Mars, allowing for abundant liquid surface water, and subsurface water or brine reservoirs. As Mars grew colder, a growing amount of permafrost water may have sequestered methane in the form of methane clathrate ice, Eq. (5.32), an inclusion compound where methane is captured in a water-ice structure (Figure 5.14) well known from the cold deep-ocean regions on Earth. From these presumed Martian subsoil deposits, methane can be occasionally released into Mars' atmosphere:

$$CH_4 + 6H_2O \rightarrow CH_4 \subset (H_2O)_6, \Delta H = -18 \text{ kJ mol}^{-1} \text{ (at 273 K and } 10^5 \text{ Pa)} \quad (5.32)$$

On Earth, most of the methane is produced via *biogenic* methanogenesis by bacteria belonging to the phylus archaea.[15] Methanogenic archaea, or methanogens, abundant in, for example, rice fields, in the rumen of ruminants, in hot springs and deep sea black smokers (thermophile methanogens), but also in salty pools in the permafrost and in Earth subsurface areas, reductively convert CO_2 or other, less oxidized carbon sources (e.g., CO, acetic acid, formaldehyde, and cellulose) to methane in a multistep process with the help of metalloenzymes, employing hydrogen as the reducing agent. With carbon dioxide, the overall reaction, Eq. (5.30), is the same as the abiotic conversion of CO_2 to CH_4. The biotic reduction of CO_2 takes place in several successive steps, summarized in Scheme 5.1. The first step, **(a)** in Scheme 5.1, is the two-electron reduction of CO_2 to the formyl cation (HCO^+) and its transfer to methanofuran to form methylmethanofuran. This reaction is catalyzed by the enzyme formylmethanofuran dehydrogenase, containing molybdenum (or, sometimes, tungsten) as the redox-active constituent of a molybdopterin cofactor. The second step, **(b)**, is the transfer of the formyl group to tetrahydropterin, followed, **(c)**, by the four-electron reduction of formyl to cationic methyl (CH_3^+), a reaction where an iron-dependent hydrogenase is involved. The methyl group is then transferred to coenzyme-M to form methylcoenzyme-M, **(d)**. This transfer is mediated by a cobalt complex, viz. methylcobalamine, a close relative to vitamin B_{12}. The final step, **(e)**, the release of methane by a two-electron reduction of CH_3^+ to CH_3^- (followed by protonation of CH_3^- to form CH_4), is catalyzed by methyl-coenzyme-M reductase, an enzyme the cofactor of which is a porphinogenic nickel complex, the factor F_{430}. All the metals involved are available on Mars.

Whether or not there are remote areas on Mars providing the conditions for prospering methanogenic archaea, in particular the availability of liquid water, remains a question which, with the present stand of knowledge, cannot concludingly be answered. Determining the isotopic signature of Martian methane, that is, the $^{13}C/^{12}C$ ratio, could reveal the origin of this gas on our neighbor planet: in methane of biogenic origin, the heavier isotope ^{13}C is depleted with respect to the lighter one. Typically, for methane of biogenic origin on Earth, the ratio $^{13}C/^{12}C$, labeled $\delta^{13}C$, is in the range of -10 to -30 per mill, which is a depletion of ^{13}C with respect to ^{12}C by an order of magnitude as compared to CH_4 of volcanic origin. For the definition of $\delta(^{13}C)$, see Eq. (5.54) in Section 5.3.2.

Methane can be oxidized to formaldehyde by, for example, oxygen, Eq. (5.33), in the presence of catalytic amounts of iron, vanadium, or molybdenum oxides. Under Martian conditions, formaldehyde is rapidly destroyed by photolysis, Eq. (5.34). The lifetime amounts to ca. 10 h (as compared to several hundred years for atmospheric methane; the lifetime drops to several hours for ground-level methane). Traces of formaldehyde are, however, present in the Martian atmos-

---

15) The phyla archaea and bacteria (sometimes also referred to as archaebacteria and eubacteria, respectively) are prokarya, which have developed independently (but from the same common ancestor) prior to the eukarya, the phylogenetically younger and more developed living beings containing a cellular nucleus (see Chapter 7 for more details).

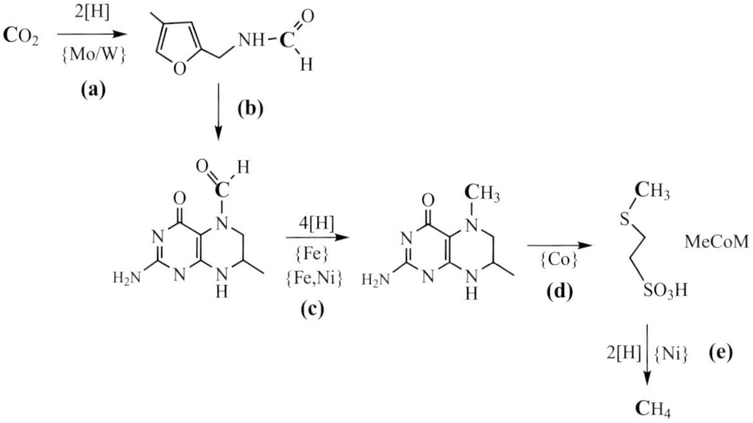

Scheme 5.1 Simplified reaction course of the reductive protonation of CO_2 to CH_4 by methanogenic archaea. [H] stands for reduction equivalent ([H] = H⁺ + e⁻). {Mo} or, less common, {W}, is formylmethanofuran dehydrogenase, containing a molybdo- or tungstopterin cofactor; {Fe} = iron-dependent hydrogenase; {Co} = (methyl)cobalamine; {Ni} = factor F_{430} of methylcoenzyme-M (MeCoM) reductase.

phere, and this has lead to speculations according to which more methane is produced on Mars than can sensibly be derived from nonbiogenic sources (such as serpentinization) alone:

$$CH_4 + O_2 \rightarrow HCHO + H_2O \tag{5.33}$$

$$HCHO + h\nu \rightarrow HCO + H;\ HCO \rightarrow CO + H \tag{5.34}$$

5.2.4.5 Carbonates, Sulfates, and Water

Carbonaceous rocks based on calcite (limestone) $CaCO_3$, magnesite $MgCO_3$, and siderite $FeCO_3$ can form as water and CO_2 react with calcium-, magnesium-, and iron-bearing minerals present in volcanic material. Calcium and magnesium carbonates are stable only in the neutral to alkaline pH range; siderite may also form under slightly acidic conditions, at a pH ≥ 5.6. In acidic media, carbonates $M^{II}CO_3$ set free CO_2, Eq. (5.35a) for M^{II} = Ca; with excess CO_2, soluble hydrogen carbonates $M^{II}(HCO_3)_2$ form, Eq. (5.35b), which can be reprecipitated at higher temperatures, Eq. (5.35c), a reaction sequence well known for the formation of boiler scales from temporarily hard water, that is, water containing hydrogen carbonates. The mobilization of Ca^{2+}, Mg^{2+}, and Fe^{2+} from basaltic rocks, followed by the formation of carbonates, suggests that the early Mars possessed a denser carbon dioxide atmosphere than today, providing a warmer climate which also allowed for the presence of liquid water. This denser and warmer atmosphere may have been provided by volcanic exhalations 4 billion years ago:

$$CaCO_3 + 2H^+ \rightarrow Ca^{2+} + H_2O + CO_2\uparrow \tag{5.35a}$$

$$CaCO_3 + H_2O + CO_2 \rightarrow Ca(HCO_3)_2 \tag{5.35b}$$

$$Ca(HCO_3)_2 \xrightarrow{\Delta} CaCO_3\downarrow + H_2O + CO_2 \tag{5.35c}$$

The Mars Reconnaissance Orbiter has recently found magnesite $MgCO_3$ along with clays [23]. Clays are phyllosilicates consisting of sheets of (alumo)silicates with hydrated cations (such as Mg^{2+}, Al^{3+}, Fe^{3+}) incorporated in-between the silicate layers. The minerals have been detected in an area, particularly rich in olivine, known as Nili Fossae at the edge of Isidis Planitia (Figure 5.9), one of the great impact basins, about 3.5 billion years old. Detection of magnesite is based on near-IR reflectance spectral features, with typical bands at, inter alia, 2.3 and 2.5 µm (4350 and 4000 cm^{-1}), which correspond to overtones and combination tones of carbonate C—O stretching and bending fundamental vibrations. An accompanying weak band at 2.4 µm (4170 cm^{-1}) has been attributed to an iron–magnesium smectite (hydrous aluminum silicate), and an electronic band at 1.1 µm originating from an electronic transition of Fe^{2+} might be attributable to siderite. Formation of all these minerals can be explained by hydrothermal serpentinization, Eq. (5.36), either by surface fluvial activity, or by groundwater percolating through fractures of the mafic rock, or by precipitation of carbonate in shallow ephemeral lakes [23]. In any case, the existence of carbonate implies that, along with an environment having experienced acidic sulfate-forming conditions (vide infra), neutral to slightly alkaline niches must also have been present on Mars, hence an environment which is generally more conductive to life than an acidic one. Carbonates have also been found, though in minor amounts only, in Mars meteorites and, by the Phoenix lander, in Mars regolith and bedrock. For the detection of CO_2, the Phoenix instruments exploited the thermal decomposition of calcite, Eq. (5.35c).[16] The decomposition of calcite to carbon dioxide and calcium oxide has an endothermic peak at an onset temperature of 675–750 °C, typical of low-pressure decomposition. Combustion of organics, and thermal decomposition of $MgCO_3$ and $FeCO_3$, occurs at lower temperature. Wet treatment of forsterite in the presence of CO_2, Eq. (5.36), yields silicate in a system efficiently buffered (pH 8.3) by calcite:

$$2Mg_2SiO_4 + 2H_2O + CO_2 \rightarrow MgCO_3 + Mg_3Si_2O_5(OH)_4 \tag{5.36}$$

Another water-bearing mineral, kieserite $MgSO_4 \cdot H_2O$, has positively been identified as an important sulfate mineral on Mars by its 2.4 µm (4170 cm^{-1}) water band. This water band is shifted toward longer wavelengths λ with respect to hydrates of other sulfates because of the stronger hydrogen-bonding interaction of the water molecule with Mg^{2+}, a small cation with a high charge density. Strong $Mg^{2+}\cdots$ H—OH interaction weakens the H—O bonds and thus shifts the corresponding IR band to longer λ. The broad distribution of sulfate minerals such as $MgSO_4$ and

---

16) The analyses have been carried out by the Phoenix TEGA analyser (Thermal and Evolved Gas Analyser), combining differential scanning calometry and mass spectrometry.

$CaSO_4$ as seen by, inter alia, the OMEGA[17] orbiter, requires water; and the deposition of these minerals affords water to evaporate. Further, an acidic environment is usually required [24]. Shallow lakes have thus been proposed to have been present on Mars.

To account for the extended regions covered by sulfate deposits, and thus the large amounts of sulfateous minerals, a powerful source of sulfur is a prerequisite, and may have been provided by volatiles (sulfur, sulfur dioxide) present in steamy volcanic exhalations. The atmospheric sulfur would have been oxidized to form sulfuric acid, Eqs. (5.11b, c) and (5.12) in Section 5.2.3.3, precipitating to the surface, and transforming other minerals [25] such as olivine and apatite. Such a transformation is exemplarily represented by Eqs. (5.37) and (5.38). Although, in this scenario (which resembles *solfataras* on Earth), water also has to be around, a watery environment as provided by a large sea or huge lake is not a necessity. On Earth, solfataras host chemolithotrophic[18] microbes such as *Thiobacillus thiooxidans*, capable of exploiting the oxidation of sulfur for sustenance, Eq. (5.39), still another issue inspiring astrobiologists:

$$(Fe, Mg)_2 SiO_4 + 2H_2SO_4 \rightarrow 2(Fe, Mg)SO_4 + SiO_2 + 2H_2O \qquad (5.37)$$

$$Ca_5(PO_4)_3 OH + 5H_2SO_4 + 9H_2O \rightarrow 5CaSO_4 + 3H_3PO_4 + H_2O \qquad (5.38)$$

$$1/n\,S_n + 1\frac{1}{2}O_2 + H_2O \rightarrow H_2SO_4 \ (n = 6 \text{ or } 8) \qquad (5.39)$$

Ancient Mars apparently hosted aqueous environments in a variety of geologic settings in which water ranged from acidic, allowing for the formation of sulfates, to (slightly) alkaline, giving rise to the formation of carbonates, phyllosilicates, and opaline silica (hydrated silicon dioxide), a diversity which bodes well for the prospect of past environments on Mars habitable for micro-organisms. The formation of carbonates, phyllosilicates, opal, and sulfates requires water; the presence of these mineral families thus indirectly proves (along with geological features such as chasms, gullies, and canals) the abundance of liquid surface water in the early days of our neighbor planet. In younger epochs, devoid of liquid water, nanosized red hematite grains formed by various oxidative weathering processes in surface and near-surface areas. Water-ice has definitely been detected by the Phoenix lander just a few inches below surface, suggesting the presence of a blanket of subsoil permafrost, along with water fixed in the polar ice-caps of about 3 km thickness (more extended in the Southern polar region), plus dust covered glaciers at Mars' mid-latitudes. Impact events can temporarily melt these deposits, locally providing relatively short-term conditions resembling those on Mars about 4 to 3.5 billion years ago. It has been estimated that the overall amount of water fixed

---

17) OMEGA is short for *Observatoire pour la Mineralogie, l'Eau, les Glace et l'Activité*, a mapping instrument of the European Space Agency's Mars Express spacecraft.

18) Chemolithotrophic bacteria sustain survival and growth by using energy and electrons from the oxidation of an inorganic (noncarbon) source (Section 7.3).

in Mars' icy deposits would cover, when liquefied, the Martian surface with an 11-m thick water layer.

Whether or not *liquid* H_2O and CO_2 can presently exist on Mars' surface is a matter of pressure and temperature. The atmospheric pressure on mean surface level is ≈600 Pa (6 mbar), the temperature varies between −87 and −5 °C, with possible peak heights above zero during summer day-time in equatorial regions, and lows down to −150 °C during night in polar regions. The triple point[19] for carbon dioxide corresponds to a pressure p of 5.1×10^2 kPa (5.1 bar) and −56.4 °C, definitely excluding liquid surface or near surface CO_2. The temperature and vapor pressure at the triple point of water are 0.01 °C and 0.612 kPa (6.12 mbar) and allow for the short-term presence of liquid water a few degrees above zero at best in depressions where the atmospheric pressure is just above that appointed by the triple point, as in the depths of Hellas Planitia ($p = 1.16$ kPa; 11.6 mbar). However, this approach applies to pure water only; for aqueous solutions of electrolytes (salts), hence brines, the vapor pressure, and freezing/melting point are well below the triple point of pure water. Another question is whether or not, some 3.5 to 2 billion years ago, outpourings of liquid water that apparently carved Mars' surface, and extended lakes and seas that allowed for the formation of sizable deposits of sulfates and carbonates, were just periodic, or have existed long enough to support the development of life. These interrogations will be picked up again in Chapter 7.

5.2.4.6 Chemistry in the Martian Atmosphere

In contrast to the two other terrestrial planets (Venus and Earth), Mars' low gravity did not allow for the retention of a dense atmosphere. The particularly thin atmosphere of Mars exerts a mean surface level pressure of 0.6 kPa (6 mbar), which compares to Earth's stratospheric pressure at an altitude of ca. 40 km. Occasional cirrus clouds of water-ice at an altitude of about 12 km, clouds of carbon dioxide ice between 40 and 50 km, and dust particles originating from sand storms and reaching up to 35 km shield off only a small part of the sunlight, which illuminates Mars' surface with an average intensity of only about 12% that on Earth. Due to the high eccentricity of Mars' orbit, there are considerable variations in Solar irradiation: at perihelion, Mars' surface receives about 40% more sunlight than at aphelion, causing heavy dust storms which temporarily obscure vast areas.

As on Venus, the main atmospheric gas constituent in the Martian atmosphere is CO_2 (Table 5.9). As much as 25% of the atmospheric CO_2 condenses at the polar caps into dry ice during the Martian winter, and sublimes back into the atmosphere during summer, leading to an atmospheric movement of CO_2 from the Northern to the Southern Hemisphere in summer, and vice versa in winter, and also creating disparities in the atmospheric distribution of other gases, such as argon: in winterly regions, the Ar/CO_2 ratio can be enhanced by a factor of up to

---

19) The triple point is the *point* where liquid, vapor, and solid coexist. Beyond this point, either the biphasic system vapor + liquid, vapor + solid or solid + liquid exists for a defined function $p = f(T)$ (p = vapor pressure, T = temperature), or just one phase (vapor, liquid, or solid) within a usually broad range of p and T.

Table 5.9 Amounts of gases in the Martian atmosphere compared to Earth.

	Mars		Earth	
	Mixing ratio c_x[a]	Column amount[b] (mol m^{-2})	Mixing ratio c_x[a]	Column amount[b] (mol m^{-2})
CO_2	95.7%	3.8×10^3	0.038%	86
N_2	2.7%	167	78.1%	2.8×10^5
Ar	1.6%	69.2	0.93%	2.3×10^3
O_2	0.13%	7	20.95%	6.5×10^4
CO	0.07%	4.3	trace	
H_2O	0.03%	2.9	0.4%[c]	2.2×10^3
NO	0.01%	0.58	0.02 ppm	67
Ne	2.5 ppm	2×10^{-2}	18.2 ppm	9
Kr	0.3 ppm	6×10^{-4}	1.14 ppm	0.14
Xe	80 ppb	1.1×10^{-4}	87 ppb	6.6×10^{-3}
CH_4[d]	10 ppb	10^{-4}	1.75 ppm	1.1
O_3	30 ppb	1.1×10^{-4}	0.0–0.07 ppm	≈7.3
H_2O_2	(30 ppb)[e]	(10^{-4})[e]	trace	
HCHO[d]	trace		trace	

a) Equals mole fraction $\times 10^{-2}$ when quoted in %, $\times 10^{-6}$ when quoted in ppm, and $\times 10^{-9}$ when quoted in ppb.
b) Cf. Eq. (5.9) in Section 5.2.3.2; $m/a = 0.173 \times 10^3$ kg m^{-2} for Mars (m = mass of the atmosphere, a = surface area). For conversion to units of cm^{-2} multiply by $10^{-4} N_A$.
c) Average over the total atmosphere.
d) For a detailed discussion on methane and formaldehyde, cf. Section 5.2.4.4.
e) Preliminary results.

10 because Ar, contrary to CO_2, does not condense out. As in the atmospheres of Venus and Earth, the Martian atmosphere is thus enriched in argon. However, the absolute amount of argon as quantified by, for example, the column amount (Table 5.9), is just 3% that on Earth and Venus.

Measurements of Ar have been performed by the Mars Odyssey orbiter, evaluating the 1294 keV γ-ray flux originating from argon-41, which forms from atmospheric argon-40 by capture of thermal neutrons, Eq. (5.40) [26], and by the X-ray spectrometer on board the Spirit and Opportunity rovers. The latter employs an alpha source (curium-244) to generate the characteristic K_α fluorescence of Ar at 2.957 keV:

$$^{40}_{18}Ar + ^{1}_{0}n \rightarrow \{^{41}_{18}Ar\}^* \rightarrow ^{41}_{18}Ar + \gamma \qquad (5.40)$$
$$\hookrightarrow ^{41}_{19}K + \beta^-$$

Stability and composition of the Martian atmosphere are controlled by photochemical processes. Since overall levels of oxygen and ozone on Mars are low, UV radiation penetrates deep into the Martian atmosphere. Chemistry with the inclusion of ice particle in Mars' cirrus clouds, and hence heterogeneous reactions, also play an important role. The size of these ice cloud particles typically is a few

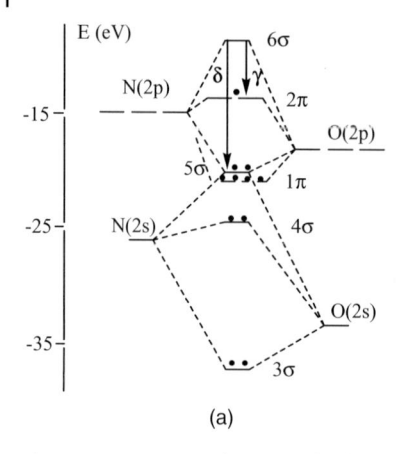

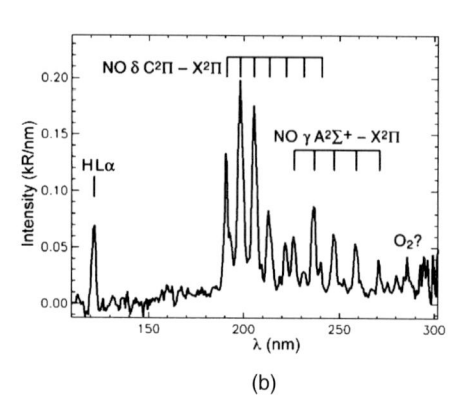

Figure 5.15 (a) Ground state molecular orbital (MO) scheme for nitric oxide NO. Only valence orbitals are shown. Dots represent electrons. The highest occupied ground state MO is the singly occupied, slightly antibonding 2π. (b) UV spectrum of the Martian nightglow, showing the transitions at 190 nm ($\pi \rightarrow \pi$), also referred to as C $\rightarrow$ X transition or δ band, and at 270 nm ($\sigma \rightarrow \pi$), also termed A $\rightarrow$ X transition or γ band. The γ and δ transitions (into empty levels of *exited* nitrous oxide, NO*) are indicated by arrows in the MO scheme (a). The multiple splittings (extension to longer wave lengths) reflect the vibrational substructure of the electronic levels. The signal at the left (121.6 nm) is the electronic Lyman-α transition of hydrogen atoms. © (spectrum) Ref. [32] Figure 2. Reprinted with permission from AAAS.

atmosphere [30, 31] (Section 5.2.3.3). Nightglow, Eq. (5.51), has also been observed on Mars [31, 32], where CO_2 and CO are the main sources for O atoms, and N atoms are formed by photolysis of N_2. Since nitrogen is clearly less abundant on Mars than CO_2, the amount of N atoms formed from N_2 is the limiting factor in NO formation and hence the intensity of nightglow. The electron distribution in the ground state of NO, $X^2\Pi$, is provided in the molecular orbital scheme in Figure 5.15a. Excited states of NO* relevant in this context correspond to a promotion of an electron either from an occupied into an unoccupied σ level, $5\sigma \rightarrow 6\sigma$ ($C^2\Pi$), or from an occupied π into an unoccupied σ level, $2\pi \rightarrow 6\sigma$ ($A^2\Sigma$). Emissions associated with the transition NO* $\rightarrow$ NO lie in the middle UV (δ and γ bands) and near IR; they are listed in Eqs. (5.52a)–(5.52c). All these transitions are doublet $\rightarrow$ doublet transitions, that is, one unpaired electron in the excited state and in the ground state. The δ and γ bands exhibit vibronic fine structures as shown in Figure 5.15b:

$$N + O \rightarrow NO^* \rightarrow NO + h\nu \tag{5.51}$$

$$C^2\Pi \rightarrow X^2\Pi, 190 \text{ nm } (\delta \text{ band}) \tag{5.52a}$$

$$C^2\Pi \rightarrow A^2\Sigma, 1224 \text{ nm (near IR)} \tag{5.52b}$$

$$A^2\Sigma \rightarrow X^2\Pi, 270 \text{ nm } (\gamma \text{ band}) \tag{5.52c}$$

Summary Section 5.2

The four inner planets, Mercury, Venus, Earth, and Mars, resemble each other in as far as they consist of an iron–nickel core, surrounded by a rocky mantle and crust. Another common feature, which includes Earth's Moon, is the presence of sizable amounts of frozen and/or liquid water. They largely differ in density and composition of the atmosphere, and surface structures related to the interaction of minerals with atmospheric gases and liquid water.

The Moon formed, shortly after the inner planets, from terrestrial material which became ejected by a collision with a giant protoplanet. The main surface features are the basaltic maria, and the highlands, a crust the composition of which is dominated by olivine and pyroxene, feldspars, and KREEP. Noteworthy is the Moon's dichotomy: the maria are almost missing on the far side, and Fe and Ti abundances are higher on the facing side. Recent findings of water-ice deposits in a permanently shaded crater have reignited the discussion on the feasibility of a Lunar station.

The innermost planet, Mercury, resembles the Moon in surface structure, although more resurfacing has occurred on Mercury due to more recent volcanism. Mercury holds an extended iron–nickel core. As on Earth, the core's outer zone is liquid, allowing for a permanent magnetic field produced by the dynamo effect. No such magnetic field is present on Venus, the rotational period of which is too long, nor on Mars, where there is no liquid interlayer. Mercury has a very thin atmosphere, termed exosphere, 35% of which are made up of sodium atoms, originating from surface minerals subjected to sputtering by the Solar wind, and by photodesorption.

Venus' surface features a couple of craters due to volcanic activity and impact events, but has been smoothed during the millennia by its thick and hot atmosphere with a ground-level pressure of 9.2 MPa (Earth: 0.1 MPa) and temperature of 750 K. The main constituent of the Venusian atmosphere, 96.5%, is supercritical CO_2, possibly stemming from thermolytical and acidic alteration of carbonates. Prominent acidic components in Venus' atmosphere are SO_2 and H_2SO_4, which form extended clouds. The second to most abundant gas is N_2 (3.5%), the oxides of which (formed from N and O generated by photodissociation of N_2 and oxygen-bearing compounds such as CO_2, CO, and H_2O in higher atmospheric levels on the dayside) can help to catalyze the oxidation of SO_2 to H_2SO_4. Sulfuric acid in turn can comproportionate with carbonylsulfide OCS (a key compound in Venus' sulfur cycle) to form elemental sulfur.

Likewise, the presence of water and methane on Mars have nourished the notion of primitive life existing or having existed on the red planet. Several surface signatures on Mars, but also the presence of carbonates, sulfates, limonite, jarosite, and goethite account for liquid surface and/or subsurface water on early Mars. Water as a constituent of the polar caps, and its presence in the form of permafrost (subsoil ice), are now well established. In addition, there may be liquid subsurface reservoirs of brines, or H_2O/H_2O_2 mixtures. The extended deposits of sulfate require a sulfur source related to volcanism (H_2S, S_n, SO_2), the oxidation of which to sulfate might have been accomplished by sulfur oxidizing bacteria. Occasional plumes of methane may reflect biogenic conversion of CO_2, the main constituent in Mars' thin atmosphere, through the activity of methanogenic archaea, albeit the inanimate production of CH_4 by serpentinization, or release from CH_4-H_2O clathrates, also explains regional and seasonal atmospheric enrichment of methane. Inclusions of crystallites of pure magnetite, found in the Martian meteorite ALH84001, are another possible clue to ancient microbial life.

5.3
Ceres, Asteroids, Meteorites, and Interplanetary Dust

5.3.1
General and Classification

Asteroids (or "small solar system bodies," as coined in 2006 by the IAU) are celestial bodies, smaller than planets but commonly larger than meteoroids. They are distinct from comets, in most cases, by their sufficiently less eccentric orbits. Basically, comets have undergone less severe processing than asteroids during the last 4.6 billion years. There is, however, a relatedness between asteroids and comets in that both derive from (and represent) vestiges from, and prior to, our Solar System. Most asteroids are located in the asteroid belt between the orbits of Mars and Jupiter at a distance between ca. 2.1 and 3.3 AU. The inner boundary of the belt, 2.06 AU, corresponds to the 4:1 orbital resonance with Jupiter: Jupiter completes one orbit in the time it takes the asteroid to complete four; this leads to the asteroid's destablization by gravitational forces and eventually to the ejection from its orbit. The outer boundary, at 3.27 AU, is related to the 2:1 orbital resonance. In addition, there are also a few objects on the facing side of the main belt, for example, the Hungaria family with a semimajor axis of 1.9 AU, and innumerous objects beyond the main belt, such as the centaurs, trojans, and trans-Neptunian (Kuiper belt) objects, the composition (and origin) of which essentially is the same as that of the belt asteroids. The Martian moons Phobos and Deimos, Section 5.2.4.2, also may be asteroids captured by Mars.

All these objects evolved from planetesimals in the early epoch of our Solar System, and most of them were ejected into the present asteroid belt when, according to the Nice model (Section 5.1), the Jovian planets had started to settle where they are today about 4.6 billion years ago. There are considerable variations in size, ranging from a few meters (and less) to 950 km in diameter for the largest object Ceres (cf. also Table 5.2 in Section 5.1), suggesting early accretion to bodies larger than 120 km in diameter on the one hand, and fragmentation by collision on the other hand. Collisions also produced asteroid families with similar orbital and composition characteristics. Many of the asteroids appear to be just rubble piles, kept together mainly by weak gravitational forces, and others are differentiated and consist of rock and ice, and sometimes have iron-based cores. Several of the asteroids also have tiny moons.

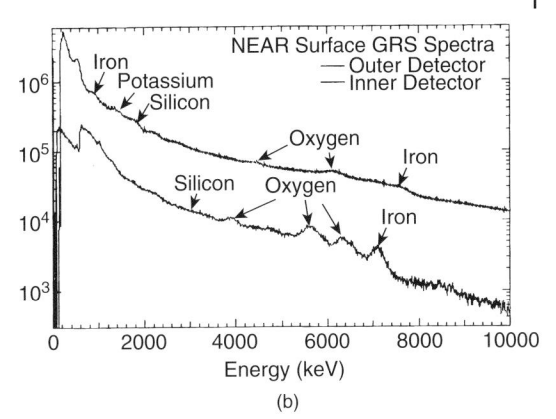

Figure 5.16 (a) The asteroid 433 Eros, photographed by the NEAR spacecraft. Credit: NASA and the Johns Hopkins University Applied Physics Laboratory. (b) Composition of Eros' surface regolith as detected by NEAR's γ-ray spectrometer at its 2001 landing site. Credit: NEAR project (Jacob Trompka), NASA.

Asteroids enjoy general interest not the least through Antoine de Saint-Exupéry's novel *Le Petit Prince* [33], acquainting the reader with the thoughts of the Little Prince living on "his" asteroid, asteroid B612. Several of the objects of the main asteroid belt have names related to this novel: the asteroid *2578 Sain-Exupéry*; the asteroid *46610 Bésixdouce* ("b-six-twelve"; 46610 is the decimal equivalent to the hexadecimal notation B612); and the moon *Petit-Prince* of the asteroid *45 Eugénie* (45 Eugenia).[20] Eugenia, measuring $305 \times 220 \times 145$ km, belongs to the large objects of the asteroid belt. The moon Petit-Prince, 12.7 km across, orbits Eugenia with a half-axis of 1184 km.

The launch of the spacecraft NEAR Shoemaker on Eros in 2001, after orbiting this asteroid for about 1 year, was celebrated as a spectacular event. NEAR stands for *N*ear *E*arth *A*steroid *R*endezvous mission; Eugene M. Shoemaker was a geologist and planetologist who died in 2000. 433 Eros is a comparatively small and irregular ($34.4 \times 11.2 \times 11.2$ km) peanut-shaped asteroid, Figure 5.16a, with a rather eccentric orbit (eccentricity = 0.223) and a semimajor axis of 1.46 AU.

The overall number of objects in the asteroid belt amounts to several million, but the overall mass is just 4% that of the Moon. Ceres makes up approximately 32% of the total mass of the asteroid belt. The dwarf planet Ceres (Figure 5.17) is almost spherical, and also the next two in size, Pallas ($580 \times 560 \times 500$ km) and Vesta ($580 \times 560 \times 460$ km; Figure 5.18a, upper right (3)) are roughly spherical, while the smaller objects, including Hygiea, the fourth in size ($500 \times 385 \times 350$ km), exhibit irregular shapes, very much like the two moons of Mars.

---

20) Discovered 1857 by Herrmann Goldschmidt and named in honour of the wife of Napoleon III. For the formal designation of an asteroid, a number is added, indicating where it came in discovery, e.g., 1 Ceres, 2 Pallas, 3 Juno, 4 Vesta, 5 Astraea, 10 Hygiea, 45 Eugenia, 433 Eros, 951 Gaspra, etc.

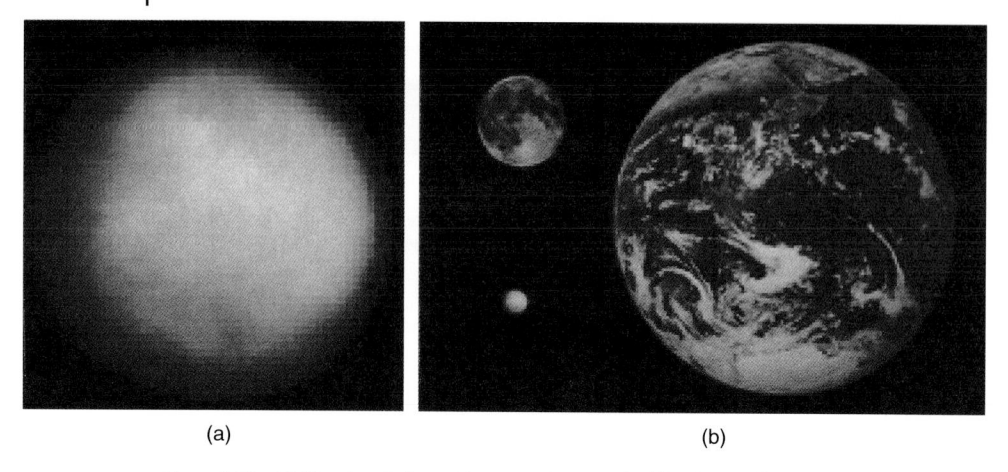

Figure 5.17 (a) The dwarf planet Ceres as photographed by the Hubble telescope. Credit: NASA/STScI. (b) The size of Ceres (lower left) in comparison to Earth and Moon. Credit: NASA (public domain).

Ceres (discovered 1801), namesake for the chemical element cerium (discovered 1803), orbits the Sun at an average distance of 2.77 AU on a near circular orbit (eccentricity 0.08) with an orbital period of 4.6 terrestrial years. The inclination of its orbit to the ecliptic is 10.6°. The dwarf planet is differentiated into a rocky core and an about 100 km thick mantle made up of water-ice and hydrated minerals [34]; the mean density, 2.08 g cm^{-3}, is thus less than that of an essentially rocky body such as the Moon (3.35 g cm^{-3}). Ceres' comparatively low albedo of 0.09 (compared to 0.12 of Mercury and 0.26 of Mars) suggests that the outer ice shell is covered by a dusty layer. Other large asteroids like Vesta are also differentiated.

Ceres is the only known gravitationally completely relaxed asteroid, that is, the only one the shape of which corresponds to thermodynamic equilibrium, which has led to its classification, along with large trans-Neptunian objects such as Pluto, as a dwarf planet. The differentiation took place soon after Ceres' accretion, when internal heating by the energy of accretion and decay of relatively short-lived radionuclides such as ^{26}Al had ceased. The decay of ^{26}Al to ^{26}Mg, Eq. (5.53a), occurs with a half-life of 0.72×10^6 a, and is accompanied by positron and γ (1.809 MeV) emission. As a result, ^{26}Mg/^{24}Mg ratios are elevated. The aluminum isotope ^{26}Al is supplied through proton capture by ^{25}Mg, Eq. (5.53b), in sufficiently hot stars, such as supernovae, novae, and/or AGB stars. Near-by core-collapse supernova activity during the very early hours of the Solar System is being anticipated. The magnesium isotope ^{25}Mg is produced by processing of nuclei formed in the Bethe-Weizsäcker cycle (Section 3.1.3), Eq. (5.53c):

$$^{26}_{13}\text{Al} \rightarrow {}^{26}_{12}\text{Mg} + \beta^+ + \gamma \tag{5.53a}$$

$$^{25}_{12}\text{Mg} + {}^{1}_{1}\text{H} \rightarrow {}^{26}_{13}\text{Al} + \gamma \tag{5.53b}$$

$$^{14}_{7}\text{N}(\alpha,\gamma)^{18}_{9}\text{F}(\beta^+\nu)^{18}_{8}\text{O}(\alpha,\gamma)^{22}_{10}\text{Ne}(\alpha,n)^{25}_{12}\text{Mg} \tag{5.53c}$$

Figure 5.18 (a) Examples for stony asteroids/meteorites. (1) 951 Gaspra (S-type achondrite; $2-3 \times 10^{13}$ kg; $18.2 \times 10.5 \times 8.9$ km); credit: NASA. (2) 253 Matilda (CI- or CM-type carbonaceous chondrite; 10^{17} kg; $66 \times 48 \times 46$ km); credit: NASA. (3) Computer model of 4 Vesta ($578 \times 560 \times 458$ km; top is North); credit: Ben Zellner (Georgia Southern University), Peter Thomas (Cornell University) and NASA. (4) Part of the Murchison meteorite (CM-type carbonaceous chondrite; ca 100 kg, from Victoria, Western Australia); © New England Meteoritic Services. (5) Millimeter-sized chondrules typical of most chondritic meteorites. (6) Cross-section of a part (ca. 2.5 cm across) of the Allende meteorite (CV-type), with spherical chondrules and irregular white fragments of CAIs. CAI stands for Ca-Al-rich inclusion, for example, $CaAl_2O_4$; Credit: The Smithsonian Institution.. (b) Metal (type) meteorites. (7) A meteorite (107.9 g; $6.4 \times 4.8 \times 1.2$ cm) from the Canyon Diablo, Northern Arizona, USA. (8) One of the Henbury meteorites (433 g), SW of Alice Springs, Australia. (9) Widmanstätten pattern, characteristic of a polished and acid-etched cross section of a metallic meteorite.

The composition of Ceres as based on its visible and near-IR spectral features is similar to that of meteorites belonging to the category of carbonaceous chondrites (class CI, vide infra): A characteristic feature at 3.05 μm (3280 cm^{-1}) can be traced back to an absorption associated with hydrated iron-rich phyllosilicates (iron-rich clays), with 4–6% contributions of carbonates ($FeCO_3$, $CaMg(CO_3)_2$ and/or $CaCO_3$) which give rise to absorptions at 3.3 and 3.85 μm (3030 and 2590 cm^{-1}). The presence of iron-rich clays such as cronstedtite $Fe_2^{II}Fe_2^{III}(SiO_5)(OH)_4$ is further based on an emission feature at 9.5 μm (1050 cm^{-1}) [35]. Cronstedtite is an aqueous alteration product of an assemblage of olivine $MgFeSiO_4$ and enstatite $Mg_2Si_2O_6$ formed in the presence of hydrogencarbonate, partway to Mg-rich phyllosilicates plus magnetite $Fe^{II}Fe_2^{III}O_4$. These findings indicate that the surface of Ceres did not experience aqueous alteration to completion.

Examples for asteroids and meteorites, including typical features, are shown in Figures 5.18a and b. Asteroids (and the large majority of meteorites) intriguingly do not only have incorporated information on the early days of our Solar System, but also hold fingerprints related to the presolar situation, originating from throw-offs by supernovae and red giants. On the other hand, despite their primordial origin, asteroids do not necessarily carry net original information from that time (except of a small fraction, about 5%, the most "primitive" ones, viz. carbonaceous chondrites): Many of them have undergone evolution since their formation, by internal heating, impacts, cosmic radiation, and, predominantly, Solar wind implantation ("space weathering") also typical for surface modifications on the Moon, on Mercury (and on Mars). Some of the "dust," or regolith, produced by space weathering escaped, and still escapes, into space, contributing to the ecliptic interplanetary cloud of dust with particle sizes around (near Earth's orbit) 40 μm, together with dust resulting from collisions, micrometeorite bombardment, ejected cometary material, and stardust from the primordial Solar nebula. This dust is responsible for the zodiacal light, an auroral glow extending from the direction of the Sun along the plane of the ecliptic. Planetesimals beyond a distance of about 2.7 AU (which is almost half-way between Mars and Jupiter) were able to accumulate water-ice, that is, asteroids beyond (as viewed from Earth) that limit can be expected to contain icy deposits along with their rocky, carbonaceous, or metallic constituents. The presence of water is an additional and important factor contributing to metamorphoses of the original material.

Classification of the asteroids is based on their orbital characteristics, or their spectral properties, that is, their color (including the IR and UV range) and albedo. Although there are links between spectral properties and composition, the correspondence often is poor. Most meteoroids, or meteorites when having plunged down to our planet, are distinct from asteroids just in size but probably not in composition, and the main part of information available for asteroids has been derived from analyses of meteorites. A secure example of a connection between an asteroid and a group of meteorites is the Vesta meteorites, also termed V or HED meteorites, where HED stands for *Howardites*, *Eucrites*, and *Diogenites*. The V meteorites consist of igneous, basaltic rock belonging to the achondrites (see below), which show distinct pyroxene spectral features also detected for the asteroid 4 Vesta and a group of smaller asteroids called Vestoids. These V meteorites and Vestoids may results from a giant impact, documented by a huge crater visible on Vesta. Factually, the greater part of about 40 000 meteorites picked up on our planet are objects stemming from the asteroid belt, with a sizable number (possibly the CI-type meteorites), however, originating from extinct comets. For about 100 meteorites, a Lunar origin has been established; for classification of the about 50 Martian meteorites (which again are achondrites), see Table 5.8 in Section 5.2.4.3.

In one specific case, a firm link between individual meteorites and their asteroidal parent has been established directly [36]: the asteroid 2008 TC_3 was discovered in October 2008, 19 h prior to its impact with our atmosphere, followed by explosion and fragmentation. The original mass of 2008 TC_3 has been estimated to $83 \pm 25 \times 10^3$ kg, which corresponds to an erratic rock of ca. 4 m diameter, assuming a density of 2.8 g cm^{-3}. 2008 TC_3 is supposed to have broken from the surface of a larger asteroid. Meteorite fragments collected shortly after atmospheric impact and shattering in the Nubian Desert allowed it to be classified as a polymict ureilite (F-type achondrite). F-type achondrites are dark gray, with a particularly low albedo. *Polymict* refers to the fact that the material is nonuniformly composed of varying lithic clasts. *Ureilites* are coarse-grained minerals consisting of magnesium-rich olivine $(Fe^{II},Mg)_2SiO_4$ and pigeonite, an inosilicate of composition $(Ca,Mg,Fe^{II})_2Si_2O_6$, plus inclusions of carbon modifications, iron–nickel, and troilite FeS. In several additional cases, the fall of a meteorite had been witnessed by the fireball and explosion on its entry into the atmosphere, and fragments have been secured prior to any major terrestrial alteration and contamination. A recent example is the Tagish Lake meteorite, which fell in the Yukon Territory, BC, Canada, in 2000 [37].

Following bulk chemistry and mineralogy patterns, the following main meteorite categories can be set up:[21]

- *Stony meteorites.* Chondrites (C) and achondrites (S, for stony). Chondrites make up about 85% of the finds of meteorites, achondrites ca. 9%. Chondrites, with a low albedo around 0.05, are named for their contents (in most cases) of chondrules (Figure 5.18a, bottom row). These are micro- to millimeter-sized spherical objects, originating from droplets of free floating molten material in the primordial space. The chondrules are usually rich in olivine $(Fe^{II},Mg)_2SiO_4$ and pyroxene (approximate composition: $M,M'(\underline{Si},Al)_2O_6$, where M and M' can be Mg^{2+}, $Fe^{2+/3+}$, Co^{2+}, Mn^{2+}, Na^+; $\underline{Si}$ emphasizes the predominance of Si over Al in these alumosilicates). Also characteristic of chondrites are irregularly shaped embeddings of refractory minerals rich in inclusions based on calcium- and aluminum-rich mineral (termed CAIs), and tiny flecks of nickel–iron alloy. *Refractory* refers to materials which are chemically and physically stable at high temperatures. While chondrites have

---

21) There is no uniformly used classification system for meteorites. The one employed here is frequently found in the literature. No comprehensive treatment of all the numerous (and constantly increasing number of) subclasses is intended.

experienced little change during their original formation, achondrites are stemming from asteroids which have experienced metamorphosis since their formation mainly by melting. Achondrites are brighter; they have albedos between 0.15 and 0.25. The Vestan, Lunar, and Martian meteorites are achondrites. A third subgroup is represented by the E-type meteorites (E for enstatite $Mg_2Si_2O_6$) with albedos of 0.4 and more. A few uncategorized objects also belong to the stony meteorites.

- Along with the stony (silicateous) C-, S-, and E-type meteorites, iron (metallic, M) meteorites occur; Figure 5.18b. The M-type meteorites, about 6% of the overall population, come from the cores of large, differentiated asteroids. They are characterized by high contents almost entirely of iron (plus some nickel and sulfur). They have a high albedo, may have a reddish or metallic luster, and show typical Widmanstätten patterns (Figure 5.18b, right (9)) when cut, polished and etched with acid at the cut face, revealing the crystalline nature of the metal phase.

- Stony-iron meteorites are a composite of achondrites (S meteorites) and metal meteorites (M meteorites).

Chondrules are present in all chondrites except in some specimens of the CI subgroup. Also typical of many chondrites, along with the refractory material and Fe–Ni flecks, is a dust matrix with embedded presolar grains (see below). Chondrites are further subdivided into ordinary chondrites (with the subgroups H, L, and LL, referring to high iron, low iron, and low iron plus low overall metal content) and carbonaceous chondrites. The latter contain around 3% of carbon in the form of carbonates, elemental carbon, carbides, and organic compounds; they represent only 5% of all of the chondrites. The famous Murchison and Allende meteorites are examples for carbonaceous chondrites. For the Murchison meteorite, see Figure 5.18a, (4). Another feature differentiating between ordinary and carbonaceous chondrites is the isotopic ratio $^{17}O/^{16}O$, which is high in ordinary, and low in carbonaceous chondrites, where high and low relate to the $^{17}O/^{16}O$ ratio in terrestrial materials. Carbonaceous chondrites are further divided into several subgroups, sometimes referred to as "clans." These are named so for characteristic specimens:

- CI (Ivuna) is friable and, in many cases, lacking chondrules and refractory inclusions, possibly as a consequence of aqueous alteration, producing phyllosilicates. In composition, they resemble the Solar photosphere (except of the volatiles). A cometary origin has been suggested.

- CM (Mighei) and CO (Ornans) contain very small (submillimeter-sized) chondrules. They are comparatively high in water contents (but less than the CI-type) and hence also exhibited some aqueous alterations.

- CV (Vigarano) contains numerous refractory inclusions particularly rich in lithophile elements and CAIs. Lithophile ("silicate-loving") elements are the group 1–6 metals, boron and aluminum.

- CK (Karoonda) materials are in a highly oxidized state. They are rich in magnetite and had been thermally metamorphosed.

- CR (Renazzo) and CH (*h*igh in Fe) are rich in metallic iron–nickel (up to 15% by weight) and iron sulfide. The CR chondrites likely contain the most primitive organic material.

- CB (Bencubbin) contains more than 50% Fe–Ni together with pyroxene-type silicates.

5.3.2
Carbon-Bearing Components in Carbonaceous Chondrites

Carbonaceous chondrites with their comparatively high contents of carbon-based compounds (commonly 2–3%) are also of interest in the context of the abiotic formation of organic matter including molecules related to life, in particular amino acids. These carbon-rich objects have remained relatively unprocessed since their formation in the early Solar System, and hence provide a record of prebiotic chemical evolution; the materials they contain can be considered to have been frozen in time shortly after the birth of the Solar System, and some compound classes even contain a record that extends back to interstellar environments before the Solar System formed [38a]. Understanding this record imprinted into carbonaceous chondrites requires that terrestrial contamination is identified and excluded. The basis for distinguishing between terrestrial contamination and primordial organic compounds contained in the meteorites can be the relative abundance of related species, the isotopic ratio R of stable isotopes ($R = {}^{2}H/{}^{1}H$, ${}^{13}C/{}^{12}C$, ${}^{15}N/{}^{14}N$, ${}^{18}O/{}^{17}O/{}^{16}O$, ${}^{30}Si/{}^{29}Si/{}^{28}Si$, and ${}^{34}S/{}^{33}S/{}^{32}S$), and the presence of racemic vs. optically enriched compounds in the case of optically active molecules. The abundance of stable isotopes is expressed as δ values (in per mill, ‰), indicating the difference between the ratio in the sample under investigation and the same ratio in a standard, Eq. (5.54). Commonly employed international standards are summarized in Table 5.10. A positive δ value means that the sample contains more of the heavier isotope than the standard:

$$\delta = 10^{3}(R_{sample} - R_{standard})/R_{standard} \text{ ‰} \quad (5.54)$$

where R = ratio of heavier to lighter isotope; for $R_{standard}$ see Table 5.10.

The main amount of carbon, about 90% and more, is present in the form of organic matter, both "free" (i.e., soluble in organic solvents; about one-third) and insoluble in the form of a polymeric macromolecular matrix resembling kerogen, a material which is formed by geological processing of hydrocarbons. The remaining about 10% carbonaceous matter are mainly carbonates, while elemental carbon and silicium carbide account for about 1% of the overall amount. The indigenous nature of the (insoluble) organic matter is confirmed by noticeable enrichments of the heavier isotopes ${}^{15}N$ (δ = +25 to +150) and ${}^{2}H$ (δ = +480 to +680 ‰ and more) bonded to carbon [38b], pointing toward a formation in a cold (about 10 K)

Table 5.10 Commonly employed standards for isotope abundances.

	Standard	Ratio $R_{standard}$
Hydrogen	Standard mean ocean water (SMOW)	$^2H/^1H = 0.156 \times 10^{-3}$
Carbon	Pee Dee belemnite (PDB)[a]	$^{13}C/^{12}C = 11.2372 \times 10^{-3}$
Nitrogen	Atmospheric air (AIR)	$^{15}N/^{14}N = 3.66 \times 10^{-3}$
Oxygen	SMOW[b]	$^{17}O/^{16}O = 0.380 \times 10^{-3}$; $^{18}O/^{16}O = 2.005 \times 10^{-3}$
Silicon	Solar composition	$^{29}Si/^{28}Si = 52.53 \times 10^{-3}$; $^{30}Si/^{28}Si = 35.99 \times 10^{-3}$
Sulfur	Cañon Diablo troilite (CDT)[c]	$^{33}S/^{32}S = 7.88 \times 10^{-3}$; $^{34}S/^{32}S = 44.16 \times 10^{-3}$

a) Belemnite is a marine fossil from the lower Jurassic (Pee Dee formation).
b) For carbonates, the PDB standard is also in use.
c) The Cañon Diablo troilite is an iron meteorite (91.6% Fe, 7.1% Ni, 1% S); Figure 5.18b left, from the Cañon Diablo in Arizona. The mineral troilite, FeS, belongs to the pyrrhotite group, $Fe_{1-x}S$ ($x = 0$–0.2).

presolar cloud or protoplanetary disk from ion–molecule reactions or reactions on grain surfaces. In either case, the frigid environment favors the survival of the molecule containing the heavier isotope, because the heavier isotopomer (the molecule containing the heavier isotope) usually reacts at a slower rate (kinetic isotope effect) and is less volatile (thermodynamic isotope effect) than the lighter isotopomer. Also, at very low temperatures, the formation of the heavier isotopomer is favored (see Section 5.3.3).

Organic matter is mainly present in those meteorites that contain a notable amount of water (CI- and CM-type chondrites), while anhydrous carbonaceous meteorites show significant depletion of organics. It was thus originally assumed that organic matter had been produced by aqueous alteration of elemental carbon. Carbon X-ray absorption near edge structure (XANES) spectroscopy and IR investigations of interplanetary dust particles (Section 4.2.5.2) have, however, revealed that the bulk of the prebiotic matter in the Solar System did not form by aqueous processing [39]. The depletion of organic matter in other than CI and CM chondrites may be the result of the high temperature (as high as 1200 °C) the parent objects of the anhydrous meteorites had been subjected to.

Elemental carbon is present in carbonaceous chondrites in the form of various polymorphs; Figure 5.19. These include amorphous carbon; nano- to microcrystalline graphite, diamond and lonsdalite; clusters such as the fullerenes C_{60} and C_{70}; and carbon monolayers such as graphene and carbon nanotubes. Diamond and lonsdalite constitute sp^3 hybridized, tetrahedral C; in the other carbon species, the carbon atoms are sp^2 hybridized (trigonal planar). Diamond and lonsdalite are distinct by their crystallographic features: in diamond, which is the more common

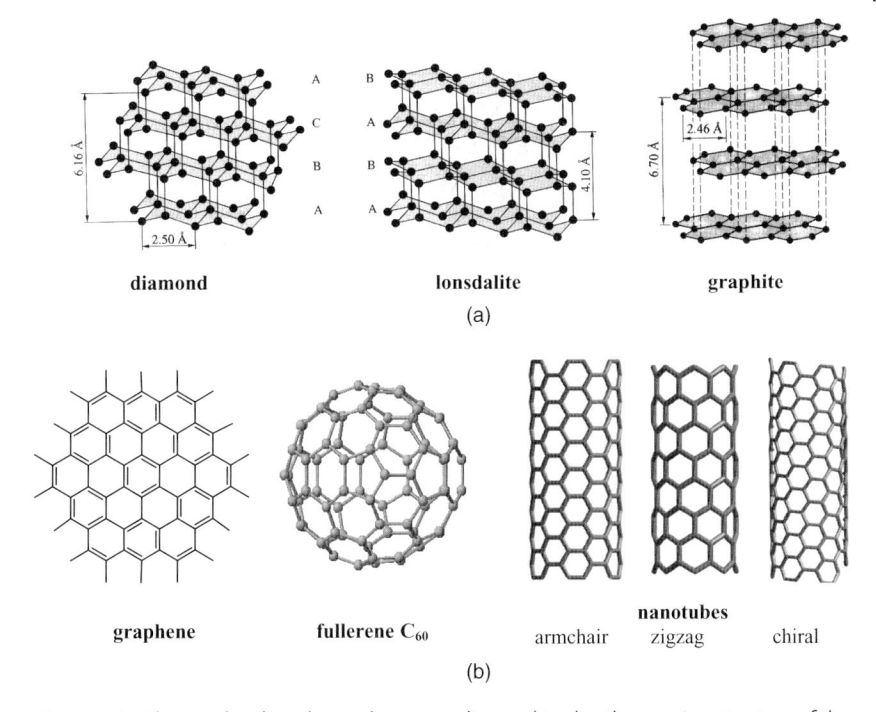

Figure 5.19 Elemental carbon detected or likely present (fullerenes and nanotubes), in carbonaceous chondrites. Single-walled carbon nanotubes (SWNT) can be capped by fullerene half-spheres. The structure of cubic diamond is also the genuine structure of the silicon carbide modification moissanite.
(a) © Hollemann-Wiberg, Lehrbuch der AnorganischenChemie, de Gruyter.

form, the carbon atoms are in a cubic arrangement (the sequence of "layers" corresponds to A,B,C,A,B,C,...), while in lonsdalite, which forms from graphite by heat and pressure transformation on occasion of impact events, the hexagonal crystal lattice (layer sequence A,B,A,B,...) of graphite is retained. Graphene represents a single sheet layer of graphite and hence is the basic structural unit for graphite, but also for nanotubes. In carbon nanotubes, three different types of connectivity for the C_6 hexagons are possible, one of which provides chirality. Of the fullerenes, the most prominent is the Buckminster fullerene C_{60}, named after the architect Buckminster Fuller, who became famous, around the fifties of the last century, for his geodesic domes based on hexagonal and pentagonal structure elements. For the laboratory synthesis of C_{60}, inspired by the chemistry in interstellar clouds, see Section 4.1. In C_{60}, an aromatic compound in a first approximation, watch-glass shaped C_6-hexagons and C_5-pentagons are linked in such a way that an ideal sphere arises, exactly in the same way a soccer ball attains its ideal spherical shape; an American football, in contrast, resembles the C_{70} fullerene. It should be cautioned that fullerenes and nanotubes detected in meteorites might also be an artifact, generated by laser ablation commonly used in combination with mass spectrometry as an analytical tool to characterize nonvolatile and nonextractable matter.

Another nonorganic carbon compound present in carbonaceous chondrites is silicon carbide SiC. Most SiC has been produced in the form of micron-sized grains in the atmosphere of carbon- and metal-rich, comparatively low mass ($<5\,m_\odot$) red giant stars, but some also represent remnants from supernovae [40]. These presolar particles became trapped in the course of asteroid formation. SiC exists as several polytypes, one of which, cubic moissanite, resembles cubic diamond in the built-up of the crystal lattice (with half of the C atoms alternately replaced by Si atoms; zincblende structure). Moissanite (or β-SiC) was discovered by Henry Moissant in the Canyon Diablo meteorite (Figure 5.18b, (7)) in 1893. Its presolar origin can be traced back to (i) the ratio of silicon isotopes $^{29}Si/^{28}Si$ and $^{30}Si/^{28}Si$, both of which are larger than those founnd in Solar material, and (ii) the presence of trace elements (Xe, Kr, Ba, Nd) formed by slow neutron capture (the so-called s-process) typical of asymptotic giant branch (AGB) stars [40] which, observationally, appear as red giants. For details of nucleosyntheses, cf. Section 3.1; for the s-process in particular, see Section 3.1.5.3.

Most information on organic compounds present in meteorites has been obtained by sampling the Murchison and the Allende meteorites and, more recently, the Tagish Lake meteorite. The Murchison meteorite (CM-type; Figure 5.18a, (4)) shut down in Western Australia in 1969, the Allende meteorite (CV-type) in the same year near the village Pueblito de Allende in the Mexican State Chihuahua. Fragments of the Tagish Lake meteorite (intermediate between CI- and CM-type) were collected immediately after the meteorite's fall was witnessed in 2000. The overall composition of these three carbonaceous chondrites is summarized in Table 5.11. Overall carbon contents are markedly different, paralleled by a corresponding trend in sulfur contents: C and S are lowest in the Allende and highest in the Tagish Lake meteorite.

In order to access the nature of the organic compounds, the material is commonly fractionated by distillation/vaporization and/or solvent extraction with solvents of increasing polarity (e.g., benzene, benzene–methanol, water, water–HCl). This fraction is commonly referred to as *free* organic matter. Macromolecular material which can be broken down by heating is termed *labile* organic matter. The remaining fraction is *refractory* organic matter, which can be degraded only by strong oxidizing agents such as perchloric acid $HClO_4$ or ruthenium tetroxide RuO_4.

Most of the organic material, about 70%, remains in the macromolecular insoluble organic matter (IOM), resembling, to some extent, kerogen. The IOM is based on di- and tricyclic aromatic[22] moieties linked by aliphatic chains with a high

---

22) In the context of meteorite chemistry, the term "aromatic" is often used in a broader sense, i.e., for any essentially planar polycyclic compound containing conjugated π systems. An authentic aromatic compound follows the Hückel rule, according to which the number of π electrons is $4n + 2$, where $n = 0, 1$ (benzene), 2, 3, etc. Thus, naphthalene (Scheme 5.2) literally is a "pseudoaromatic compound."

Table 5.11 Composition (selection of main elements; H, N, and O excluded) in three carbonaceous chondrites.

	Tagish Lake (CI/CM)	Murchison (CM)	Allende (CV)
Fe	19.3%	22.13%	23.6%
Si	11.4%	13.59%	15.9%
Mg	10.8%	12.0%	14.9%
S	3.9%	3.0%	1.96%
C	3.6%	2.2%	0.5%
Ni	1.16%	1.4%	1.2 %
Ca	0.99%	1.35%	1.9%
Al	0.99%	1.1%	1.7%
Cr	0.28%	0.33%	0.35%
Mn	0.14%	0.15%	0.15%
P	930 ppm	0.10%	0.11%
Co	517 ppm	630 ppm	800 ppm
Ti	520 ppm	780 ppm	900 ppm
V	54 ppm	61 ppm	93 ppm

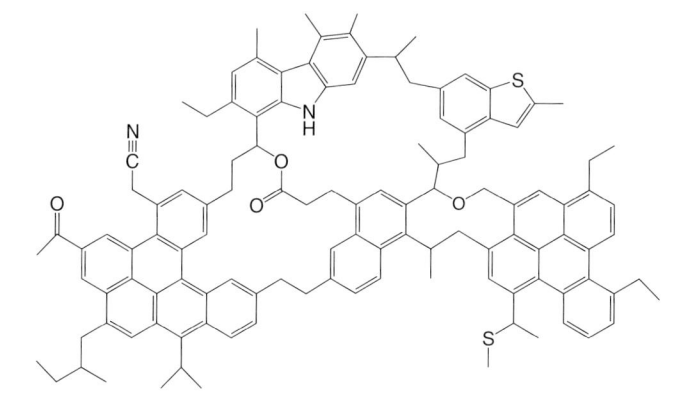

Figure 5.20 Model for the structure and composition of the insoluble organic matter (IOM) in CM-type meteorites such as Murchison; modified from Ref. [41].

level of branching/cross-linking, and containing a substantial amount of the heteroatoms oxygen (in ester and ether functions), sulfur (in the form of aliphatic sulfides and thiophene), and nitrogen (as N-heterocycles and some nitrile). A further characteristic feature of IOM is the presence of heterogeneously distributed radicals. A summary of molecular information on the chemical composition of IOM in the Murchison meteorite [41] is provided by the idealized structure in Figure 5.20. A selection of aryls (aromatic compounds) exclusively based on carbon, and either detected directly in volatile fractions and solvent extracts by MS, or by a combination of laser desorption and MS, are collated in Scheme 5.2. The

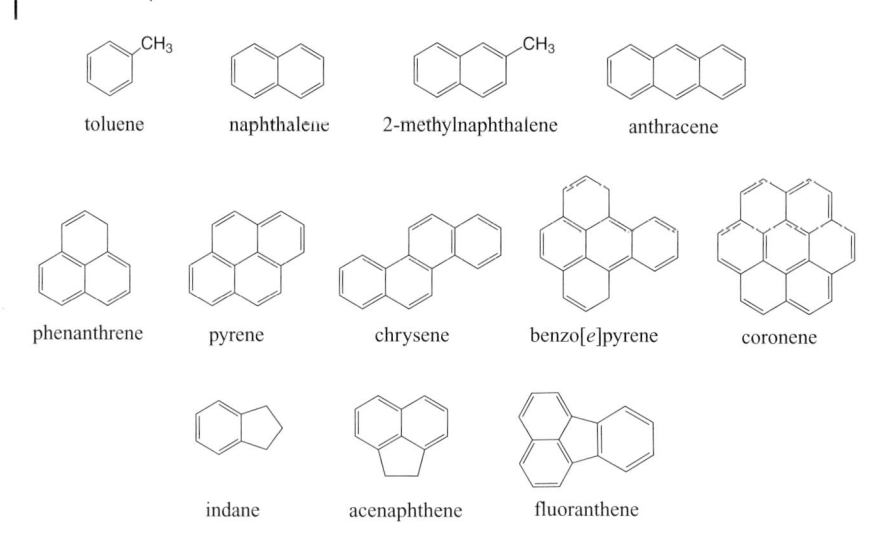

Scheme 5.2 Selection of aryls ("polyaromatic hydrocarbons" PAH) detected in meteorites.

PAHs are produced in the high-temperature high-density ejecta of asymptotic red giants (AGB stars), and have hence been readily available when our Solar System formed. Assignment of the individual species is usually based on ^{13}C NMR (nuclear magnetic resonance) and MS, often coupled with GC (gas chromatography) or other chromatographic methods for fractionation and separation. Laser desorption/laser ionization MS (L²MS) takes advantage of the abrasion, or desorption, of material from a surface by a laser beam which, concomitantly, ionizes the vaporized material to make it accessible for mass-spectrometric detection. In volatiles and solvent extractions, saturated and unsaturated straight-chain and branched aliphatic compounds ranging from C_1 to C_7 were also detected. Whether or not these aliphatic compounds, at least as far as *n*-alkanes are concerned, are of primordial origin, or terrestrial contaminants, is still under debate [38b].

Of special interest in the frame of the origin of "life-molecules" are amino acids, which can be abundant constituents in the organic equipment of meteorites. The total amino acid concentration in some CR-type meteorites collected in Antarctica amounts to up to 250 ppm, with about 10% represented by isovaline (Scheme 5.3). Most of the just under 100 amino acids found in meteorites do not occur on Earth, and those of the meteoritic amino acids also occurring on our planet, essentially as building blocks of proteins, viz. glycine, alanine, aspartic acid, glutamic acid, valine, leucine, isoleucine, and proline, usually (but not always; see below) are racemic, another indication of their nonterrestrial origin. Chiral amino acids that play a role in life attain, in most cases, the L-configuration at their asymmetric carbon (the C adjacent to the carboxylate function, Cα); in a few rare cases, D-configurated amino acids are also used. D- and L-form (see frame in Scheme 5.3 for a definition), so-called enantiomers, are related to each other in terms of image and mirror image, and this phenomenon is referred to as chirality or handedness.

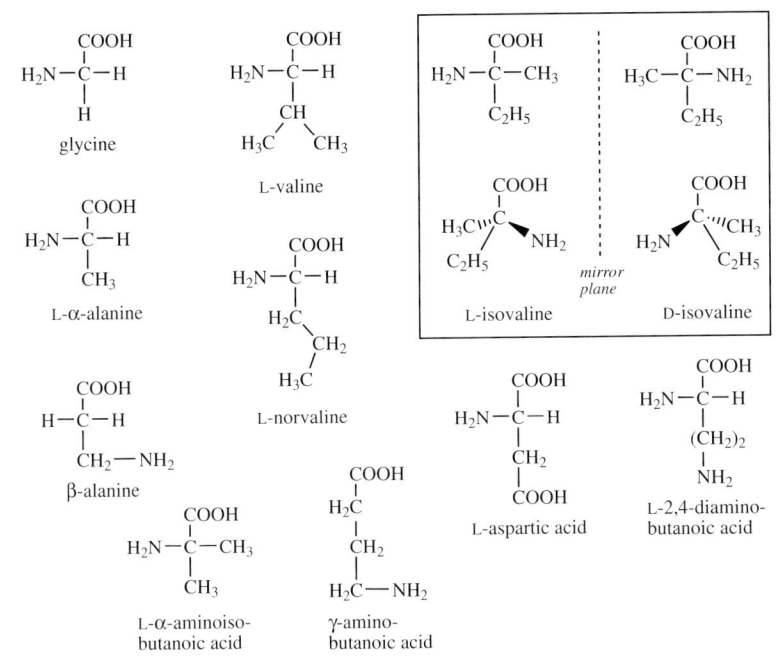

Scheme 5.3 Selection of amino acids found in carbonaceous chondrites. For isovaline (framed), the two optical isomers (enantiomers) are shown. Glycine, β-alanine, and γ-aminobutanoic acid do not have an asymmetric carbon center and hence are optically inactive. For the remaining amino acids, only the L-enantiomer is shown, which is the common biotic form on Earth. Meteoritic amino acids are of abiotic origin, and most of them are present in the form of racemates or with a slight excess of one of the enantiomers.

Chiral compounds are "optically active," that is, they are able to rotate the plane of polarized light either to the left (anticlockwise) or to the right (clockwise).

Living organisms can synthesize, and utilize, only one of the chiral versions, mostly the L-version in the case of amino acids. Among the few exceptions is isovaline (Scheme 5.3), a rare nonprotein amino acid on Earth, found in its D-configuration in fungal peptides.[23] If L- and D-form are present in equal amounts – as a *racemate* – this hints toward an *abiotic* formation, at least in the case of amino acids such as isovaline which, because there is no hydrogen on the asymmetric carbon atom, cannot undergo racemization by aging. A rather high enantiomeric excess (*ee*) of 18.5% for L-isovaline has been found in the Murchison meteorite [42]. Gradual destruction preferentially of one of the enantiomers in a racemate can be the result of irradiation with circularly polarized UV light in

---

23) An example is the fungus *Tolypocladium niveum*, which contains several peptide antibiotics with 16 amino acids as building blocks, including D-isovaline.

the presolar cloud. Homochirality, that is, the presence of just one of the optical isomers can, however, also be effectuated by an autocatalytic reaction in which the reaction product serves as the catalyst to produce more of itself, suppressing production of the enantiomer [43], a process which is referred to as *asymmetric amplification*: amino acids such as proline, alanine, and serine, forming racemic crystal lattices, have recently been shown to propagate an initially small chiral imbalance toward a homochiral system when in a coupled solid–liquid (H_2O) phase system; cf. also the more detailed discussion and Eq. (7.53) in Section 7.4.2. The comparatively high *ee* of isovaline in meteorites (such as Murchison) may thus reflect extended aqueous alteration on the parent body of CM- (and CI-) type meteorites. In any case, the presence of an excess of the L-enantiomer in Murchison, as compared to the exclusive use of the D-enantiomer in terrestrial fungal peptides, clearly points to an extraterrestrial origin of meteoritic isovaline.

For the prebiotic formation of amino acids, three main pathways have been proposed, all of which start from simple molecules (ammonia, hydrogen cyanide, hydrogen, water, carbon monoxide, methane, formaldehyde, methanol) present in the interstellar medium [44, 45] (Section 4.2):

1) In the Strecker cyanohydrin synthesis, Eq. (5.55), condensation of aldehyde, NH_3, and HCN results in the formation of aminonitrile and cyanohydrin. Hydrolysis of cyanohydrin yields α-amino acids (Eq. (5.55a)), and α-hydroxy acids (Eq. (5.55b)), respectively. Along the reaction path (5.55a), glycine forms from formaldehyde (R = H), and α-alanine from acetaldehyde (R = CH_3). The reaction products obtained along reaction path (5.55b) are glycolic acid (R = H), and lactic acid (R = CH_3). Replacing the aldehyde for acetone, α-amino-*iso*butanoic acid is produced via the aminonitrile intermediate, (Eq. (5.55c)):

$$\text{(5.55a)}$$
$$\text{(5.55b)}$$
$$\text{α-amino-}iso\text{butanoic acid (5.55c)}$$

2) The Strecker synthesis does not allow for the unambiguous formation of β-amino acids such as β-alanine. A convenient way to generate β-amino acids is the Michael addition (of ammonia) to α,β-unsaturated nitriles with subsequent hydrolysis. This is shown for the reaction between acrylonitrile (formed from ethyne and HCN) and ammonia in Eq. (5.56). The β-aminopropionitrile thus generated is hydrolyzed to β-alanine. Cyanoacetylene (cyanoethyne HC≡C–C≡N) can also be the starting compound for the generation of β-alanine. Cyanoacetylene is a common interstellar species.

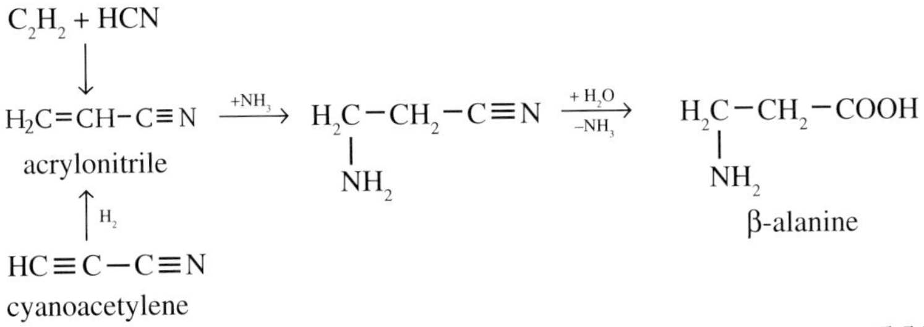

$$C_2H_2 + HCN$$
$$\downarrow$$
$$H_2C=CH-C\equiv N \xrightarrow{+NH_3} H_2C-CH_2-C\equiv N \xrightarrow[-NH_3]{+H_2O} H_2C-CH_2-COOH$$

acrylonitrile ... NH_2 ... NH_2

$\uparrow H_2$ ... β-alanine

$HC\equiv C-C\equiv N$
cyanoacetylene

(5.56)

3) Simple amino acids such as glycine and alanine may also be produced through processing of compounds generated by HCN polymerization. Hydrogen cyanide forms, along with acetylene (ethyne), from methane and ammonia by, for example, spark discharge, Eq. (5.57a). HCN readily trimerizes to aminomalodinitrile and further polymerizes as shown in Eq. (5.57b); the structural unit provided for the polymer is idealized. Hydrolysis and partial decarboxylation of aminomalodinitrile yields glycine, Eq. (5.57c), together with some alanine

$$CH_4 + NH_3 \xrightarrow{h\nu} HC\equiv N \text{ and } HC\equiv CH \ (+ H_4) \quad (5.57a)$$

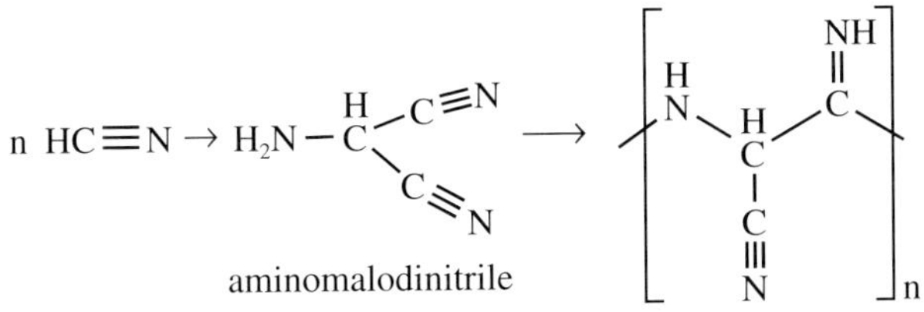

$$n\ HC\equiv N \rightarrow H_2N-C\underset{C\equiv N}{\overset{H\diagdown C\equiv N}{\diagup}} \rightarrow \left[\underset{N}{\overset{NH}{\underset{\|}{\underset{C}{\overset{H}{\diagdown}}}}} \underset{C\equiv N}{\overset{H}{\diagdown C \diagup}} \right]_n \quad (5.57b)$$

aminomalodinitrile

$$H_2N-C\underset{C\equiv N}{\overset{H\diagdown C\equiv N}{\diagup}} \xrightarrow[\text{2. decarboxylation}]{1.\ \text{hydrolysis}} H_2N-CH_2-COOH \quad (5.57c)$$

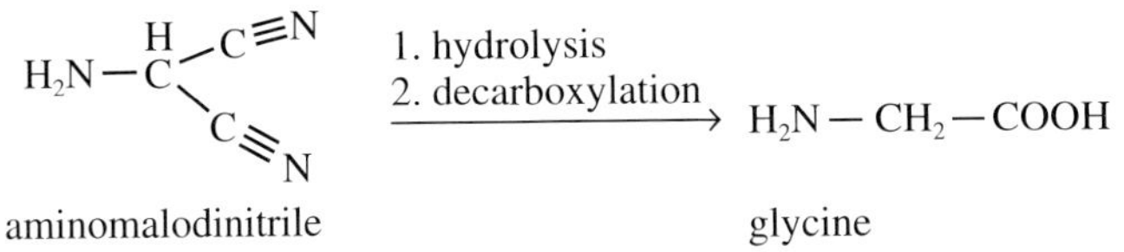

aminomalodinitrile ... glycine

In contrast to the CM-type Murchison meteorite, isovaline is low and β-alanine is high in the CI-type meteorites Orgueil and Ivuna. This points to a predominance of Michael addition (Eq. (5.56)) and HCN polymerization (Eq. (5.57)) as the synthetic pathways in the case of these two meteorites. The necessary precursor compounds for β-alanine, HCN, cyanoacetylene, NH$_3$, and H$_2$O, all are typical ingredients of comets, which has lead to the assumption that Orgueil and Ivuna are remnants of an extinct comet rather than of an asteroid [45].

Apart from amino acids, carboxylic acids, hydroxycarboxylic acids, and dicarboxylic acids have been found in meteoritic organic material and ascribed an indigenous origin based on their $\delta(^{13}C)$ and $\delta(^2H)$ values, which are clearly elevated with respect to the corresponding compounds on Earth. Particularly abundant is propanonic acid, CH$_3$CH$_2$COOH. The formation of carboxylic acids resembles the Strecker cyanohydrin synthesis, but proceeds under reducing conditions, Eq. (5.58). Possible sources for hydrogen in this reaction are aqueous oxidative alterations of iron minerals. Examples are provided by Eqs. (5.59a)–(5.59c). General features of all the organic acids encompass (i) a drastic decline in amount with increasing number of carbon atoms, (ii) dominance of branched chain isomers, (iii) decrease of $\delta(^{13}C)$ with increasing number of carbons, and (iv) a distinct deuterium enrichment [38b]. Other classes of organic compounds have been detected, among these sulfonic and phosphonic acids, aldehydes and ketones, amines and amides, O-, S- and N-heterocycles (including the nucleic acid bases adenine, guanine and uracil), and dihydroxyacetone HOCH$_2$–C(=O)–CH$_2$OH, a triose sugar:

$$H_2O + RCHO + HCN + H_2 \rightarrow RCH_2\text{-}CO_2H + NH_3 \tag{5.58}$$

R=H: acetic acid

R=CH$_3$: propanoic acid

$$Fe + FeS + 2H_2O \rightarrow FeS \cdot Fe(OH)_2 + H_2 \tag{5.59a}$$

$$2Fe_3O_4 + H_2O \rightarrow 3Fe_2O_3 + H_2 \tag{5.59b}$$

$$3Fe_2SiO_4 + 2H_2O \rightarrow 2Fe_3O_4 + 3SiO_2 + 2H_2 \tag{5.59c}$$

Related to carbonaceous chondrites on the one hand, and to interstellar/interplanetary dust particles on the other hand, are ultracarbonaceous Antarctic micrometeorites with sites up to 1 mm and masses up to a few μg. Contrasting carbonaceous chondrites, these micrometeorites contain up to 85% organic matter, highly enriched in deuterium, and mineral assemblages (Fe, NiS, Mg-rich silicates) embedded within the organic matter.

5.3.3
Interplanetary Dust Particles (Presolar Grains)

Interplanetary dust particles [46] (<100 μm; typically 0.1–20 μm) are recognized by the rather unusual isotopic compositions of their hydrogen, carbon, nitrogen, and silicon contents, but also by elevated $^{26}Mg/^{24}Mg$ ratios, indicating that they contained radioactive ^{26}Al; cf. Eq. (5.53). Interplanetary dust has been (and still is) the main source for the supply of extraterrestrial matter to Earth. The overall mass delivered to early Earth amounted to about 10^9 kg per year (i.e., about 1 million tons), including ca. 10^7 kg of carbon-based matter; cf. also Section 7.4.4. This is two orders

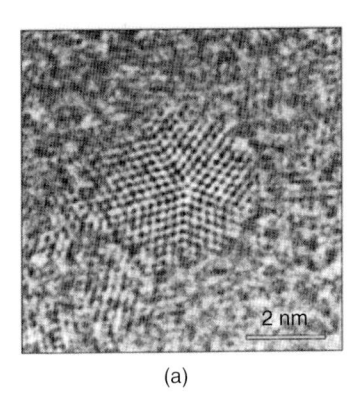

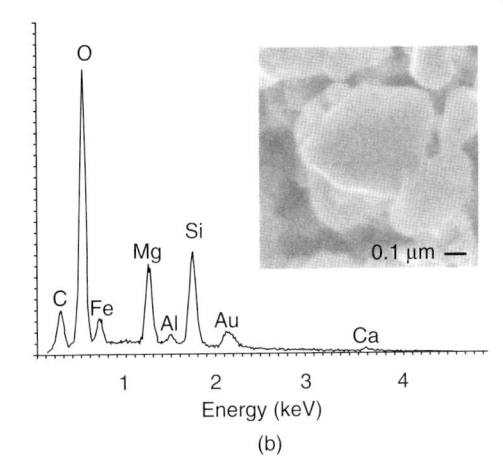

Figure 5.21 (a) TEM (transmission electron micrograph) lattice image ([111] plane) of a nano-scaled diamond crystallite from the Allende meteorite. © Ref. [47a] Figure 1 inset, Elsevier. (b) A micrometer-sized silica grain from the CO/CM-type meteorite Acfer 094 and its EDX spectrum (EDX = energy dispersive X-ray spectroscopy, a variant of XRF; cf. footnote 12), showing the main components (Si and O) and small admixtures of the elements Fe, Mg, Al, and Ca. The C signal is a contamination, and the Au signal is from the substrate. © Ref. [47b] Figure 3. Reprinted with permission from AAAS.

of magnitude more than delivered by grains in the size range 100 μm to 1 mm ("micrometeorites"); meteorite-sized objects contribute just 10^{-5} of that mass. Stratospherically collected micrometer-sized grains have revealed signatures resembling those of meteorites, that is, the grains (can) contain graphite and diamond (Figure 5.21a), silicon carbide SiC, carbide mineral solid solutions of Ti, V, Fe, Zr, Mo, and Ru, silicon nitride Si_3N_4, corundum Al_2O_3, spinels $Mg(Cr,Al)_2O_4$, silica SiO_2 (Figure 5.21b), silicate minerals such as olivine $(Fe,Mg)_2SiO_4$, forsterite Mg_2SiO_4 and pyroxene (general composition $MM'[(Si,Al)_2O_6]$, M for example, Na, Mg, Ca, Fe^{II}; M' = Cr, Al, Fe^{III}; see also Tables 5.4 and 5.8), hibonite (a particularly alumina-rich mixed oxide of empirical formula $Ca_{0.8}Ce_{0.1}La_{0.1}Al_{10.4}Ti_{0.5}Fe^{II}_{0.7}Mg_{0.05}O_{19}$, roughly $CaAl_{12}O_{19}$), rutile TiO_2, and Fe–Ni. Traces of noble gases (Ne, Kr, Xe) are frequently incorporated. Further, PAHs were found with a more pronounced degree of alkylation and higher nitrogen contents (in the form of amines, nitriles, and isonitriles) than PAH samples from carbonaceous chondrites. Also, the overall amount of PAH is clearly lower than in meteorites, likely as a consequence of volatilization from these tiny particles.

An about spherical grain of 2 μm diameter and a density of 3.1 g cm^{-3} (the approximate density of SiC) contains ca. 4×10^{11} molecules. For comparison, 1 ml of water contains ca. 3×10^{22} molecules of H_2O. The tiniest, nanometer-sized grains are represented by diamond; they contain just about 1000 C atoms. The formation of grains affords efficient association of atoms and molecules out of the gaseous phase, often in competition with chemical reactions, for example, the – kinetically favored – association of carbon atoms to graphite as opposed to the chemical reaction between C and O atoms to form CO. The grains condensed during cooling of gases in ancient stellar outflows; they can originate from different sources. Main sources are ejecta from supernovae (in particular for diamond, graphite, SiC, Si_3N_4) and AGB stars (for presolar oxide grains from O-rich stars, and carbon-rich phases from C-rich AGB stars <5 $m_\odot$). Supernovae and AGB stars (Section 3.2) are thus an important source in providing material for stellar evolution. Different groups of grains having comparable $^{13}C/^{12}C$, $^{15}N/^{14}N$, $^{30}Si/^{29}Si$, and $^{26}Mg/^{24}Mg$ patterns can often be assigned to a specific stellar source [46]:

- The *main stream* particles, accounting for about 90%, are stemming from carbon-rich AGB red giant stars: silicon-bearing grains from this source have a higher ^{13}C and a lower ^{15}N abundance than the Sun; $^{12}C/^{13}C$ ratios typically vary between 40 and 80 (Solar: 89), $^{14}N/^{15}N$ ratios between 500 and 5000 (Solar: 272). The $\delta(^{29}Si)$ and $\delta(^{30}Si)$ amount to about +200 ‰. In oxidic grains (silicates, corundum, spinel, hibonite), ^{17}O levels are enhanced and ^{18}O levels moderately to strongly depleted.

- In SiC and Si_3N_4 grains originating from supernovae (so-called *X-type grains*), both ^{15}N ($^{14}N/^{15}N$ = 7 to 19 000) and ^{13}C ($^{12}C/^{13}C$ = 2 to 7000) can be enhanced or depleted. In most X-type grains, however, ^{13}C is depleted and ^{15}N is enhanced, contrasting the trend in main-stream grains. Oxidic grains have elevated ^{16}O levels. The $\delta(^{29}Si)$ and $\delta(^{30}Si)$ are around −600‰. An additional intrinsic feature is the high abundance of ^{26}Mg, stemming from extinct ^{26}Al, and the presence of neon almost exclusively as ^{22}Ne, which is the isotope formed by the s-process, ending by β-decay of ^{22}Na.

- In the few grains which have been associated with novae, both ^{13}C and ^{15}N are enriched.

These different isotope ratios for grains from different types of stars reflect differences in nucleosyntheses (and secondary processes such as neutrino spallation), along with the downstream isotopic fractionation in the interstellar cloud; see below.

The Tagish Lake meteorite addressed in the preceding section (Table 5.11 and accompanying text) has also turned out to carry interstellar grains which apparently have been picked up by the meteorite's parent asteroid and ferried to Earth. Interplanetary dust particles are remnants of the protosolar cloud, or have formed at the fringes of the Solar nebula before it started to coalesce. Thus, meteorites, along with grains collected in Earth's stratosphere by high-flying planes, and grains incorporated in objects from the outer reaches of the Solar System can be, together with remnants of comets, important indigenous witnesses from the time prior to the formation of our Solar System.

Along with nanosized diamond and silicon carbide crystals, hollow organic globules mainly consisting of amorphous carbon-dominated organic matter have been traced in the Tagish Lake meteorite [48]. These globules, with a diameter of 140 nm to 1.7 μm, may have originated from ice grains rich in H_2O, CO_2, CH_3OH, HCHO, and NH_3, converted by UV radiation to refractory organic material. The interiors of the about spherical objects were protected from processing by UV, and later escaped by volatilization, leaving cavities of several tens to several hundred nanometer in diameter. An intriguing feature of the organic matter in these

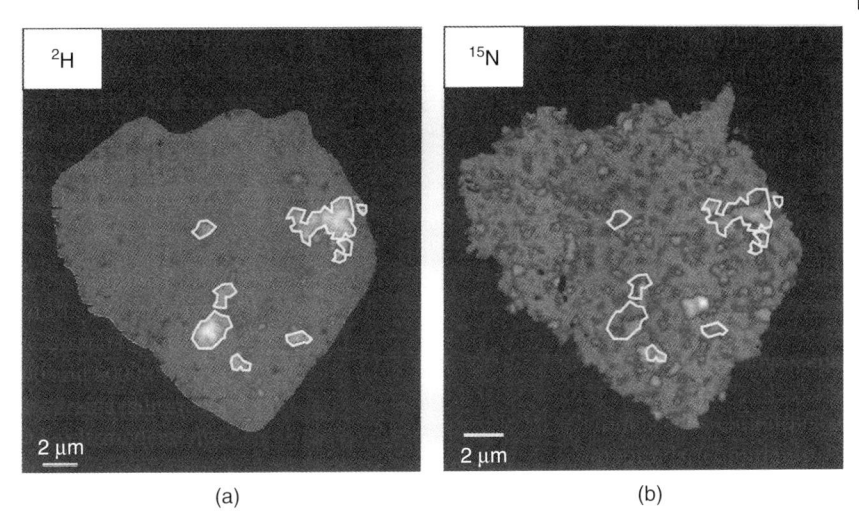

Figure 5.22 Mapping of (a) $\delta(^2H)$ ($\equiv \delta(D)$) and (b) $\delta(^{15}N)$ hotspots in a sample of insoluble organic matter (IOM) of the carbonaceous chondrite EET 92042. © Ref. [49] Figure 1. Reprinted with permission from AAAS.

globules is the particularly high enrichment of the isotopes ^{15}N and ^2H. The isotopic excess, quantified by the δ value (see Eq. (5.54) for definition), amounts to $\delta(^{15}N) = 1000$ and $\delta(^2H) = 8100‰$, exceeding the "normal" δ values of carbonaceous matter in meteorites by an order of magnitude. On the other hand, even higher ^{15}N and ^2H anomalies have been found in IOM of the CR-type meteorite EET 92042 [49], Figure 5.22, with an isotope excess $\delta(^{15}N)$ up to 1770, and $\delta(^2H)$ up to 16 300 ‰, referred to as "hotspots." As far as deuterium (D, ^2H) enrichment is concerned, these particularly high δ values reflect chemical fractionation of the isotopes in ion–molecule reactions, Eq. (5.60), characteristic of temperatures as low as 10–20 K. These reactions took place in the interstellar medium either under homogeneous conditions or on fridge surfaces of dust particles. Bonds X-^2H (X = C, N, O, ...) are somewhat stronger than X-^1H bonds, and hence the formation of the X-^2H bond is favored, in particular close to absolute zero. Reaction (5.60b), for example, is exergonic. A second process leading to enrichment in deuterium is isotopic exchange of protium (^1H) for deuterium in hydrocarbons, represented symbolically by Eq. (5.61), in a deuterium-rich Solar System reservoir. This mechanism is supported by findings according to which deuterium enrichment increases with decreasing C–H dissociation energy of an organic compound:

$$D^+ + CH_2 \rightarrow CHD^+ + H \tag{5.60a}$$

$$H_3^+ + HD \rightarrow H_2D^+ + H_2 \tag{5.60b}$$

$$-\overset{|}{\underset{|}{C}}-H + D \rightleftharpoons -\overset{|}{\underset{|}{C}}-D + H \tag{5.61}$$

For the enrichment of ^{15}N in organics, vacuum-UV ($\lambda < 200$ nm) triggered dissociation of interstellar N_2 has been proposed as a source for fractionation of nitrogen isotopes: as compared to $^{14}N^{14}N$, the heavier isotopomer $^{14}N^{15}N$ is split more easily into nitrogen atoms, followed by incorporation of ^{15}N into an organic compound, Eq. (5.62). An additional mechanism for the enrichment of a heavier isotope, "isotopic self-shielding," which will be described below in the context of $^{17,18}O$ enrichment in silica-rich grains, does not seem to have played a decisive role for ^{15}N enrichment:

$$^{14}N \equiv ^{15}N \xrightarrow{h\nu} {}^{14}N + {}^{15}N$$
$$\downarrow h\nu, -e^-$$
$$^{15}N^+ \xrightarrow{H_2} {}^{15}NH_3 \xrightarrow{C^+, e^-} HC^{15}N \text{ and } H^{15}NC \tag{5.62}$$

Silica-rich grains embedded in the IOM of the Murchison meteorite contain particularly high amounts of the heavy oxygen isotopes, with ratios $^{17}O/^{16}O$ of ca. 7×10^{-2} and $^{18}O/^{16}O$ of ca. 11×10^{-2}, corresponding to $\delta(^{17}O)$ of 180 000 and $\delta(^{18}O)$ of 54 000‰. These oxygen isotope fortification could have been produced during a period of intense X-ray activity of the nascent Sun [50], by high-energetic ionized particles (^1H, ^3He, ^4He) in Solar flares targeting N, O, and Ne atoms in the circumsolar gas, see for example, Eqs. (5.63a)/(5.63b), and followed by interception through SiO, Eq. (5.64):

$$^{20}_{10}Ne + {}^3_2He \rightarrow {}^{18}_9F + {}^1_1H + {}^4_2He \tag{5.63a}$$
$$\hookrightarrow {}^{18}_8O + \beta^+$$

$$^{14}_7N + {}^4_2He \rightarrow {}^{17}_8O + {}^1_1H \tag{5.63b}$$

$$^{17}O + SiO_{(gas)} \rightarrow SiO^{17}O_{(solid)} (= \text{silica}) \tag{5.64}$$

An alternative process for the enrichment of heavy isotopes arises from *self-shielding* in interstellar clouds of gas and dust subjected to vacuum-UV irradiation from nearby stars. This irradiation breaks up molecules such as the abundant interstellar CO into atoms, C and O, at wavelengths between 90 and 110 nm. Different isotopes, that is, $C^{17}O$ or $C^{18}O$ vs. the far more abundant $C^{16}O$, absorb the photons at slightly different energies. Near the edge of the cloud, $C^{16}O$ absorbs the major part of the photons, thus shielding $C^{16}O$ in deeper areas in the cloud. This shielding is not effective for $C^{17}O$ and $C^{18}O$ so that inside the cloud more of the heavier isotopomers will become dissociated. Independent of self-shielding, wavelength-dependent isotopic fractionation through photodissociation of CO can also lead to an atomic oxygen reservoir enriched in ^{17}O and ^{18}O [51]. In any case, the resulting oxygen atoms are then built in into OH, H_2O, SiO_2, and so forth.

A conspicuous feature of SiC grains from AGB stars is the nearly pure s-process composition of traces of associated heavy elements such as ^{96}Mo [40b]. The s-process refers to nucleosynthesis by slow capture of neutrons by an element ^mE to form the isotope ^{m+1}E. These processes are typical of comparatively low neutron densities ($<10^8\,cm^{-3}$) and medium temperature conditions in stars. Main neutron sources are the processes $^{13}C(\alpha,n)^{16}O$ and $^{22}Ne(\alpha,n)^{25}Mg$. *Slow* means that the neutron capture process is slow with respect to successive β^- decay of ^{m+1}E. Equation (5.65) illustrates a sequence by which ^{96}Mo is formed, starting from ^{88}Sr:

$$^{88}_{38}Sr(n,\gamma)^{89}_{38}Sr(\beta^-)^{89}_{39}Y(n,\gamma)^{90}_{39}Y(\beta^-)^{90}_{40}Zr(5n,\gamma)^{95}_{40}Zr(\beta^-)^{95}_{41}Nb(\beta^-)^{95}_{42}Mo(n,\gamma)^{96}_{42}\textbf{Mo} \tag{5.65}$$

On the other hand, xenon present in trace amounts in diamond crystallites originating from supernovae has inscribed the signature of *p*- and *r*-processes taking place in core-collapse supernovae: this so-called xenon-HL consists of the light (L) isotopes ^{124}Xe and ^{126}Xe originating from *p*-processes, and the heavy (H) isotopes ^{134}Xe and ^{136}Xe created in *r*-processes. The *p*-process produces *proton*-rich isotopes by knocking out, with γ rays, neutrons or α-particles (helium nuclei), while the *r*-process corresponds to *rapid* neutron capture, rapid with respect to the competing β⁻ decay of the daughter isotope. For the *s*- and *r*-processes, see also Section 3.1.5.

5.4
Comets

5.4.1
General

There is a close relation between asteroids and comets in that both are remnants from the early time of the Solar System and even contain information from the time before. CI-type meteorites have in fact been suggested to be remnants of comets, or "extinct" comets. A substantial amount of comets has formed by aggregation of material from the Solar nebula in the giant planets region of the Solar System (and are thus remnants of planetary building blocks), and either been ejected toward the inner domain where they impacted with the Sun and the inner planets, or gravitationally scattered out to great distances where they formed the Oort cloud at a distance of around 50 000 AU, Figure 5.2 in Section 5.1. From the Oort cloud, with an estimated reservoir of 10^{13} cometary objects ("cometisimals"), they are occasionally thrown out by gravitational forces exerted by the Sun's close neighbors such as Proxima Centauri and α Centauri, or by by-passing stars. These long-period comets have extremely elliptical orbits, and sometimes the "orbit" can even be parabolic or hyperbolic, that is, the comet is prone to leave the Solar System for good. Orbits may also be retrograde, and have inclinations toward the ecliptic up to 90°. Examples are the comet Halley (inclination 162.3°), and the recently discovered object 2008KV$_{42}$ (inclination 103.3°, eccentricity 0.508, orbital period 277 a, semimajor axis 42.5 AU) [52a]. Short-period comets in turn have originally been native in the Kuiper Belt/Scattered Disk (Section 5.5), which is at a distance of roughly 50–100 AU, or they originate from the centaurs,[24] an unstable orbital class of celestial bodies with characteristics of asteroids and comets.

---

24) Not to mistaken with the constellation of Centaurus with Proxima Centauri, the star nearest to our Sun. In the Greek mythology, centaurs are creatures composed of part human and part horse.

Centaurs have transient orbits extending to the Kuiper belt and often crossing the orbits of the giant planets. The short-period comets frequently have mean orbital radii similar to those of the asteroids.

Due to their different region of formation, the giant planet region on the one hand, and the sufficiently colder Kuiper belt on the other, comets of different populations can be expected to show diversity in chemistry, which will further be influenced by different evolutionary histories. Most of the information on the chemical constituents present in the comae and tails has been obtained from UV, visible (Vis), IR, and far-IR (FIR), characteristics gathered by Earth-based instruments. Additional data have been collected on occasion of rendezvous with spacecrafts, such as Giotto and the International Comet Explorer (Halley flybys 1986), Deep Space 1 (comet Borrelly 2001), Deep Impact (impact encounter with the comet Temple 1 in 2005), and Stardust which visited the comet Wild 2 in 2004, collecting dust particles in its coma, and returning these to Earth in 2006. Studies of these cometary grains provide evidence of mixing between the Solar System and interstellar matter, and extensive radial mixing of matter across the Solar System at an early stage of comet formation, also casting light on the early stages of the formation of our Solar System.

Typical comets are conglomerates of ice (mainly water, methane and carbon oxides), dust, some carbonaceous material, and small rock particles, and therefore popularly referred to as "dirty snowballs." Their density commonly is around $0.4\,\mathrm{g\,cm^{-3}}$, demonstrating (i) the predominance of ice and (ii) their fluffy porous texture. In their perihelion, that is as they approach the point closest to the Sun, part of the volatiles evaporates and forms a tenuous atmosphere (termed coma) plus, together with submicrometer dust particles, the typical bright (through reflection of sunlight) cometary tail, sometimes with a reddish hint. The direction of the tail follows the thermal gradient and motion of the comet.

In addition, there is a plasma tail, generated through ionization by the Solar wind, and this tail is oriented along the field lines of the Sun's magnetic field and thus points in the direction opposite the Sun. This ion tail often appears blue (CO^+) or green (CN^+, C_2^+) by fluorescence of the positively charged ions recombining with electrons. The tails can span several hundred million kilometers and thus stretch over half the sky. The inner part, or solid nucleus of the comet, typically around 1–50 km in diameter, is surrounded by the coma extending to about 10^5 km, and a hydrogen envelop, extending to 10^7 km. In Figure 5.23, the general built-up of a comet is shown. Figure 5.24 pictures the long-period comets Hale-Bopp and Lulin with the typical bluish and greenish glows of the plasma tails.[25] In Figure 5.25, the outer appearance of the nuclei of the comets Temple 1 and Wild 2 prior to their approach to the inner region of the Solar System are shown, along with the core of Halley's comet close to its perihelion. Temple 1, photographed by Deep Impact a few minutes before the spacecraft's impactor intentionally smashed into

---

25) When Earth crosses the plane of a comet's orbit, part of the dust tail may be seen edge-on as an "anti-tail," formed like a spike and pointing toward the Sun. This phenomenon, caused by the viewing geometry, is thus physically nonexistent.

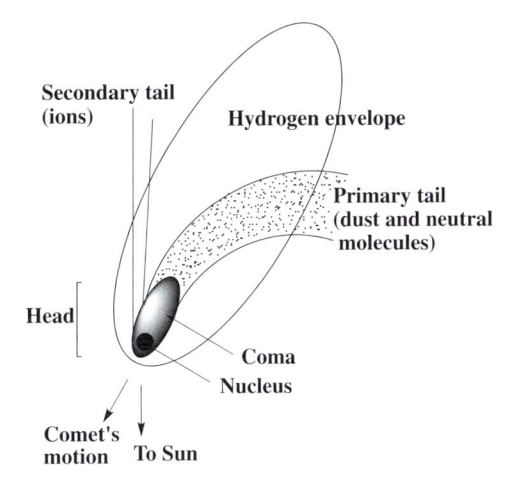

Figure 5.23 The general architecture of a comet (not to scale).

Figure 5.24 (a) The comets Hale-Bopp (1997; with permission of E. Kolmhofer and H. Raab (Linzer Astronomische Gemeinschaft), Johannes-Keppler-Observatory, Linz, Austria). (b) Comet Lulin (2009; with permission of Richard Richins, New Mexico State University). Both comets belong to the long-period category. For Hale-Bopp, see also Table 5.12. The colorful glows of the tails come about by fluorescence of CO (blue) and CN/C_2 (green). A third tail of Hale-Bopp, in-between the dust and the plasma tails and not visible here, constitutes sodium atoms.

the comet, liberated about 10^6 tons of debris. Temple 1 displays the low albedo typical of a comet's nucleus. The dark appearance is due to an envelope, or shell, of carbonaceous matter surrounding the nucleus which, by itself, supposedly is structured. The very low albedo of a comet's nucleus is also apparent for Halley's comet (Figure 5.25b), which also shows areas of high activity, namely outbursts of dust and gas where the comet faces the Sun. Orbital characteristics of four comets and a rough account of their chemical composition are provided in Table 5.12.

(a) (b) (c)

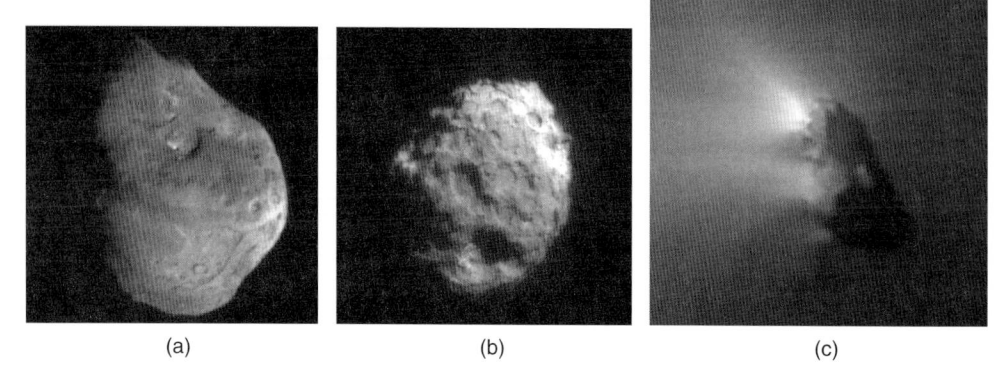

Figure 5.25 (a) Shape and surface topography of comet Tempel 1 (2005), photographed by the spacecraft Deep Impact a few minutes prior to impact. Dimensions: 7.6 × 4.9 km. Credit: NASA/JPL-Caltech/UMD. (b) Comet Wild 2 (2004), photographed by Stardust. Dimensions: 5.5 × 4.0 × 3.3 km. Credit: NASA/JPL Stardust Team. (c) The nucleus of comet Halley (1986), imaged on occasion of an encounter by the spacecraft Giotto, and showing dust jets streaming away from the active surface facing the Sun. Dimensions: 15.3 × 7.2 × 7.2 km. Credit: Halley Multicolor Camera Team, Giotto, ESA; © Max Planck Institut für Aeronomie. Cf. also Table 5.12.

Table 5.12 Characteristic data of the comets Temple 1, Wild 2, Halley, and Hale-Bopp.

Name and type	Aphelion/peri-helion; period	Eccentricity; inclination	Main molecular chemical constituents in the coma[a] (%)
Halley, short period (Halley family)	35.1/0.586 AU; 76.1 a	0.976; 162.3° (retrograde)	H_2O (80), CO (10), CO_2 (2.5), CH_4 + NH_3 (2.5), HC, Fe, Na
Hale-Bopp, long-period	370.8/0.914 AU; 2533 a	0.9951; 89.4°	H_2O (80), CO (10), CH_3OH (1.6), CH_4 (1), NH_3 (0.6), C_2H_6 (0.5), HCN (0.3), C_2H_2
Temple 1, short-period (Jupiter family)	4.74/1.51 AU; 5.5 a	0.516; 10.5°	H_2O (80), CO (3.4), CH_3OH (0.9), CH_4 (<0.7), H_2CO (0.6), C_2H_6 (0.2), HCN, C_2H_2
Wild 2,[b] short-period (Jupiter family)	5.31/1.59 AU; 6.4 a	0.538; 3.2°	PAH, HC, alcohols, carbonyls, amines, amides, nitriles.

a) An about equal amount of water (80%) is assumed for the four comets. HC = hydrocarbons, PHA = Polyaromatic hydrocarbons. For more details, see Section 5.4.2.
b) Pronounced "Vilt."

Along with the comet families mentioned in Table 5.12 (Halley family and long period comets; Jupiter family), there is a third family, the main belt comets [52b]. These objects, which are confined to the asteroid belt, develop dust tails when close to perihelion. Physically, they are thus comets; dynamically, they are asteroids in character. At least two of these comets belong to the asteroid 24 Themis family, i.e., they were chipped off 24 Themis.

Comets leave a trail of debris behind them, giving rise to meteor showers as Earth passes through the debris. Halley's comet, for example, is the source of the Orionid shower. When all the volatiles are used up, a lump of rubble remains, not unlike an asteroid. Comets may also break apart under the gravitational influence exerted by the Sun. Illustrative examples are Schwasmann-Wachmann 3 (orbital period 5.34 a), the continuing fragmentation of which has been followed on occasion of repeated apparitions, and Shoemaker-Levy, which broke apart and collided with Jupiter in 1994. For this event, see also Section 5.6.1.

5.4.2
Comet Chemistry

The chemical composition, and abundance, of species found in the comae (and the tails) of comets depends on temperature and radiation intensity, and hence on the heliocentric distance at which analyses are performed. Similar considerations apply to many of the potential precursor compounds of the detected chemical species. Short-period comets have been processed considerably by multiple passes near the Sun, and their chemistry does not necessarily reflect the pristine situation.

The discovery of comet Hale-Bopp in 1995 and the spectacular appearance when it approached the Sun in 1997 has been exploited to derive the versatile chemistry taking place in its coma. In addition to the species listed in Table 5.12, the molecules and molecular ions shown in Scheme 5.4 have been detected during observations close to perihelion (0.91–1.4 AU) [53].

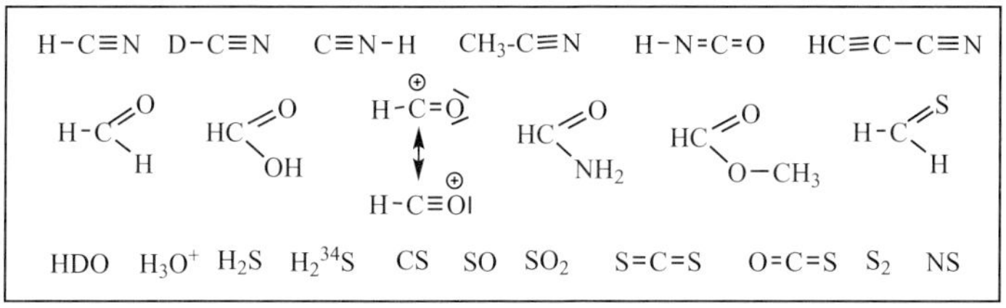

$H-C\equiv N$ $D-C\equiv N$ $C\equiv N-H$ $CH_3-C\equiv N$ $H-N=C=O$ $HC\equiv C-C\equiv N$

$H-\overset{O}{\underset{H}{C}}$ $HC\overset{O}{\underset{OH}{}}$ $H-C=\overset{\oplus}{O}\updownarrow H-C\equiv \overset{\oplus}{OI}$ $HC\overset{O}{\underset{NH_2}{}}$ $HC\overset{O}{\underset{O-CH_3}{}}$ $H-C\overset{S}{\underset{H}{}}$

HDO H_3O^+ H_2S $H_2^{34}S$ CS SO SO_2 $S=C=S$ $O=C=S$ S_2 NS

Scheme 5.4 Molecules which have been detected in the comae of comets. Most of the molecules shown have been found in Hale-Bopp.

Formic acid can form directly by gas phase reaction between OH radicals and formaldehyde, Eq. (5.66a), or by an ion–molecule pathway from water and CHO$^+$, Eq. (5.66b), followed by proton abstraction or transfer. Reactions leading to cyanoacetylene HC$_3$N are depicted in Eqs. (5.67a) and (5.67b). Equations (5.66b) and (5.67b) represent ion–molecule radiative recombinations, that is, the kinetic energy of the two reactants is dissipated in the form of electromagnetic radiation ($h\nu$). Model reactions of the gas phase chemistry toward formic acid, cyanoacetylene, methylformate HCOOCH$_3$, dimethylether (CH$_3$)$_2$O, and acetonitrile CH$_3$CN failed, however, to reproduce the abundances of these species found in the comet's coma, suggesting that they did not primarily form by gas phase reaction in the coma, but have already been present in the nucleus [54a]:

Table 5.13 Column amounts N and production rates Q of cyano species in the coma of Hale-Bopp [54b].

Species	N (cm^{-2})	N (mol m^{-2})	Q (s^{-1})	Q (mol s^{-1})
HCN	$1-7.6 \times 10^{13}$	$\sim 10^{-6}$	$\sim 5 \times 10^{27}$	$\sim 10^4$
H^{13}CN	7×10^{11}	10^{-8}	1.2×10^{26}	0.2×10^3
HC^{15}N	2.3×10^{11}	4×10^{-9}	4.2×10^{25}	0.7×10^2
HNC	3.4×10^{12}	6×10^{-8}	5×10^{26}	10^3
CN	$0.8-1.5 \times 10^{12}$	$\sim 2 \times 10^{-8}$	$\sim 2 \times 10^{27}$	$\sim 3 \times 10^3$
[HCNH]$^+$	$\leq 2 \times 10^{12}$	$\leq 3 \times 10^{-8}$	–	–

$$\mathrm{OH + H_2CO \rightarrow HCOOH + H} \tag{5.66a}$$

$$\mathrm{H_2O + [CHO]^+ \rightarrow [HCOOH_2]^+ + h\nu}$$

$$\mathrm{[HCOOH_2]^+ \rightarrow HCOOH + H^+} \tag{5.66b}$$

$$\mathrm{C_2H_2 + CN + M \rightarrow HC_3N + M} \tag{5.67a}$$

$$\mathrm{HCN + CH_3^+ \rightarrow [CH_3CNH]^+ + h\nu} \tag{5.67b}$$

In terms of the origin of molecules relevant for life, cyanide chemistry is of particular interest because molecules containing the unit CN can be considered building panels for N-heterocycles present in the nucleo-bases, and for amino acids (formed by condensation of formaldehyde and HCN under hydrolytic conditions). Table 5.13 summarizes cyano species located in the coma of Hale-Bopp close to perihelion, together with the column amounts N and production rates Q [54b]. Production rates Q are a measure for the amount of molecules of a specific species produced in the coma and/or released to the coma per second. The ratio hydrocyanic acid/isohydrocyanic acid (HCN/HNC) is approximately 7. Both molecules can originate from the protonated form [H–C≡N–H]$^+$ by dissociative electron recombination; Eq. (5.68); see also the reaction sequence (4.35) in Section 4.2.2. Since C–H bonds are somewhat stronger (bond energy 413 kJ mol^{-1}) than N–H bonds (391 kJ mol^{-1}), the preferential formation of the isomer HCN is to be expected. The HCN isotopomers H^{13}CN and HC^{15}N have also been observed. The ratios ^{12}C/^{13}C (109 ± 22) corresponding to $\delta(^{13}$C$) \approx -184‰$, and ^{14}N/^{15}N (330 ± 98) corresponding to $\delta(^{15}$N$) \approx -180‰$ (cf. Eq. (5.54) in Section 5.3.2 for the definition of δ), are close to the local interstellar ratio, underlining the contemporaneous formation of Hale-Bopp with the Solar System:

$$\mathrm{[H-C\equiv N-H]^+ + e^- \rightarrow (H-C\equiv N + C\equiv N-H) + H} \tag{5.68}$$

In addition to the species quoted in Table 5.13 and Scheme 5.4, the molecules C$_2$, C$_3$, CN, and NH$_2$ have been discovered in comets at distances >3 AU afar the Sun [55]. Ethane (C$_2$H$_6$) and cyanoacetylene (HC≡C–C≡N, HC$_3$N) have been proposed as the parent molecules for the formation of C$_2$ by photodissociation or electron-impact dissociation. The corresponding reaction paths are provided by Eqs. (5.69) and

(5.70). Reaction (5.70) depicts a possible path to the formation of the cyan radical CN, which otherwise can also derive from HCN by H abstraction. Ethane (C_2H_6) can either originate from the catalytic hydrogenation of ethyne on grain (M) surfaces, Eq (5.71), or in the gas phase from ethyne and water (or ethyne and methane) by UV- and γ-radiation. The direct formation of C_2 from pristine ethyne also comes in: C_2H_2 forms through ion–neutral reaction in interstellar clouds (Scheme 4.4 in Section 4.2.2), and hence will also have been available in the cloud from which our Solar System condensed. As far as C_3 is concerned, model calculations have shown that C_3H_4 is an appropriate precursor for the photodissociative formation of tricarbon. C_3H_4 can either be propyne (methylacetylene) $H_3C-C\equiv CH$, or allene $H_2C=C=CH_2$. In either case, an allenyl intermediate is involved in its production, Eq. (5.72):

$$C_2H_6 \xrightarrow{h\nu} H, CH_2, C_2H_4, C_2H_5$$
$$\xrightarrow{h\nu} C_2H_2$$
$$\xrightarrow{h\nu} C_2H \xrightarrow{h\nu} C_2 + H \quad (5.69)$$
$$\xrightarrow{e^-} C_2 + H_2$$

$$HC_3N \xrightarrow{h\nu} H, CN, C_2H, C_3N$$
$$\xrightarrow{h\nu} C_2 + CN \quad (5.70)$$

$$HC\equiv CH + 2H_2 + M \rightarrow H_3C-CH_3 + M \quad (5.71)$$

$$\left.\begin{array}{l}H_3C-C\equiv CH \\ H_2C=C=CH_2\end{array}\right\} \xrightarrow{h\nu} H_2C=C=CH \xrightarrow{h\nu} H, H_2, C_3 \quad (5.72)$$
$$(C_3 \text{ is } |C=C=C|)$$

The occurrence of a multitude of sulfur species in comets (bottom line in Scheme 5.4) [53, 56a] has prompted model calculations on sulfur chemistry in cometary comae [56b] with the aim to establish which of these species is either originally present in the nucleus, or formed by gas-phase chemical reactions in the coma, and hence an approach similar to that for the generation of cyano species. A key reaction in sulfur chemistry is the formation of the HS radical (which then readily reacts to form other sulfur species) either by neutral exchange, Eq. (5.73), or by photodissociation, Eq. (5.74). The modest activation energy necessary for reaction (5.73) to occur is overcome by the large amount of suprathermal H atoms produced by photodissociation of abundant H_2O, and by H atoms formed according to Eq. (5.74). As is the case in cyanide chemistry, the model cannot explain the abundances of molecules such as SO, SO_2, or NS in the coma, again suggesting that they have already been present in the nucleus:

$$H_2S + H \rightarrow HS + H_2 \quad (5.73)$$
$$H_2S + h\nu \rightarrow HS + H \quad (5.74)$$

The enrichment $^2H/^1H$ of $\approx 3 \times 10^{-4}$ in water present in the comets Halley, Hale-Bopp, and Hyakutake corresponds to an enrichment by an order of magnitude with respect to the protosolar value (3×10^{-5}), and a factor of 2 relative to terrestrial water, indicating that cometary water cannot be the only source of water in Earth's

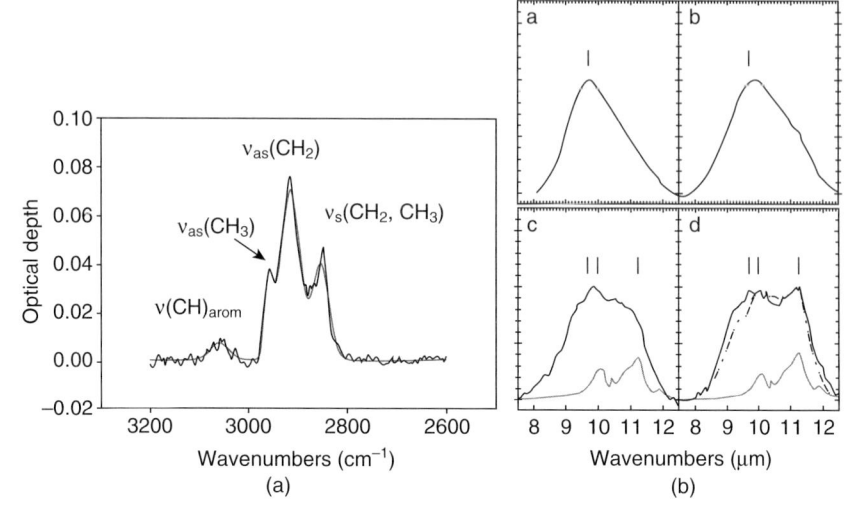

Figure 5.26 (a) The C–H stretching regime of the IR spectrum (continuum corrected) of organic matter found in grains from the coma of comet Wild 2, collected by the spacecraft Stardust © Ref. [58a] Figure 2. Reprinted with permission from AAAS. The labels v_s and v_{as} stand for the symmetric and antisymmetric stretching vibrations of the aliphatic constituents, respectively. (b) Silicate emissions in the 8–12 μm (1250–833 cm^{-1}) range. Panels **a** (interstellar grains) and **b** (EX Lupi; March 2005) show amorphous silicate, panels **c** (EX Lupi, April 2008) and **d** (comet Halley; full line) the superimposed features of crystalline silicate. The dashed line in panel **d** features ejecta from Temple 1; the gray bottom lines in panels **c** and **d** correspond to fosterite Mg_2SiO_4 [61]. © Ref. [61a] Figure 1. Reprinted with permission from Nature.

oceans (see also Section 7.4.4), as sometimes hawked. The $^2H/^1H$ ratio in HCN is seven times the value in water. Such a ratio is common for interstellar clouds and should indicate that cometary materials retain a memory of the interstellar matter [57]. Gas-grain surface reactions and ion–molecule reactions in the gas phase both lead to species-dependent deuterium enrichment typical of ion–molecule reactions at a temperature of ca. 30 K. This temperature for comet formation has also been inferred from the ratio of *ortho*- to *para*-H_2O. In o-H_2O, the nuclear spins (the nuclear magnetic moments) of the two hydrogen atoms are parallel, while in p-H_2O they are antiparallel. Re-equilibration of the *ortho* and *para* states does not easily occur (it is, in fact, not allowed). The canonical ratio o-H_2O/p-H_2O is 3 for temperatures >60 K. In comet Hale-Bopp, a ratio of 2.5 has been found, which goes along with to a temperature of 30 K at the time where the comet accreted.

Organic material detected in dust particles in the 1–300 μm size range, collected by the spacecraft Stardust in the coma of comet Wild 2 at a distance of 234 km from the comet's nucleus, has features similar to material from meteorites and interstellar dust particles (Section 5.3.3). Species identified by IR (Figure 5.26a) encompass aromatic compounds (naphthalene, phenanthrene, pyrene; cf. Scheme 5.2 in Section 5.3.2), alkylated (methyl, ethyl, propyl, butyl) derivatives of

Table 5.14 Minerals, detected by their IR features, in ejecta of Temple 1 [59b], and by analyses of dust grains collected from the coma of Wild 2 [60].

Mineral group	Temple 1	$\underline{N}$[a]	Wild 2
Orthosilicates	Olivine MgFeSiO$_4$	0.35	Olivine MgFeSiO$_4$[b]
	Forsterite Mg$_2$SiO$_4$	0.70	Forsterite Mg$_2$SiO$_4$
	Fayalite FeSiO$_4$	0.18	Enstatite MgSiO$_3$
Chain silicates	Pyroxene (Mg,Fe)$_2$Si$_2$O$_6$	0.06	Pyroxene (Mg,Ca,Fe)SiO$_3$[c]
	Ferrosilite Fe$_2$Si$_2$O$_6$	0.50	Diopside (Mg,Ca)SiO$_3$
	Diopside (Mg,Ca)$_2$Si$_2$O$_6$	0.18	
	Enstatite Mg$_2$Si$_2$O$_6$	0.16	
Clay	Smectite nontronite [Na$_{0.3}$Fe$_2^{III}$(Al, Si)$_4$O$_{10}$(OH)$_2$]·3H$_2$O	0.07	Not detected
Carbonates	Magnesite MgCO$_3$	0.11	Not detected
	Siderite FeCO$_3$	0.17	
Sulfides	Niningerite (Mg,Fe^{II}(Mn))S	0.92	Troilite/pyrrhotite Fe$_{1-x}$S
			Pentlandite (Fe,Ni)$_9$S$_8$
			Cubanite CuFe$_2$S$_3$

a) $\underline{N}$ (compositional abundance) is defined here [59a] as $\underline{N}$ = (density/molecular mass) × (weighted surface area [of dust particles]); the dimensions thus are mol cm^{-1}. $\underline{N}$(H$_2$O$_{ice}$) ≈ 0.27 mol cm^{-1} (≈3% by surface area).
b) Broad compositional range: the mole fraction f_{Mg} varies between 0.04 and 1; and there are also variations in contents of Mn, Al, and Cr.
c) Low (dominant) and high Ca varieties have been found.

these aromatic compounds, and derivatives thereof, functionalised by —OH, —C=O, —C≡N, —NH$_2$, and —C(O)NH$_2$ [58]. But there are also distinctive differences to organics in meteorites and IDPs, such as superior relative oxygen and nitrogen abundance, a more abundant aliphatic component, and a larger fraction of volatiles. PAHs have also been identified, as inferred from their spectral signatures in the IR, in the ejecta of comet Temple 1 [59].

Some reprocessing of the matter constituting the dust particles from Wild 2 may have occurred during capture, and by interaction with the capture frame, a silica-based aerogel. This reserve also applies to the silicates detected and analyzed in the grains. The majority of these silicates is amorphous and occurs in the form of so-called GEMS also typical of interstellar dust particles. GEMS stands for glass with embedded metal and sulfide, where the metal fraction is Fe–Ni, and the sulfide fraction is mainly FeS, dispersed in a magnesium silicate matrix. Crystalline silicates comprise pyroxene (predominating), enstatite, Fe-rich olivine, fosterite, and diopside [60]; see Table 5.14. Crystalline silicates present in comets must have been formed by annealing, that is, tempering of amorphous silicates at temperatures of at least 800 K, suggesting their formation in the inner Solar region, or an origin in the Solar nebula. Thermal annealing is a fast process, which has recently been detected in the outburst spectrum of the young sun-like star EX Lupi[26] [61]; see Figure 5.26b. In one single grain, a calcium- and aluminum-rich inclusion (CAI, mainly Ca-aluminates and -alumosilicates) has been found. CAIs are products of gas–solid separations at high temperature, and they are thought to represent the oldest samples of the Solar System. Apparently, these high temperature materials either formed near the protosun (or the Sun) and became transported as far out as the Kuiper belt, the presumed place of origin of Wild 2, or they were supplied as remnants of the presolar nebula.

Clays (phyllosilicates) and carbonates do not appear to be present in grains of Wild 2, excluding appreciable aqueous alteration otherwise typical of, for example, meteorites and their parent asteroids. This is in contrast with findings for Temple 1, where silicate hydrates and magnesite $MgCO_3$ have been traced. A possible explanation for the presence of silicate hydrates in Temple 1, and thus of products which had to be formed by hydrothermal activity, are impact events. The overall mineral configuration in the debris plume of Temple 1, as provided by the IR and FIR emission signatures and modeled by fitting to the spectra of a mix of Earth-bound minerals [59b], is assorted in Table 5.14 together with mineralogical findings in grains of Wild 2 [60].

5.5
Kuiper Belt Objects

The Kuiper belt [62], a circumstellar accretion disk of icy remnants from the Solar System, extends from just beyond Neptune's orbit (about 30 AU) to approximately 55 AU. Pluto's orbit, with a semimajor axis of 39.5 AU, lies in the inner, sun-ward part of this belt of objects mainly consisting of frozen volatiles such as water, nitrogen, ammonia, and carbon oxides. The dwarf planets[27] compiled in Table 5.2 of Section 5.1 are the more prominent representatives of the thousands of objects constituting the Kuiper belt, and the partly overlapping Scattered Disk and Extended Scattered Disk. The largest of these dwarf planets is Eris, exceeding the mass of Pluto (who is next in size and mass) by almost 14%. The Scattered Disk, extending well beyond 100 AU, and the Extended Scattered Disk reaching out as far as 1000 AU and even further are distant regions populated, as the Kuiper belt, by icy objects, most of which in unstable, highly eccentric, and strongly inclined orbits. An example for a Scattered Disk object is Sedna with an aphelion of 976 and a perihelion of 76 AU.

Many of the comparatively short-term comets, plus the trojans and the irregular satellites of the giant planets, Jupiter's in particular, mainly originate from the Scattered Disk, from which they have been perturbed by Neptune. The Kuiper belt

---

26) EX Lupi, in the constellation of Lupus, is the proto-type of the so-called EXor class of the pre-main sequence variables (cf. Section 3.1.1). EXor refers to the generally low luminosity.

27) Ceres is the only dwarf planet not being accommodated in the Kuiper belt/Scattered Disk. For Ceres, the largest object in the Asteroid belt, see Section 5.3.

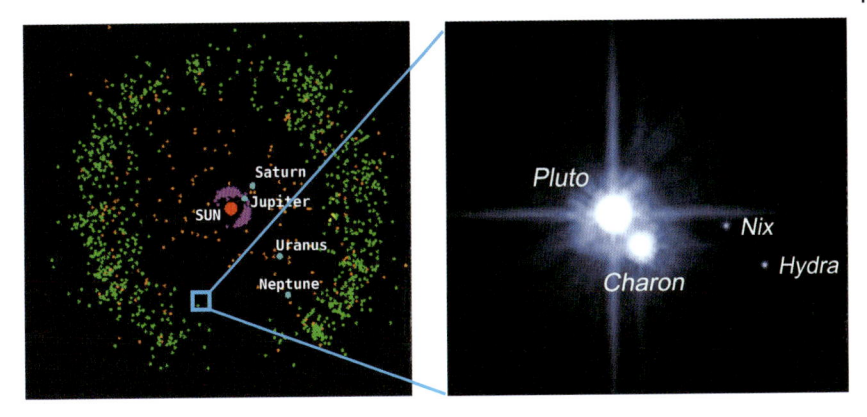

Figure 5.27 The Kuiper belt (green objects; "errands" [comets, Centaurs, Trojans] in orange) as related to the Sun and the four giant planets. The enlarged area is the Pluto system. Source: Wikimedia Commons; credit: NASA, ESA, H. Weaver, A. Stern.

objects are more stable in their orbits, which in many cases are resonance-locked to Neptune. Those which are not in orbital resonance with Neptune are also referred to as "classical Kuiper type objects" or "cubewanos." The QB1os, or cubewanos, are named after the first trans-Neptunian object, discovered 1992: (15760) 1992 QB_1, with a semimajor axis of 43.74 AU and an eccentricity of 0.0654. The Centaurs, a population of icy bodies between Jupiter and Saturn, are also believed to have been dislocated from a dynamic Scattered Disk and/or the Kuiper belt. On the other hand, Scattered Disk object may, in part, also represent planetesimals which have originally been formed in the region of the giant planets, and flung outward to orbits beyond Neptune. Neptune's big moon Triton (see the next section) is considered a perturbed and finally captured Kuiper belt object.

Figure 5.27 provides an impression of the relation of Pluto (named after the Greek God of the Underworld) and its moons to the Kuiper belt. Discovered in 1930, Pluto was treated as the ninth planet until it became reclassified as dwarf planet in 2006. Pluto is just a fifth of the mass of Earth's moon, and a third of its volume. Its orbit is highly eccentric and somewhat inclined with respect to the ecliptic. The eccentricity of its orbit brings it closer to the Sun than Neptune every ca. 250 years. The last such situation occurred between 1979 and 1999. The planet has three moons. The largest moon is Charon ("the ferryman of the dead"), with about half the diameter of Pluto and 12% of its mass. Since the conjoint center of gravity of the Pluto–Charon system lies beyond Pluto's surface, the Pluto–Charon ensemble is sometimes considered a binary dwarf planet system. Some of the characteristics of Pluto are summarized below:

- Aphelion: 49.3 AU; perihelion: 29.7 AU; eccentricity: 0.245; orbital period: 248 a
- 2 : 3 (Pluto : Neptune) orbital resonance
- Mean radius: 1153 km
- Mean density: $2.03\,g\,cm^{-3}$

Table 5.15 Characteristics, related to the surface composition, of large Kuiper belt objects (except of Pluto and Triton)[a].

Name, identification	Mean density (g cm^{-3})	Albedo[b]	Color	Color index[b] B–V, V–R	Surface composition
Ixion, 2001 KX$_{76}$	(2)	0.15–0.37	Moderately red	1.03, 0.61	Hydrocarbons, tholins, H$_2$O?
Orcus, 2004 DW	1.6	≈2	Neutral	0.68, 0.37	Cryst. H$_2$O, tholins
Haumea, 2003 EL$_{61}$	ca. 3	0.7 ± 0.1	–	–	2/3–4/5 cryst. H$_2$O, HCN?, kaolinite?
Makemake, 2005 FY$_9$	(2)	0.78	–	–	CH$_4$, C$_2$H$_6$, N$_2$
Eris, 2003 UB$_{313}$	ca. 2	0.86 ± 0.07	Gray	–	CH$_4$
Varuna, 2000 WR$_{106}$	0.99	0.04–0.26	Moderately red	0.93, 0.64	H$_2$O
Quaoar, 2002 ML$_{60}$	4.2	0.1–0.2	Moderately red	0.94, 0.65	Cryst. H$_2$O, NH$_3$·H$_2$O, ≈5% CH$_4$ + C$_2$H$_6$
Sedna, 2003 VB$_{12}$	(2)	0.16–0.30	Red	1.24, 0.78	CH$_4$ (33%), CH$_3$OH, tholins (24%), C(amorphous) (7%)

a) For size, mass, and orbital characteristics cf. Table 5.2 in Section 5.1.
b) See text for details.

- Core: Rock; mantle: water-ice; surface: dinitrogen frost (+CH$_4$, CO, C$_2$H$_6$, tholins)
- Mean surface temperature: 44 K; upper atmosphere: ~100 K
- Atmospheric surface pressure [63]: 0.65–2.4 Pa (depending on distance to Sun)
- Atmospheric composition [63]: >98% N$_2$, methane (0.5%), CO, ethane.

Tholins (literally "mud"), which are constituents of the surface, are polymeric organic compounds formed by UV irradiation from simple precursors such as CH$_4$, N$_2$, C$_2$H$_6$, and HCN. For more details, cf. Section 5.6.3.

Table 5.15 provides an overview of specific features, related to their surface chemistry, of the other trans-Neptunian dwarf planets. Eris and Sedna are noteworthy because their orbits are particularly eccentric (0.855 in the case of Sedna) and inclined toward the ecliptic (44.2° in the case of Eris); cf. Table 5.2 for details. Varuna and Quaoar on the other hand have almost spherical orbits. Some of the dwarf planets are accompanied by small moons (Eris: one; Haumea: two; Quaoar: one satellite). Information on the build-up and composition naturally is quite restricted. Since there are substantial differences in the mean density, ranging from ca. $1\,\text{g}\,\text{cm}^{-3}$ (Varuna) to ca. $4.2\,\text{g}\,\text{cm}^{-3}$ (Quaoar), the overall composition will also greatly differ, suggesting, together with the strikingly differing orbital characteristics, a different origin and/or history of development. The high density of Haumea and Quaoar indicates that these two objects almost entirely consist of silicate rock such as olivine $(\text{Fe,Mn,Mg})_2\text{SiO}_4$ and pyroxene (Mg,Fe,Mn-alumosilicates).

Despite of its rocky nature, Haumea has a rather high albedo of 0.7, a feature which it shares with Makemake (0.78) and Eris (0.86), while the other Kuiper belt dwarf planets have low albedos. The albedo of an object describes the extent to which it diffusely reflects light from an external light source; the albedo ranges between 0 for no reflection and 1 for total reflection. A high albedo is correlated with a surface of ice. Such a surface is generally present for all of the dwarf planets, but may have been "contaminated" with carbonaceous materials, such as condensed hydrocarbons, tholins, or amorphous carbon, to give rise to a low albedo and a more or less reddish appearance of the object, quantified by color indices $B–V$ around 1 and $V–R$ of 0.6–0.8. B, V, and R stand for the apparent magnitude in the blue, visible (green-yellow), and red, respectively; see also Section 3.1. A reddish star, for example, is characterized by a comparatively large $B–V$ index. In the case of the high albedo objects Orcus, Haumea, and Quaoar, absorptions in the near IR, in particular a band at $1.65\,\mu\text{m}$, is indicative of crystalline water-ice [64]. *Crystallinity* of water-ice is unexpected at the low temperatures typical of this distant region of the Solar System, because cosmic rays and Solar wind will cause rearrangement of crystalline to amorphous ice on the million year time scale. Some resurfacing is, therefore, required. One option for resurfacing is the production of internal heat produced by radioactive decay and transported to the surface. Model investigations suggest an admixture of HCN and phyllosilicates such as kaolinite $\text{Al}_4[\text{Si}_4\text{O}_{10}(\text{OH})_8]$ to the surface layer of crystalline water-ice. Methane, ethane, and dinitrogen have been detected as ice constituents on other dwarf planets.

Summary Sections 5.3–5.5

Between Mars and Jupiter, in a zone spanning 2.06–3.27 AU, the asteroid belt extends, a conglomerate of innumerous rocky and icy chunks, including its largest object, Ceres, classified as a dwarf planet. Asteroids and asteroidal fragments impacting Earth as meteorites serve as witnesses from the time where our Solar System accreted out of a protosolar gas and dust nebula. But meteorites also contain imprinted signatures related to processing in the 4.5 billion years that followed. The main constituents of the meteorites are silicateous chondrites and achondrites, the former named so for inclusions of small spherical objects, the chondrules. About 6% of the overall population are metallic meteorites, which almost exclusively consist of metallic iron plus some nickel and sulfur. A subgroup of chondritic meteorites, the carbonaceous chondrites, contain up to 3% of carbon in the form of carbonates, carbides (SiC), and organic compounds; they thus provide a valuable record of prebiotic chemical evolution. Organic matter in meteorites is distinct from that of terrestrial origin by an elevated fraction of the heavier isotopes ^{2}H, ^{13}C, ^{15}N, and $^{29/30}$Si. Among the organic compounds found in meteorites are amino acids, including eight amino acids which are essential for proteins of terrestrial life forms. Most of these meteoritic amino acids are racemic. An exception is isovaline in the Murchison meteorite, with an 18.6% ee of the L-enantiomer. Reaction paths toward the formation of amino acids on asteroids and their fragments start from simple interstellar precursor molecules such as formaldehyde, ethyne, hydrogen cyanide, ammonia, and water. The Strecker cyanohydrin synthesis is an example.

Meteorites can also be remnants of comets. Comets are "dirty snowballs," rubble piles consisting of ice (mainly water) and "rock" (silicates, carbonates, sulfides), and covered with regolith, but also containing a rich supply of complex molecules. Many of the comets have highly eccentric orbits; they are errants from the Oort cloud and the Kuiper belt/Scattered Disk, and hence carry essentially unprocessed information from the mergence of the Solar System. When approaching the Sun, part of the material constituting the comet's parent body evaporates and forms an extended coma plus tail, with a chemical composition reminiscent of that in interstellar clouds, including exotic species such as C_3.

The by far main amount of extraterrestrial material is carried to Earth by interplanetary dust particles, typically between 0.1 and 20 μm in size. Their chemical signature is similar to that of meteoritic material, with the exception that, due to their small size, volatiles are under-represented. Interplanetary dust grains are remains of the protosolar cloud. The two more important categories are the "main stream particles" and the "X-type grains." Main stream particles represent about 90% of the grain population; they originating from AGB stars, that is, developed stars blowing off the main part of their envelopes. X-type grains have been supplied by supernovae; these grains are enriched, by two to three orders of magnitude with respect to Solar values, in the heavy silicon isotopes ^{29}Si and ^{30}Si.

The Kuiper belt, a circumstellar accretion disk of icy remnants of the Solar System, extends from beyond Neptune's orbit to approximately 55 AU. Several dwarf planets, among these Pluto, Eris, and Haumea, reside in the Kuiper belt. Typical Kuiper belt objects consist of frozen volatiles, mainly H_2O, N_2, NH_3, CO, and CO_2, plus CH_4 and other simple as well as polymeric hydrocarbons (tholins). Exceptions are Haumea and Quaoar, which mainly are composed of silicate rock. Beyond the Kuiper belt, the Scattered Disk extends. Sedna is a dwarf planet which, on its very eccentric orbit, plunges deep into the Scattered Disk.

5.6
The Giant Planets and Their Moons

5.6.1
Jupiter, Saturn, Uranus, and Neptune

The four planets Jupiter, Saturn, Uranus, and Neptune are distinct from the inner planets by, inter alia, their larger mass and hence gravitational force, the predomi-

Table 5.16 Characteristics of the four giant planets.

	Jupiter	Saturn	Uranus	Neptune
Distance from the Sun (AU)	5.2	9.6	19.2	30.1
Rotational period (h)	9.8	10.5	17.9	192
Mass relative to Earth, $m_\oplus$	318	95	14.5	17
Mean density (g cm^{-3})	1.33	0.69	1.29	1.64
Mean surface temp. at the 1 bar level (K)	165	134	76	72
Number of moons	>63[a]	>61[b]	>27	>13[c]
Planetary ring system	faint; dust	Very complex[d]; 93% water ice (+ tholins[e]), 7% amorphous C; dust and meter-sized boulders	Thin rings; organics (?); meter-sized boulders, low dust fraction	Few thin rings; ice layered with silicates and organics; high dust fraction

a) Including the four large Galilean moons Europa, Io, Ganymede, and Callisto.
b) Including particularly large Titan.
c) Including the large Triton.
d) Saturn's outermost ring, with a core radius of about 200 r_{Saturn}, is associated with the outer Saturnian moon Phoebe, who keeps the ring populated with dust particles <100 μm. With a density of 20 particles km^{-3}, the ring is extremely tenuous.
e) Tholins are brownish organic polymers made up of different C- and N-based monomers.

nance of hydrogen, and the presence of a plethora of moons and moonlets, as well as icy and rocky lumps and dust swaying in the equatorial plane. Table 5.16 provides an overview of several of the planets' properties; see also Table 5.1a in Section 5.1 for orbital characteristics. Uranus' axis of rotation is almost in the plane of the planet's orbit; the tilt is 97.8°. The largest of the giant planets, Jupiter, has attained particular interest during the last few years since it serves as an analog for many of the newly discovered extrasolar planets ("hot Jupiters"; Chapter 6). Jupiter's mass is slightly less than 1/1000 that of the Sun, and about 1/15 that of the lowest mass brown dwarf stars (more in Chapter 6). Jupiter, Saturn, and Neptune (but not Uranus) produce more heat than they receive from the Sun. This outflow of extra energy is predominantly due to slow gravitational compression.

The two gas giants (Jupiter and Saturn) and the two ice giants (Uranus and Neptune) formed by what is known as the "nucleation model." According to this model, these planets accreted in the protosolar system from icy planetesimals (with volatiles such as methane and noble gases trapped and deposited in the form of water-ice clathrates), followed by gravitational capture of the surrounding gaseous nebula in the case of Jupiter and Saturn. The masses of Uranus and Neptune are just at the limit where this gravitational accretion of gas can occur; they are lacking the extended gaseous envelops of the two other giant planets for this reason. In the case of Jupiter, the mass of the core formed from planetesimals is only 5.7% of the planet's overall mass (Saturn: 15%); these planets thus mostly consist of protosolar gas. The nucleation model is supported by the enrichment of heavy elements such as C, N, S, Ar, Kr, and Xe in the planets, enriched with respect to Solar abundances by a factor of 3 on Jupiter, and ca. 40 on Neptune and Uranus. The C:H ratio here is particularly meaningful. Oxygen, helium, and nitrogen are depleted. Further, while the D/H ratio of Jupiter and Saturn compares to the Solar value, deuterium is enriched as compared to protium by a factor of 2 in Uranus and Neptune as a consequence of ion–molecule and molecule–molecule reactions, which favor the formation of deuterated molecules in the icy bodies of these two planets. A comparable enrichment has been discussed in the context of deuterium fractionation in the interstellar medium (Sections 4.2.3.2 and 4.2.5.2) and long period comets (Section 5.4.2). The deuterium enrichment factor in the latter amounts to 3.

The depletion of helium (and neon) in the atmospheres of Jupiter and Saturn with respect to the Solar abundance is thought to be due to the condensation of helium in the form of droplets absorbed in the "ocean of metallic hydrogen" in the inner regions of these planets, overlaying the dense rocky core. In the case of Jupiter, metallic hydrogen supposedly extends outward to about ¾ of the planet's radius. Liquid metallic hydrogen is formed at pressures of ca. 3 Mbar (300 GPa) at ambient temperatures [65a], and is present in a degenerate state, in which the hydrogen atoms are devoid of their electron and hence at inter-"atomic" distances clearly below the Bohr radius (the radius of a hydrogen atom) of 53 pm. The "freely floating" electrons provide conductivity comparable to metals. Metallization of hydrogen is even facilitated, that is, the necessary pressure goes down by an order of magnitude, in the presence of silane SiH_4, which forms a $SiH_4 \cdot (H_2)_2$ adduct with particularly weak hydrogen–hydrogen bonds [65b]. So far, SiH_4 has not been found on Jupiter, but the corresponding hydride of the higher homolog germanium, germane GeH_4, has been detected in the troposphere and may act accordingly. The metallic hydrogen is layered by liquid nonmetallic hydrogen that smoothly transits to hydrogen gas beyond the critical temperature (33.2 K) and pressure (13.1 bar = 1.31 MPa). On Uranus and Neptune, the internal pressure is probably insufficiently high to enable formation of metallic hydrogen. Here, the rocky core is surrounded by a mantle consisting of ices (H_2O, NH_3, CH_4), which gradually pass into the surface layer and atmosphere.

While most of the atmospheric hydrogen is present as H_2, the molecular cation H_3^+, also an important interstellar molecule (Section 4.2.3), has also been detected in the outer reaches of the atmospheres of the giant planets. Equation (5.75) represents one way by which H_3^+ can form under stratospheric conditions, where ionization of molecular hydrogen is anticipated:

Table 5.17 Main atmospheric constituents (%) of the four giant planets.

	Jupiter	Saturn	Uranus	Neptune
H_2	90	94	83	82; 0.02 HD
He	9	5	15	17
CH_4 and water	0.3, water 0.1	0.2, water 0.1	2.3	1.5
NH_3	0.026	0.01	0.01	0.01

$$H_2 + H_2^+ \rightarrow H + H_3^+ \tag{5.75}$$

Swirling movements of conducting materials (so-called eddy or Foucault currents) within the metallic hydrogen are responsible for the generation of a magnetosphere. The magnetic field of Jupiter is particularly strong, ranging from 0.4 mT at the equator to 1.4 mT at the poles; it is thus up to ca. 20 times as strong as the terrestrial magnetic field (30 and 60 µT at the equator and poles, respectively). Interaction of Jupiter's magnetic field (and plasma) with the Solar wind creates a magneto-tail extending anti-sun-ward to distances of 2500 Jovian radii. This magneto-tail is supplied with H_3^+ and H^+ from Jupiter's ionosphere, and with S^{2+} and O^+ originating from volcanic exhalations of Jupiter's moon Io. The magnetic field of Saturn, 20 µT, is slightly weaker than Earth's magnetic field.

The composition of the atmosphere (see also Table 5.17) is clearly distinct for the troposphere and the stratosphere [66, 67] (with the tropopause usually set at 0.1 bar = 10 kPa; the temperature at this level is <110 K). The main species present in the troposphere are methane and, in condensed form, ammonia NH_3, hydrogen sulfide either as such (H_2S) or in the form of ammonium hydrogen sulfide NH_4SH, and water. These molecules are processed from H_2, C, N, S, and O in the deep atmosphere where temperature and pressure are high, and subsequently convectively transported upward into cooler regions. The species are equilibrium species, that is, they are stable under the planets' atmospheric conditions. They are not necessarily uniformly distributed; rather, they appear to be zoned. On Jupiter, for example, NH_3 is confined to the upper troposphere (0.5–1 bar), NH_4SH to 2–4 bar, and H_2O to 3–10 bar. Along with the equilibrium species, there are minor amounts of disequilibrium species present in the atmospheres of the gas giants, viz. phosphane PH_3, arsane AsH_3, and germane GeH_4. Disequilibrium refers to the fact that these species should *not* be present according to thermodynamic models. Arsane, for example, is supposed to react with water to form arsenic As_2O_3 according to Eq. (5.76), and germane should yield germanium dioxide, Eq. (5.77a), and/or germanium sulfide, Eq. (5.77b). Chemical kinetics can, however, be slow at low temperatures, and adjustment to thermodynamic equilibrium then is hampered by tropospheric dynamics, such as strong upward motions in the troposphere, leading to quenching of the nonequilibrium composition:

$$AsH_3 + 1\frac{1}{2}H_2O \rightarrow \frac{1}{2}A_2O_3 + 3H_2 \tag{5.76}$$

$$GeH_4 + 2H_2O \rightarrow GeO_2 + 4H_2 \tag{5.77a}$$

$$GeH_4 + H_2S \rightarrow GeS + 3H_2 \tag{5.77b}$$

The chemistry in the upper layers of the troposphere beyond clouds, and in particular in the stratosphere, is dominated by photochemistry. Since methane is the major constituent in the stratosphere, methane photolysis initiates the production of more complex hydrocarbons, Scheme 5.5, and a few other carbon-bearing compounds. The starting reaction is the formation of the methyl radical CH_3 (framed in Scheme 5.5), followed by further photolysis to form methylene CH_2 and methyne CH. CH_3 has been detected as a minor species on all giant planets except of Uranus. Various condensation reactions can then take place, either by molecule–molecule reactions and release of H, or by mediation through dust particles. The major species thus formed are ethyne (acetylene C_2H_2; $^{12}C^{13}CH_2$ has also been detected), which is present in the stratospheres of all the giant planets, and ethane C_2H_6, which appears not to be present on Uranus. C_2H_6 is shielded from photolysis by methane; its photochemical lifetime amounts to more than 100 years. Propyne (CH_3C_2H), ethene (ethylene C_2H_4), and benzene (and possibly higher condensed aromatic compounds, responsible for the formation of haze) are present as minor species on Jupiter and Saturn only. Figure 5.28 shows the IR emissions for C_6H_6, C_2H_2, and C_2H_6 from the upper Jovian stratosphere, where irregular heating by energetic electrons and ions gives rise to "hot spots," and hydrocarbons can be produced by ion–molecule reactions in a manner similar to those in diffuse interstellar clouds.

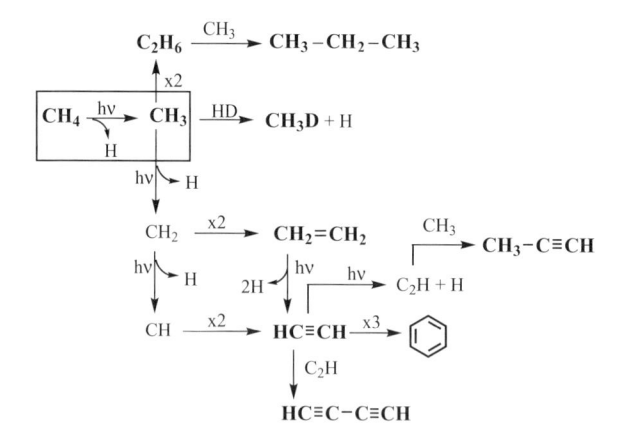

Scheme 5.5 Reaction scheme for the formation of hydrocarbon species detected (in bold) in the stratospheres of the giant planets. The starting reaction (the photolysis of methane) is framed. Most of the condensation reactions afford the presence of a matrix (dust particles).

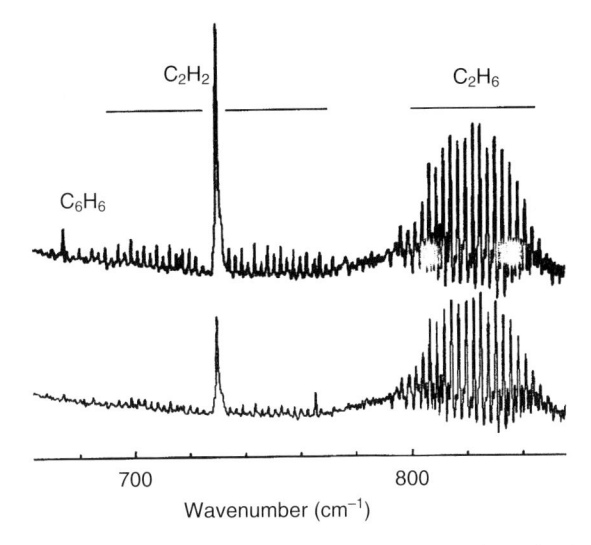

Figure 5.28 IR emissions (CH bending modes) observed in the upper (auroral) stratosphere of Jupiter inside and outside "hot spots" (top and bottom, respectively). The emissions *within* the hot spot are clearly enhanced with respect to those *outside*. The rotational fine structures are also resolved, in particular for the ethane band. © Ref. [64] Figure 4A. Reprinted with permission from AAAS.

A second candidate (along with aromatic compounds) for haze in the planets' atmospheres is hydrazine N_2H_4, which is generated by photolysis of ammonia via the NH_2 radical. This is represented by Eqs. (5.78a) and (5.78b), where M is a matrix such as a dust grain. Basically, the amine radical can also react with the methyl radical to form hydrogen cyanide (HCN), Eq. (5.79). The spatial separation of ammonia (in the troposphere) and methane (in the stratosphere) apparently prevents the generation of HCN – except on Neptune[28] where the abundance of HCN is particularly high. Interestingly, this HCN abundance is correlated to a high abundance of carbon monoxide. Alternatively to the generation of HCN according to Eq. (5.79), this molecule may also have formed in the Neptunian atmosphere by reaction between methyl radicals and nitrogen atoms, Eq. (5.80), the latter originating from the dissociation of N_2 ejected from Neptune's large moon Triton, which does have a faint but distinct N_2 atmosphere:

$$NH_3 + h\nu \rightarrow NH_2 + H \qquad (5.78a)$$

$$2NH_2 + M \rightarrow N_2H_4 + M \qquad (5.78b)$$

$$NH_2 + CH_3 \rightarrow HCN + 2H_2 \qquad (5.79)$$

$$N + CH_3 \rightarrow HCN + H_2 \qquad (5.80)$$

28) HCN has also been detected on Jupiter after an impact event (comet Shoemaker-Levy 9) in 1994. See also text.

For the stratospheric oxygen species CO, CO_2, and H_2O, a partial external origin has been implemented, because the troposphere acts as a cold trap at least for CO_2 and H_2O. External sources can be debris from the rings (in particular in the case of Saturn), moonlets, micrometeorites and interplanetary dust, and cometary impacts. The impact, in 1994, by several fragments of comet Shoemaker-Levy 9 supplied large quantities of N-, O-, and S-bearing molecules to the Jovian stratosphere, and HCN, CO, CO_2, and CS as secondary products. The formation of carbon monoxide may also occur via the reaction between water and photolytically produced methyl radicals; Eq. (5.81). Further reaction of CO, for example, with OH radicals originating from the photolysis of H_2O, can produce CO_2, Eq. (5.82):

$$H_2O + CH_3 \rightarrow CO + 2\frac{1}{2}H_2 \tag{5.81}$$

$$CO + OH \rightarrow CO_2 + H \tag{5.82}$$

5.6.2
The Galilean Moons

The four large moons of Jupiter have already been discovered in early 1610 by Galileo Galilei (see Chapter 1) and later conjointly named Galilean moons for this reason. For an overview, see Table 5.18. The radii range from 1569 km (Europa) to 2634 km (Ganymede), which compares to 1737 km for Earth's Moon. Ganymede is the largest moon in the Solar System, and even exceeds Mercury in size ($r = 2440$ km) – though not in mass. While Ganymede is differentiated into a liquid metal core, a rock mantle, and ice layer, Callisto, although very similar in size and mean density, remained essentially undifferentiated. This dichotomy reflects the fact that Ganymede, who is closer to Jupiter, was more vigorously subjected to the "late heavy bombardment" (4.2–3.9 Ga ago), thus receiving sufficiently more energy to thaw the moon's material and thus to enable differentiation [68]. The innermost of the Galilean moons, Io, is remarkably distinct from the three others by its high mean density, its volcanism, the apparent lack of all but traces of water, its sulfur-rich surface, and its coupling to the Jovian upper atmosphere along magnetic flux lines. While the two outer moons' overall composition is dominated by water-ice, Io and Europa have extended iron or Fe–FeS cores and a silicate mantle. Io's mean density, $3.53 \mathrm{g\,cm^{-3}}$, even exceeds that of Earths' Moon ($3.346 \mathrm{g\,cm^{-3}}$), and makes Io the densest object in the Solar system beyond Mars. The pronounced volcanism (and the lack of water) is a consequence of internal heating. The main source of this energy production is "tidal heating," going back to the gravitational pull by Jupiter, overlaid and thus enhanced by Io's orbital resonance with Europa and Ganymede. The enormous internal friction thus caused generates enough heat to account for Io's volcanism.

Due to the volcanism with its exhalations and eruptions of sulfur, sulfur dioxide, and silicate magmas, Io's surface is permanently restructured. The abundance of

Table 5.18 Overview of Jupiter's four Galilean moons.

Moon	Io	Europa	Ganymede	Callisto
Origin of name	Priestess of Hera, and Zeus' lover	Phoenician noblewoman, courted by Zeus	Trojan prince and hero, and Zeus' beloved	A nymph, and Zeus' lover
Mean orbit radius (km)	421 700	670 900	1 070 400	1 882 700
Radius (km)	1737	1569	2634	2410
Mean density (g cm^{-3})	3.53	3.01	1.94	1.83
Mean surface temperature (K)	130	102	110	124
Core	Fe–FeS (10% S)	Fe–FeS	Fe–FeS	Rock and ice
Mantle[a]	Silicates	Silicates	Silicates	Silicates and ice
Crust/outer layers/surface	Elemental S, SO_2-frost, basalts	Water-ice (120 km thick)[c], H_2O_2	Water-ice (900 km thick)[c]	Water-ice (120 km thick)[c] + NH_3, Fe, Mg-silicates
Atmosphere,[b] pressure (mPa)	10^{-1}	10^{-4}	10^{-3}	7.5×10^{-4}
Atmospheric composition	SO_2 (90%) SO, S_2, SO^+, NaCl, Na, Cl, Cl^+, O^+, S^{n+}, (CO, CO_2)	O_2	O_2, H_2O, H_2, SO_2, CO_2, CN, CH, OH	O_2 (98%), CO_2 (2%), SO_2, CN, CH

a) Magnesium-rich mafic (orthopyroxenes) and ultramafic silicates (such as komatiite) are likely candidates.
b) Atmospheres are extremely tenuous (with densities of 10^8–10^{10} molecules cm^{-3}) and thus rather exo- than atmosphere; cf. the discussion of the exosphere of Mercury, Section 5.2.2.
c) Salty subsurface oceans may be present; see text.

sulfur and sulfur compounds [69, 70] provides a yellow-orange appearance of the surface, with interspersed reddish and greenish specs, very much reminiscent of a pizza, Figure 5.29a. Sulfur species identified or strongly suggested for Io are pictured in Figure 5.29b.

Elemental sulfur is present in the form of S_2 (mauve), some S_3 (blue), S_4 (red), S_8 (light yellow), and S_∞ (yellow). Among the sulfur compounds present as surface constituents, sulfur dioxide, in the form of SO_2 frost, is the most prominent one. Io's surface, and hence the sulfur dioxide frost, is subjected to the bombardment by high energetic electrons stemming from Jupiter's ionosphere, initiating, together with photolytic interactions, reaction cascades such as those represented by Eqs. (5.83) and (5.84): sulfur monoxide formed by photolysis of SO_2 disproportionates to regenerated SO_2 together with disulfur monoxide S_2O, which further

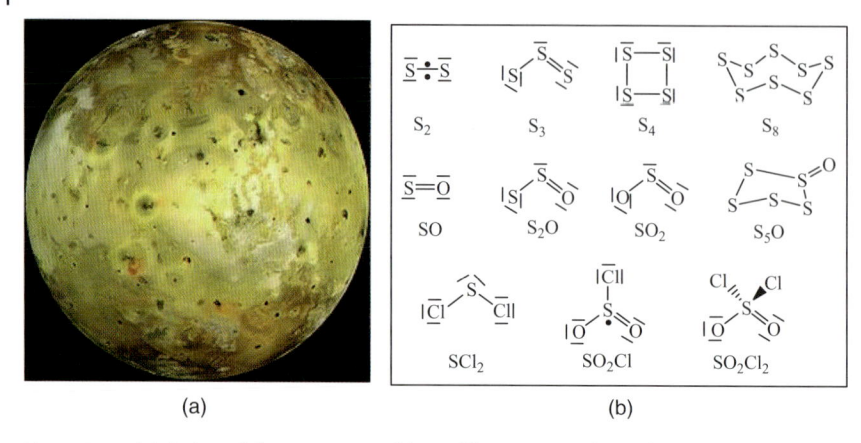

Figure 5.29 (a) A view of the innermost of the Galilean moons, the sulfur moon Io. Credit: NASA (public domain). (b) Structural formulae (formal charges omitted) of a selection of sulfur allotropes and sulfur compounds in Io's surface areas.

degrades to SO_2 and S_3. The trisulfur thus generated reacts with S_2O to form S_5O, the decay of which yields SO_2, S_3, S_4, and other species. Radiolysis of SO_2 by electrons (reaction sequence (5.84)) yields sulfur trioxide SO_3 and sulfur atoms. SO_3 is unstable in an electron flux and disintegrates to form SO_2 and O_2. The net reaction is a break-up of SO_2 into sulfur and oxygen, and at least part of the surface sulfur on Io should originate from this reaction. The instability of SO_3 on the one hand and the lack of any appreciable amounts of water on the other apparently prevent the formation of sulfuric acid, a prominent molecule in the clouds of Venus:

$$SO_2 + h\nu \rightarrow SO + O$$

$$3SO \rightarrow SO_2 + S_2O$$

$$2S_2O \rightarrow SO_2 + S_3$$

$$S_3 + S_2O \rightarrow S_5O$$

$$S_5O \rightarrow \rightarrow SO_2, S_nO, S_4, \ldots \tag{5.83}$$

$$SO_2 + e^- \rightarrow \rightarrow SO_3 + S$$

$$SO_3 + e^- \rightarrow \rightarrow SO_2 + O_2 \tag{5.84}$$

Photodissociation of volcanic NaCl deposited on the surface delivers sodium and chlorine atoms to the atmosphere. This chlorine, together with chlorine supplied by volcanic plumes, can react with SO_2 to form sulfuryl chloride SO_2Cl_2 via SO_2Cl, Eq. (5.85). Photo-induced reaction of chlorine with sulfur produces sulfur dichloride SCl_2, Eq. (5.86), the red color of which may be responsible for reddish flecks on Io's surface:

$$NaCl + h\nu \rightarrow Na + Cl$$

$$SO_2 + Cl \rightarrow SO_2Cl \rightarrow \rightarrow SO_2Cl_2 \quad (5.85)$$

$$S_n + Cl + h\nu \rightarrow \rightarrow SCl_2 \quad (5.86)$$

Surface sulfur chemistry is a specific attribute for Io. For the outer three moons, a characteristic feature is the partially fissured thick crust of water-ice which, on Ganymede, amounts to 47% of the overall mass. Water-ice is detectable through the low-temperature water vibrational absorption bands in the approximate 1.5–3.1 μm (6800–3200 cm^{-1}) range. These bands are usually "distorted," and the distortion is attributed to the presence of "non-ice" constituents. According to a common assumption, part of the water is present in the form of water of hydration, such as hydrated sulfates of sodium, magnesium, and calcium, and/or $H_2SO_4 \cdot H_2O$.

From an astrobiological point of view, the three outer moons are of interest mainly for two reasons:

- There is evidence for extended salty water oceans deep underneath the ice crust (this is also being suggested for Saturn's moon Enceladus);
- The exposed surface of the ice crust contains hydrogen peroxide H_2O_2, up to 0.13% in the case of Europa.

H_2O_2 has a characteristic IR feature at 3.5 μm (2850 cm^{-1}) corresponding to a combination band (of the symmetric and the antisymmetric deformation modes), which has also been detected in the rings of Saturn, on other icy moons and in interplanetary grains. H_2O_2 can be considered both beneficial and detrimental for life: due to its high oxidative power, it tends to oxidatively alter or even destroy organic material. On the other hand, its formation depends on an energy source; its presence hence proves that such an energy source – which may also be employed to power life – is present. Further, decay of H_2O_2 delivers oxygen, an essential molecule for aerobic organisms, or H_2O_2 may be used directly as an electron acceptor in respiration (see also Section 7.5).

The formation of hydrogen peroxide from frozen water can proceed by (i) photolysis, (ii) cleavage through bombardment with ions [71a], and (iii) through interaction with low-energy electrons. The reaction sequence (5.87) demonstrates the impact of photons with $\lambda < 240$ nm, or ions such as H$^+$ and, to a lesser extent, S$^+$ and O$^+$, subsumed as X in Eq. (5.87). Ions with a suitable energy of around 100 keV are available from Jupiter's magnetosphere. The interaction of these energetic ions with water-ice exposed to the surface at a temperature of ca. 100 K can produce water molecules in an excited state, H_2O^*, or split water into water cations H_2O^+ and electrons. The former decays to H atoms and OH radicals, the latter can interact with a second water molecule in the ice lattice to form hydroxonium ions H_3O^+ and hydroxyl radicals. Bimolecular combination of two hydroxyl radicals in a solid matrix produces H_2O_2. The bimolecular reaction between two OH radicals may also result in the formation of water and oxygen atoms, Eq. (5.88):

$$h\nu/X + H_2O \begin{cases} \to H_2O^* \to H + OH \\ \to H_2O^+ + e^- \end{cases} \quad OH + OH \to H_2O_2$$

$$H_2O^+ \xrightarrow{H_2O} H_3O^+ + OH$$

$$e^- \to H_3O^+ + H \tag{5.87}$$

$$OH + OH \to H_2O + O \tag{5.88}$$

An alternative, though less favored, path for the decay of excited water molecules leads to hydrogen molecules and oxygen atoms, and the latter can combine to produce molecular oxygen; Eq. (5.89). Oxygen production also occurs via hydroperoxyl[29] radicals HO_2, formed in the reaction between H_2O_2 and H, Eq. (5.90):

$$H_2O^* \to H_2 + O$$
$$O + O \to O_2 \tag{5.89}$$

$$H_2O_2 + H \to H_2 + HO_2$$
$$HO_2 + HO_2 \to H_2O_2 + O_2 \tag{5.90}$$

The production of H_2O_2 and subsequent reactions, as initiated by dissociative electron attachment to water, is summarized in the reaction sequence (5.91) [71b]. High fluxes of low-energy (<100 eV) electrons are produced as icy surfaces of the moons are subjected to bombardment by ions from the magnetospheric plasma, Solar photons, and protons (plus other ions) originating from the Solar wind and cosmic rays. The electron energy needed to generate the reactive intermediate H_2O^- at a temperatures of ca. 90 K is around 5 eV. Further electron association by H_2O_2 results in the formation of the hydroperoxyl radical HO_2 or, alternatively, oxygen and hydrogen:

$$H_2O + e^- \to H_2O^- \to H^- + OH$$
$$\xrightarrow{2x} H_2O_2$$
$$\downarrow (e^-)$$
$$H_2O_2^* \begin{cases} \to H + HO_2 \\ \to H_2 + O_2 \end{cases} \tag{5.91}$$

At slightly higher electron energies (~7.3 eV), water molecules in the ice matrix can also become excited, followed by dissociation in a similar way as outlined above for the ion impact excitation. The reaction sequence (5.92) summarizes some of the consecutive paths: H_2O^* may dissociate into H and OH (which again generates H_2O_2 in a bimolecular reaction), or into molecular hydrogen and oxygen atoms, the latter either in the singlet state (^1D state, all electrons paired, hence a valence electron configuration $s^2p_x^2p_y^2$), or in the triplet state (^3P, two electrons unpaired, valence electron configuration $s^2p_x^2p_y^1p_z^1$). Singlet oxygen readily

---

29) Hydro-*superoxyl* is a more appropriate denomination for the HO_2 radical, since the oxygen is in the oxidation state –½, as in superoxide, and not in the –1 state, as in peroxide.

adds to water to form H_2O_2, while triplet oxygen reacts with OH radicals to form HO_2 which, via dissociative electron attachment, produces hydride ions and molecular oxygen:

$$H_2O \xrightarrow{(e^-)} H_2O^* \begin{cases} \rightarrow H + OH \\ \rightarrow H_2 + O(^1D) + O(^3P) \end{cases} \quad (5.92)$$

$$\downarrow H_2O \qquad \xrightarrow{OH} HO_2$$
$$H_2O_2 \qquad \qquad \xrightarrow{e^-} H^- + O_2$$

5.6.3
The Moons Enceladus, Titan and Triton

Of the Saturnian moons, the comparatively small Enceladus and the large Titan (second largest moon in the Solar System, next to Ganymede) have attained special interest in the context of prebiotic chemistry, and even the possibility of the presence of primitive life forms: Enceladus for its subsurface ocean, and Titan for its dense atmosphere, hydrocarbon lakes on its surface, and the versatility of its organic chemistry. Like the Jovian moons Europa, Ganymede, and Callisto, Enceladus and Titan have a massive crust predominantly consisting of water-ice. Triton, the largest Neptunian moon, is different in that the main component of its ice coverage is frozen dinitrogen. Triton has been captured from the Kuiper belt and thus is a "Kuiper belt object," rather than an object resulting from the accretion of planetesimals in the circumplanetary disk in the last phase of Neptune's growth. Triton's capture from the Kuiper belt is also strongly supported by the D/H ratio [72], which is the same ($\sim 3 \times 10^{-4}$) as found in comets and hence in authentic remains from the infancy of the Solar System. Table 5.19 provides an overview of characteristics of the three moons.

Evidence for a subsurface ocean on Enceladus comes from the emission of supersonic plumes [73] (Figure 5.30a), eruptions of water vapor and ice particles from tectonically active, and relatively warm[30] fractures ("tiger stripes") near its South Pole region. The driving force for these eruptions can be provided by release of CO_2 dissolved in liquid subsoil water (as in the case of geyser eruptions on Earth), and by the abrupt decomposition of N_2 and CH_4 clathrate hydrates[31] when these are exposed, by tectonic action, to low-pressure conditions. Figure 5.30b is a schematic of how the process of plume ejection can evolve.

The *ice* particles with radii typically between 0.1 and 1 µm are the major supply to one of Saturn's faint outer rings, the E-ring. They contain minor amounts of ammonia, organic, and siliceous components, the latter indicative of some interaction of liquid water with Enceladus' rocky core, and up to 2% by mass of sodium

---

30) The "heat" in the cracks, temperature in excess of 180 K (the mean surface temperature of Enceladus is 75 K) may come about by "tidal heating" as Enceladus circles its giant mother planet on a slightly elliptical orbit.

31) The composition of the methane clathrates is close to $(CH_4)(H_2O)_6$; see Figure 5.14 in Section 5.2.4.4. The dissociation into water and methane is slightly *endo*thermic ($+18$ kJ mol^{-1} at 273 K and 100 kPa).

Table 5.19 Overview of Saturn's moons Enceladus and Titan, and Neptune's moon Triton.

Moon	Enceladus	Titan	Triton[a]
Origin of name	A giant in Greek mythology	The Titans were a race of powerful deities	A Greek sea god, son of Poseidon
Mean orbit radius (km)	23 800	1 222 000	354 759
Radius (km)	252	2 576	1 353
Mean density (g cm^{-3})	1.61	1.88	2.06
Mean surface temperature (K)	75	94	38
Core	Silicates	Silicates and Fe[c]	Silicates and Fe
Mantle	H_2O-rich ice	High-pressure ice layers	Silicates and water ice
Crust/outer layers/surface	Clean water-ice; subsurface liquid water with NaCl, $NaHCO_3$, CO_2, NH_3	Water-ice (ca. 75 km) on liquid H_2O/NH_3 layer (ca. 300 km), hydrocarbon lakes	N_2 ice (55%), water ice (ca. 30%), CO_2 ice (ca. 15%), traces of CH_4 and CO
Atmosphere, pressure (Pa)	Trace	1.47×10^5, opaque	1.4
Atmospheric composition	H_2O (91%), N_2, CO_2, CH_4, organics[b], ^{40}Ar and H_2S in minor amounts	N_2 (98.4%), CH_4 (1.6%), many organics (see text)	N_2, CH_4

a) A captured Kuiper belt object.
b) Organic molecules detected in the plumes include H_2CO, CH_3OH, C_2H_2, C_3H_6, C_4H_2, C_4H_4, C_4H_6, C_4H_8, and C_6H_6 [72]; cf. also Scheme 5.6 for structural formulae of these molecules.
c) Alternatively, a core consisting of rock + ice, or silicate hydrates, has been proposed [77].

salts. For the possible formation of these salt-loaded particulates, see Figure 5.30b. The *vapor* of the plumes in contrast is free of sodium. Sodium is readily detected by its intense electronic doublet transition centered at 589.29 nm, the $2s(^2S_{1/2}) \rightarrow 3p(^2P_{1/2})$ and $2s(^2S_{1/2}) \rightarrow 3p(^2P_{3/2})$ transitions, responsible for the bright orange-yellow glow of sodium vapor lamps, well known to chemistry students from the flame test. Potassium salts contribute to a minor degree. Sodium salts (chloride and carbonates) are dissolved from rock in contact with liquid water, and remain in the liquid phase as a water reservoir begins to freeze out. On the other hand, as water slowly evaporates in pressurized chambers, the water vapor remains essentially salt-free. According to model calculations, the sodium salt contents in Enceladus' ocean range from 0.3–1.3% NaCl plus 0.2–0.8% $NaHCO_3$, with an inferred pH of 8.5–9 [74].

The comparatively high percentage of CH_4 in the Titanian atmosphere (1.6%), as well as the presence of methane (and ethane) lakes on Titan's surface, give rise

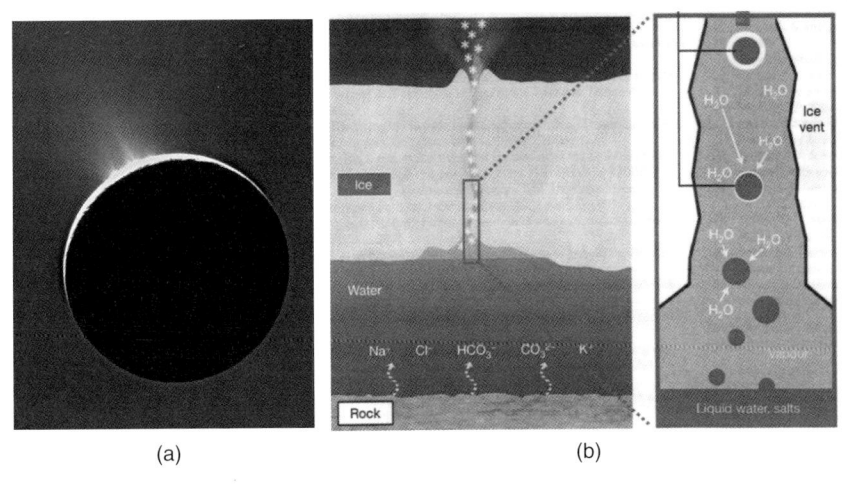

Figure 5.30 (a) Plumes ejected from cracks in the South Pole area of Enceladus. Credit: Cassini Imagine Team, SSI, JPL, ESA, NASA. (b) Schematic presentation of the expulsion of icy grains from a liquid salt water reservoir through a vent in the ice cover, and (enlargement) the formation of salt-bearing H_2O droplets/ice-grains in the vent. © Ref. [74] Figure 3. Reprinted with permission from Nature.

to a rich chemistry on Titan. The remarkable methane abundance raises the question of its origin. A possible scenario is serpentinization, the generation of methane through the reduction of water by mineralized ferrous iron, Eq. (5.93a), and subsequent reaction of the hydrogen thus formed with carbon oxides, Eq. (5.93b), catalyzed by appropriate minerals. Serpentinization has already been discussed in the context of methane plumes observed on Mars in Section 5.2.4.4. Equation (5.93c) in a way summarizes this process, based here on the mineral fayalite. Hydrogen can also be produced by radiolysis of water in Titan's interior:

$$3H_2O + 2Fe^{2+} \rightarrow H_2 + 4H^+ + Fe_2O_3 \tag{5.93a}$$

$$4H_2 + CO_2 \rightarrow CH_4 + 2H_2O \tag{5.93b}$$

$$6Fe_2SiO_4 + CO_2 + 2H_2O \rightarrow 4Fe_3O_4 + 6SiO_2 + CH_4 \tag{5.93c}$$

Serpentinization alone cannot fully explain the relatively high D/H ratio of 1.3×10^{-4} observed for methane in Titan's atmosphere, an enrichment by a factor of 5–7 relative to the protosolar value. Direct capture of CH_4 in the satellite's planetesimals in Saturn's feeding zone at the time of their formation in the Solar nebula appears to provide an appropriate explanation both for the D/H ratio and the appreciable amounts of methane trapped on Titan [75].

Methane and dinitrogen are the starting molecules for the versatile chemistry observed in the Titanian atmosphere at a height of around 1000 km. The initiating reaction is photolysis of N_2 and CH_4 to form neutral radicals and cations, Eq. (5.94), from which more complex molecules are generated. Contrasting the atmospheres of Jupiter and Saturn, reactions between neutral radicals, such as depicted in

Scheme 5.5, are essentially ineffective in Titan's atmosphere due to the low temperature of ca. 125 K. Rather, ion–molecule (ion–neutral) reactions account for the formation of more complex molecules. The inventory of molecules and molecular cations detected in Titan's upper atmosphere [76] is provided in Scheme 5.6. Simple cations such $C_2H_5^+$, from which benzene can form, are generated either by starting from methane and the methyl cation CH_3^+, Eq. (5.95a), or from ethene C_2H_4 and the methanium cation CH_5^+, Eq. (5.95b). The cation CH_5^+ forms by radiative association between CH_3^+ and H_2; cf. Eq. (4.18) in Section 4.2.1. Both reactions (5.94a) and (5.94b) have rate constants k close to $10^{-9}\,\mathrm{cm}^3\,\mathrm{s}^{-1}$. The reaction paths (5.96) and (5.97) illustrate the formation of benzene from successive ion–molecule reactions, finalized by dissociative electron capture:

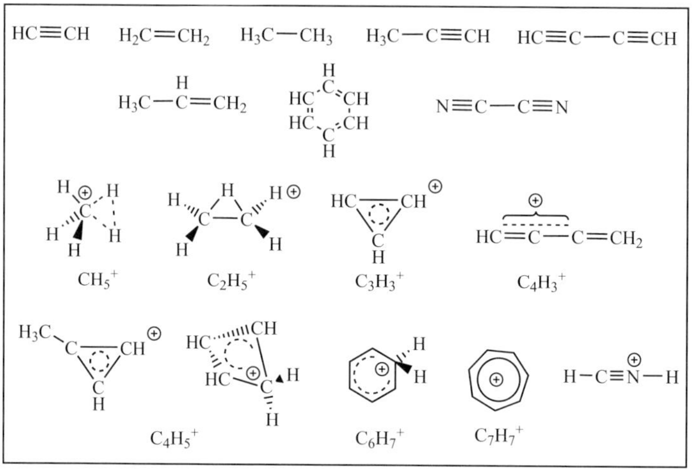

Scheme 5.6 Trace organic molecules, which have been found in Titan's atmosphere at high altitudes [76]. For the cations, detected by mass spectrometry installed on the Cassini spacecraft, the more stable structural arrangement of several possible structures is shown.

$$\left. \begin{array}{c} N_2 \\ CH_4 \end{array} \right\} \xrightarrow[e^-]{h\nu} \begin{array}{c} N,\ CH_n\ (n=1\text{--}3),\ H \\ N^+,\ CH_n^+\ (n=1\text{--}3) \end{array} \quad (5.94)$$

$$CH_4 + CH_3^+ \rightarrow C_2H_5^+ + H_2 \quad (5.95a)$$

$$C_2H_4 + CH_5^+ \rightarrow C_2H_5^+ + CH_4 \quad (5.95b)$$

$$C_2H_2 + C_2H_5^+ \xrightarrow[H_2]{} C_4H_5^+ \xrightarrow[H_2]{C_2H_4} C_6H_7^+ \xrightarrow{e^-} C_6H_6 + H \quad (5.96)$$

$$C_4H_2 + H^+ \xrightarrow[h\nu]{} C_4H_3^+ \xrightarrow[H_2]{C_2H_4} C_6H_5^+ \xrightarrow[h\nu]{H_2} C_6H_7^+ \xrightarrow{e^-} C_6H_6 + H \quad (5.97)$$

Along with benzene and condensed aromatic and pseudo-aromatic hydrocarbons (in the molecular mass range 80–350 g mol^{-1}, among these anthracene and naphthalene), the nitrogen containing organic molecule dicyan C_2N_2 and protonated hydrogen cyanide [HCNH]$^+$ have been detected, which can give rise to the formation of *hetero*cyclic polyaromatic compounds. Efficient reactions leading to [HCNH]$^+$ are represented in Eqs. (5.98) ($k = 2.7 \times 10^{-9}$ cm^3 s^{-1}) and (5.99) ($k = 21.1 \times 10^{-9}$ cm^3 s^{-1}). Negatively charged massive molecules with molecular masses (M) up to 8000 g mol^{-1} have also been found, and these can further condense to form hydrocarbon-nitrile aerosols with $M \approx 40000$ g mol^{-1}, corresponding to particle sizes with radii of 260 nm. These so-called tholin aerosols move to lower altitudes; they are responsible, at least in part, for the orange coloration of the hazes and clouds so characteristic of Titan. Clouds (of methane and/or ethane) on Titan are typically confined to altitudes below 26 km, but can eventually reach altitudes of up to 45 km:

$$HCN + C_2H_5^+ \rightarrow HCNH^+ + C_2H_4 \tag{5.98}$$

$$HCN^+ + CH_4 \rightarrow HCNH^+ + CH_3 \tag{5.99}$$

Several of the chemical processes taking place in Titan's upper atmosphere, and the molecular as well as condensed chemical species found there, are reminiscent of the chemistry that happens in dense interstellar clouds, a chemistry by which molecules are synthesized which subsequently became conserved in the remnants of the protosolar nebula, and in the comets and asteroids and their messengers to Earth, the meteorites. The density of the Titanian atmosphere at an altitude of 1000 km is, however, still higher by four orders of magnitude than in dense interstellar clouds: 10^{10} vs. $<10^6$ atoms/molecules per cubic centimeter.

Summary Section 5.6

The four giant planets accreted from icy and rocky planetesimals in the protosolar disk. In the case of the particularly large planets Jupiter and Saturn, accretion was followed by gravitational capture of surrounding gases, mainly hydrogen and helium: these two planets thus developed into gas giants, while the smaller Uranus and Neptune remained on the level of ice giants. A major part of Jupiter's hydrogen is present in the form of liquid metallic hydrogen. The troposphere is dominated by CH_4, NH_3, NH_4SH, and H_2O; minor amounts of PH_3, AsH_3, and GeH_4 are also present. The main constituent in the Jovian stratosphere is methane, starting point for a versatile photochemistry.

Among the large moons of the giant planets, Jupiter's innermost moon Io is remarkably distinct from the other moons insofar as its surface layer contains a variety of sulfur allotropes (S_n; $n = 2, 3, 4, 8, \infty$) and sulfur compounds (e.g., S_2O, S_5O, and SO_2Cl). The three other Galilean moons (Europa, Ganymed, and Callisto), as well as Saturn's Enceladus and Titan, carry a bulky ice crust and possibly extended salty water oceans underneath this crust. In the case of Enceladus, this is evidenced by violent eruptions into space of water vapor, ice particles, and salt-bearing water droplets. The icy layers also contain some H_2O_2, generated from water either by photolysis, or by high-energy cations, or via dissociative electron attachment to H_2O. The largest Neptunian moon, Triton, a captured Kuiper belt object, contains frozen dinitrogen as the main constituent of its crust.

Titan is the only moon to have a dense atmosphere (the surface pressure is about 1.5 times that of the terrestrial atmosphere). The main atmospheric constituents, N_2 (98.4%) and CH_4 (1.6%), give rise to a rich, photolytically induced chemistry in the upper atmosphere, including the formation of high molecular mass tholins. These tholins, organic compounds based on aromatic component parts, are co-responsible for the orange hazes typical of Titan's atmosphere. An additional feature specifically for Titan is the existence of CH_4/C_2H_6 lakes on its surface.

References

1 van Dishoeck, E.F. (2006) Chemistry in low-mass protostellar and protoplanetary regions. *Proc. Natl. Acad. Sci. U. S. A.*, **103**, 12249–12256.
2 Kleine, T., Palme, H., Mezger, K., and Halliday, A.N. (2006) Hf-W Chronometry of Lunar metals and the age and early differentiation of the Moon. *Science*, **310**, 1671–1674.
3 (a) Lee, D.-C., and Halliday, A.N. (1995) Hafnium-tungsten chronometry and the timing of terrestrial core formation. *Nature*, **378**, 771–774.; (b) Lee, D.-C., Halliday, A.N., Snyder, G.N., and Taylor, L.A. (1997) Age and origin of the Moon. *Science*, **278**, 1098–1103.; (c) Righter, K., and Shearer, C.K. (2003) Magmatic fractionation of Hf and W: constraints on the timing of core formation and differentiation in the Moon and Mars. *Geochim. Cosmochim. Acta*, **67**, 2497–2507.; (d) Rutherford, M.J., and Papale, P. (2009) Origin of basalt fire-fountain eruptions on Earth versus the Moon. *Geology*, **37**, 219–222.
4 (a) Denevi, B.W., Robinson, M.S., Solomon, S.-C., Murchie, S.L., Blewett, D.T., Domingue, D.L., McCoy, T.J., Ernst, C.M., Head, J.W., Watters, T.R., and Chabot, N.L. (2009) The evolution of Mercury's crust: a global perspective from messenger. *Science*, **324**, 613–621.; (b) Solomon, S.C., McNutt, R.L., Jr., Watters, T.R., Lawrence, D.J., Feldman, W.C., Head, J.W., Krimigis, S.M., Murchie, S.L., Phillips, R.J., Slavin, J.A., and Zuber, M.T. (2009) Return to Mercury: a global perspective on MESSENGER's first Mercury flyby. *Science*, **321**, 59–62.
5 (a) Domingue, D.L., Koehn, P.L., Killen, R.M., Sprague, A.L., Sarantos, M., Cheng, A.F., Bradley, E.T., and McClintock, W.E. (2007) Mercury's atmosphere: a surface-bounded exosphere. *Space Sci. Rev.*, **131**, 161–186.; (b) Zurbuchen, T.H., Raines, J.M., Gloeckler, G., Krimigis, S.M., Slavin, J.A., Koehn, P.L., Killen, R.M., Sprague, A.L., McNutt, R.L., Jr., and Solomon, S.C. (2008) Messenger observations of the composition of Mercury's ionized exosphere and plasma environment. *Science*, **321**, 90–92.
6 Glassmeier, K.-H. (2009) Magnetic twisters on Mercury. *Science*, **324**, 597–598.
7 (a) Vogt, A.E. (1952) *Destination: Universe*, The New American Library of World Literature [Signet Book], New York.; (b) Bradbury, R.D. (1950) *The Long Rain*, Love Romances Publishing Co, Stamford, Conn.; (c) Asimov, I. (1950) *Lucky Starr and the Oceans of Venus*, Doubleday & Co., NY.
8 Mikkola, S., Brasser, R., Wiegert, P., and Innanen, K. (2004) Asteroid 2002 VE68,

a quasi-satellite of Venus. *Mon. Not. R. Astron. Soc.*, **351**, L63–L65.

9 Basher, R.E. (2006) Units for column amounts of ozone and other atmospheric gases. *Quart. J. R. Meteorol. Soc.*, **108**, 460–462.

10 Mills, F.P., and Allen, M. (2007) A review of selected issues concerning the chemistry in Venus' middle atmosphere. *Planet Space Sci.*, **55**, 1729–1740.

11 Krasnopolsky, V.A. (2007) Chemical kinetic model for the lower atmosphere of Venus. *Icarus*, **191**, 25–37.

12 Pernice, H., Garcia, P., Willner, H., Francisco, J.S., Mills, F.P., Allen, M., and Yung, Y.L. (2004) Laboratory evidence for a key intermediate in the Venus atmosphere: peroxychloroformyl radical. *Proc. Natl. Acad. Sci. U. S. A.*, **101**, 14007–14010.

13 Piccioni, G., Drossart, P., Zasova, L., Migliorini, A., Gérard, J.-C., Mills, F.P., Shakun, A., García Muñoz, A., Ignatiev, N., Grassi, D., Cottini, V., Taylor, F.W., and Erard, S. (2008) First detection of hydroxyl in the atmosphere of Venus. *Astron. Astrophys.*, **483**, L29–L33.

14 Johnson, N.M., and Fegley, B., Jr. (2003) Tremolite decomposition on Venus II. Products, kinetics, and mechanism. *Icarus*, **164**, 317–333.

15 Klingelhöfer, G., and Fegley, B., Jr. (2000) Iron mineralogy of Venus' surface investigated by Mössbauer spectroscopy. *Icarus*, **147**, 1–10.

16 (a) Weinbaum, S.G. (1934) *A Martian Odyssey*, Wonder Stories, NY; (b) Bradbury, R. (1950) *The Martian Chronicles*, Doubleday & Co., NY.; (c) Clarke, A.C. (1951) *The Sands of Mars*, Sidgewick & Jackson, London, UK; (d) Asimov, I. (1957) *I'm in Marsport without Hilda*, Venture Science Fiction, Fantasy House, NY; (e) Pesek, L. (1970) *Die Erde ist nah – Die Marsexpedition (The Earth is Near)*, Georg Bitter Verlag, Recklinghausen, Germany.

17 Wells, H.G. (1898) *The War of the Worlds*, William Heinemann Publ., London.

18 Catling, D.C., Claire, M.W., Zahnle, K.J., Quinn, R.C., Clark, B.C., Hecht, M.H., and Kounaves, S. (2010) Atmospheric origin of perchlorate on Mars and in the Atacamo. *J. Geophys. Res. [Planets]*. **115**, E00E11.

19 Bogard, D.D., and Johnson, P. (1983) Martian gases in an Antarctic meteorite. *Science*, **221**, 651–654.

20 (a) Raoult, D., Drancourt, M., Azza, S., Nappez, C., Guieu, R., Rolain, J.-M., Fourquet, P., Campagne, B., La Scola, B., Mege, J.-L., Mansuelle, P., Lechevalier, E., Berland, Y., Gorvel, J.-P., and Renesto, P. (2008) Nanobacteria are mineralo fetuin complexes. *PLoS Pathog.*, **4**, e41. doi: 10.1371/journal. ppat.0040041.; (b) McKay, C.P., Friedmann, E.I., Frankel, R.B., and Bazylinski, D.A. (2003) Magnetotatic bacteia on Earth and Mars. *Astrobiology*, **3**, 263–270.

21 Formisano, V., Atreya, S., Encrenaz, T., Ignatiev, N., and Giuranna, M. (2004) Detection of methane in the atmosphere of Mars. *Science*, **306**, 1758–1761.

22 Oze, C., and Sharma, M. (2005) Have olivine, will gas: serpentinization and the abiogenic production of methane on Mars. *Geophys. Res. Lett.*, **32**, L10203.

23 Ehlmann, B.L., Mustard, J.F., Murchie, S.L., Poulet, F., Bishop, J.L., Brown, A.J., Calvin, W.M., Clark, R.N., Des Marais, D.J., Milliken, R.E., Roach, L.H., Roush, T.L., Swayze, G.A., and Wray, J.L. (2008) Orbital identification of carbonate-bearing rocks on Mars. *Science*, **322**, 1828–1832.

24 Bibring, J.-P., Langevin, Y., Mustard, J.F., Poulet, F., Arvidson, R., Gendrin, A., Gondet, B., Mangold, N., Pinet, P., Forget, F., Berthé, M., Bibring, J.-P., Gendrin, A., Gomez, C., Gondet, B., Jouglet, D., Poulet, F., Soufflot, A., Vincendon, M., Combes, M., Drossart, P., Encrenaz, T., Fouchet, T., Merchiorri, R., Belluci, G.C., Altieri, F., Formisano, V., Capaccioni, F., Cerroni, P., Coradini, A., Fonti, S., Korablev, O., Kottsov, V., Ignatiev, N., Moroz, V., Titov, D., Zasova, L., Loiseau, D., Mangold, N., Pinet, P., Douté, S., Schmitt, B., Sotin, C., Hauber, E., Hoffmann, H., Jaumann, R., Keller, U., Arvidson, R., Mustard, J.F., Duxbury, T., Forget, F., and Neukum, G. (2006) Global mineralogical and aqueous Mars history derived from

OMEGA/Mars express data. *Science*, **312**, 400–404.

25. Hurowitz, J.A., and McLennan, S.M. (2007) A ~3.5 Ga record of water-limited, acidic weathering conditions on Mars. *Earth Planet Sci. Lett.*, **260**, 432–443.

26. Sprague, A.L., Boynton, W.V., Kerry, K.E., Nelli, S., Murphy, J., Reedy, R.C., Metzger, A.E., Hunten, D.M., James, K.D., and Crombie, M.K. (2005) Distribution and abundance of Mars' atmospheric argon. *Lunar Planet. Sci.*, **36**, 2085.

27. Lelieveld, J. (2008) A reverse ozone hole on Mars. *Angew. Chem. Int. Ed.*, **47**, 9804–9807.

28. Matthiesen, J., Wendt, S., Hansen, J.Ø., Madsen, G.K.H., Lira, S., Galliker, P., Vestergaard, E.K., Schaub, R., Lægsgaard, E., Hammer, B., and Besenbacher, F. (2009) Observation of all the intermediate steps of a chemical reaction on an oxide surface by scanning tunnelling microscopy. *ACS Nano*, **3**, 517–526.

29. Lefèvre, F., Bertaux, J.-L., Clancy, R.T., Encrenaz, T., Fast, K., Forget, F., Lebonnois, S., Montmessin, F., and Perrier, S. (2008) Heterogeneous chemistry in the atmosphere of Mars. *Nature*, **454**, 971–975.

30. Gérard, J.-C., Cox, C., Saglam, A., Berteaux, J.-L., Villard, E., and Nehmé, C. (2008) Limb observations of the ultraviolet nitric oxide nightglow with SPICAV on board Venus Express. *J. Geochem. Res.*, **113**, E00B03.

31. (a) Muñoz, A.G., Mills, F.P., Piccioni, G., and Drossart, P. (2009) The near-infrared nitric oxide nightglow in the upper atmosphere of Venus. *Proc. Natl. Acad. Sci. U. S. A.*, **106**, 985–988.; (b) Krasnopolsky, A. (2006) A sensitive search for nitric oxide in the lower atmospheres of Venus and Mars: detection on Venus and upper limit on Mars. *Icarus*, **182**, 80–91.

32. Berteaux, J.L., Leblanc, F., Perrier, S., Quemerais, E., Korablev, O., Dimarellis, E., Reberac, A., Forget, F., Simon, P.C., Stern, S.A., and Sandel, B. (2005) Nightglow in the upper atmosphere of Mars and implications for atmospheric transport. *Science*, **307**, 566–569.

33. de Saint-Exupéry, A. (1943) *The Little Prince/Le Petit Prince*, Reynal & Hitchcock, Inc, New York.

34. Thomas, P.C., Parker, J.W., McFadden, L.A., Russel, C.T., Stern, S.A., Sykes, M.V., and Young, E.F. (2005) Differentiation of the asteroid Ceres as revealed by its shape. *Nature*, **437**, 224–226.

35. Rivkin, A.S., Volquardsen, E.L., and Clark, B.E. (2006) The surface composition of Ceres: discovery of carbonates and iron-rich clays. *Icarus*, **185**, 563–567.

36. Jenniskens, P., Shaddad, M.H., Numan, D., Elsir, S., Kudoda, A.M., Zolensky, M.E., Le, L., Robinson, G.A., Friedrich, J.M., Rumble, D., Steele, A., Chesley, S.R., Fitzsimmons, A., Duddy, S., Hsieh, H.H., Ramsay, G., Brown, P.G., Edwards, W.N., Tagliaferri, E., Boslough, M.B., Spalding, R.E., Dantowitz, R., Kozubal, M., Pravec, P., Borovicka, J., Charvat, Z., Vaubaillon, J., Kuiper, J., Albers, J., Bishop, J.L., Mancinelli, R.L., Sandford, S.A., Milam, S.N., Nuevo, M., and Worden, S.P. (2009) The impact and recovery of asteroid 2008TC$_3$. *Nature*, **458**, 485–488.

37. Brown, P.G., Hildebrand, A.R., Zoensky, M.E., Grady, M., Clayton, R.N., Mayeda, T.K., Tagliaferri, E., Spalding, R., MacRae, N.D., Hoffman, E.L., Mittlefehldt, D.W., Wacker, J.F., Bird, J.A., Campbell, M.D., Carpenter, R., Gingerich, H., Glatiotis, M., Greiner, E., Mazur, M.J., McCausland, P.J.A., Plotkin, H., and Mazur, T.R. (2000) The fall, recovery, orbit, and composition of the Tagish Lake meteorite: a new type of carbonbaceous chondrite. *Science*, **290**, 320–325.

38. (a) Sephton, M.A. (2005) Organic matter in carbonaceous meteorites: past, present and future research. *Philos. Trans. R. Soc. A*, **363**, 2729–2742.; (b) Sephton, M.A. (2002) Organic compounds in carbonaceous meteorites. *Nat. Prod. Rep.*, **19**, 292–311.

39. Flynn, G.J., Keller, L.P., Feser, M., Wirick, S., and Jacobsen, C. (2003) The

origin of organic matter in the Solar System: evidence from the interplanetary dust particles. *Geochim. Cosmochim. Acta*, **67**, 4791–4806.

40 (a) Clayton, D.D. (1997) Placing the Sun and mainstream SiC particles in galactic chemodynamic evolution. *Astrophys J.*, **484**, L67–L70.; (b) Clayton, D.D., and Nittler, L.R. (2004) Astrophysics with presolar stardust. *Annu. Rev. Astron. Astrophys.*, **42**, 39–78.

41 Remusat, L., Robert, F., and Derenne, S. (2007) The insoluble organic matter in carbonaceous chondrites: chemical structure, isotope composition and origin. *Comt. Rend. Geosci.*, **339**, 895–906.

42 Glavin, D.P., and Dworkin, J.P. (2009) Enrichment of the amino acid L-isovaline by aqueous alteration on CI and CM meteorite parent bodies. *Proc. Natl. Acad. Sci. U. S. A.*, **106**, 5487–5492.

43 (a) Blackmond, D.G. (2004) Asymmetric autocatalysis and its implications for the origin of homochirality. *Proc. Natl. Acad. Sci. U. S. A.*, **101**, 5732–5736.; (b) Klussmann, M., Iwamura, H., Mathew, S.P., Wells, D.H., Jr, Pandya, U., Armstrong, A., and Blackmond, D.G. (2006) Thermodynamic control of asymmetric amplification in amino acid catalysis. *Nature*, **441**, 621–623.

44 Botta, O. (2006) Organic chemistry in meteorites, comets and the interstellar medium, in *Astrochemistry: Recent Successes and Current Challenges* (eds D.C. Lis, G.A. Blake, and E. Herbst), Int. Astronom. Union, Paris, France, pp. 479–487.

45 Ehrenfreund, P., Glavin, D.P., Botta, O., Cooper, G., and Bada, J.L. (2001) Extraterrestrial amino acids in Orgueil and Ivuna: tracing the parent body of CI type carbonaceous chondrites. *Proc. Natl. Acad. Sci. U. S. A.*, **98**, 2138–2141.

46 Ott, U. (2007) Presolar grains in meteorites and their compositions. *Space Sci. Rev.*, **130**, 87–95.

47 (a) Daulton, T.L., Eisenhour, D.D., Bernatowicz, J.T., Lewis, R.S., and Busek, P.R. (1996) Genesis of presolar diamonds: Comparative high-resolution transmission electron microscopy study of meteoritic and Terrestrial nano-diamonds. *Geochim. Cosmochim. Acta*, **60**, 4853–4872.; (b) Nguyen, A.N., and Zinner, E. (2004) Discovery of ancient silicate stardust in a meteorite. *Science*, **303**, 1496–1499.

48 Nakamura-Messenger, K., Messenger, S., Keller, L.P., Clemett, S.J., and Zolensky, M.E. (2006) Organic globules in the Tagish Lake meteorite: remnants of the protosolar disk. *Science*, **314**, 1439–1442.

49 Busemann, H., Young, A.F., O'D. Alexander, C.M., Hoppe, P., Mukhopadhyay, S., and Nitter, L.R. (2006) Interstellar chemistry recorded in organic matter from primitive meteorites. *Science*, **312**, 727–730.

50 (a) Aléon, J., Robert, F., Duprat, J., and Derenne, S. (2005) Extreme oxygen isotope ratios in the early Solar System. *Nature*, **437**, 385–388.; (b) Marty, B. (2006) The primordial porridge. *Science*, **312**, 706–707.

51 Chakraborty, S., Ahmed, M., Jackson, T.L., and Thiemens, M.H. (2008) Experimental test of self-shielding in vacuum ultraviolet photodissociation of CO. *Science*, **321**, 1328–1331.

52 (a) Kavelaars, J.J., Petit, J.M., Gladman, B., and Marsden, B.G. (2009) 2008 KV42. Minor Planet Electronic Circ. C75.; (b) Hsieh, H.H. (2010) A frosty finding. *Nature*, **464**, 1286–1287.

53 Crovisier, J., Bockelée, D., Colom, P., Biver, N., Despoi, D., and Lis, D.C. (2004) The composition of ices in comet C/1995 O1 (Hale-Bopp) from radio spectroscopy. *Astron. Astrophys.*, **418**, 1141–1157.

54 (a) Rodgers, S.D., and Charnley, S.B. (2001) Organic synthesis in the coma of comet Hale-Bopp? *Mon. Not. R. Astron. Soc.*, **320**, L61–L64.; (b) Ziurys, L.M., Savage, C., Brewster, M.A., Apponi, A.J., Pesch, T.C., and Wyckoff, S. (1999) Cyanide chemistry in comet Hale-Bopp (C/1995 O1). *Astrophys. J.*, **572**, L67–L71.

55 Helbert, J., Rauer, H., Boice, D.C., and Huebner, W.F. (2005) The chemistry of C2 and C3 in the coma of comet C/1995 O1 (Hale-Bopp) at heliocentric distances. *Astron. Astrophys.*, **442**, 1107–1120.

56 (a) Woodney, L.M., A'Hearn, M.F., McMullin, J., and Samarasinha, N.

(1997) Sulfur chemistry at millimetre wavelengths in C/Hale-Bopp. *Earth Moon Planets*, **78**, 69–70.; (b) Rodgers, S.D., and Charnley, S.B. (2006) Sulfur chemistry in cometary comae. *Adv. Space Res.*, **38**, 1928–1931.

57 Meier, R., and Owen, T.C. (1999) Cometary deuterium. *Space Sci. Rev.*, **90**, 33–43.

58 (a) Keller, L.P., Bajt, S., Baratta, G.A., Borg, J., Bradley, J.P., Brownlee, D.E., Busemann, H., Brucato, J.R., Burchell, M., Colangeli, L., d'Hendecourt, L., Djouadi, Z., Ferrini, G., Flynn, G., Franchi, I.A., Fries, M., Grady, M.M., Graham, G.A., Grossemy, F., Kearsley, A., Matrajt, G., Nakamura-Messenger, K., Mennella, V., Nittler, L., Palumbo, M.E., Stadermann, F.J., Tsou, P., Rotundi, A., Sandford, S.A., Snead, C., Steele, A., Wooden, D., and Zolensky, M. (2006) Infrared spectroscopy of comet 81P/Wild 2 samples returned by Stardust. *Science*, **314**, 1728–1731.; (b) Sandford, S.A., Aléon, J., O'D. Alexander, C.M., Araki, T., Bajt, S., Baratta, G.A., Borg, J., Bradley, J.P., Brownlee, D.E., Brucato, J.R., Burchell, M.J., Busemann, H., Butterworth, A., Clemett, S.J., Cody, G., Colangeli, L., Cooper, G., D'Hendecourt, L., Djouadi, Z., Dworkin, J.P., Ferrini, G., Fleckenstein, H., Flynn, G.J., Franchi, I.A., Fries, M., Gilles, M.K., Glavin, D.P., Gounelle, M., Grossemy, F., Jacobsen, C., Keller, L.P., Kilcoyne, A.L.D., Leitner, J., Matrajt, G., Meibom, A., Mennella, V., Mostefaoui, S., Nittler, L.R., Palumbo, M.E., Papanastassiou, D.A., Robert, F., Rotundi, A., Snead, C.J., Spencer, M.K., Stadermann, F.J., Steele, A., Stephan, T., Tsou, P., Tyliszczak, T., Westphal, A.J., Wirick, S., Wopenka, B., Yabuta, H., Zare, R.N., and Zolensky, M.E. (2006) Organics captured from comet 81P/Wild 2 by the Stardust spacecraft. *Science*, **314**, 1720–1724.

59 (a) Mumma, M.J., DiSanti, M.A., Magee-Sauer, K., Bonev, B.P., Villanueva, G.L., Kawakita, H., Dello Russo, N., Gibb, E.L., Blake, G.A., Lyke, J.E., Campbell, R.D., Aycock, J., Conrad, A., and Hill, G.M. (2005) Parent volatiles in comet 9P/Temple 1 before and after impact. *Science*, **310**, 270–274.; (b) Lisse, C.M., VanCleve, J., Adams, A.C., A'Hearn, M.F., Fernández, Y.R., Farnham, T.L., Armus, L., Grillmair, C.J., Ingalls, J., Belton, M.J.S., Groussin, O., McFadden, L.A., Meech, K.J., Schultz, P.H., Clark, B.C., Feaga, L.M., and Sunshine, J.M. (2006) Spitzer spectral observations of the deep impact ejecta. *Science*, **313**, 635–640.

60 Zolensky, M.E., Zega, T.J., Yano, H., Wirick, S., Westphal, A.J., Weisberg, M.K., Weber, I., Warren, J.L., Velbel, M.A., Tsuchiyama, A., Tsou, P., Toppani, A., Tomioka, N., Tomeoka, K., Teslich, N., Taheri, M., Susini, J., Stroud, R., Stephan, T., Stadermann, F.J., Snead, C.J., Simon, S.B., Simionovici, A., See, T.H., Robert, F., Rietmeijer, F.J.M., Rao, W., Perronnet, M.C., Papanastassiou, D.A., Okudaira, K., Ohsumi, K., Ohnishi, I., Nakamura-Messenger, K., Nakamura, T., Mostefaoui, S., Mikouchi, T., Meibom, A., Matrajt, G., Marcus, M.A., Leroux, H., Lemelle, L., Le, L., Lanzirotti, A., Langenhorst, F., Krot, A.N., Keller, L.P., Kearsley, A.T., Joswiak, D., Jacob, D., Ishii, H., Harvey, R., Hagiya, K., Grossman, L., Grossman, J.N., Graham, G.A., Gounelle, M., Gillet, P., Genge, M.J., Flynn, G., Ferroir, T., Fallon, S., Ebel, D.S., Dai, Z.R., Cordier, P., Clark, B., Chi, M., Butterworth, A.L., Brownlee, D.E., Bridges, J.C., Brennan, S., Brearley, A., Bradley, J.P., Bleuet, P., Bland, P.A., and Bastien, R. (2006) Mineralogy and petrology of comet 81P/Wild 2 nucleus samples. *Science*, **314**, 1735–1739.

61 (a) Ábrahám, P., Juhász, A., Dullemond, C.P., Kóspál, Á., van Boekel, R., Bouwman, J., Henning, T., Moór, A., Mosoni, L., Sicilia-Aguilar, A., and Sipos, N. (2009) Episodic formation of cometary material in the outburst of a young sunlike star. *Nature*, **459**, 224–226.; (b) Li, A. (2009) Cosmic crystals caught in the act. *Nature*, **459**, 173–176.

62 Gladman, B. (2005) The Kuiper belt and the Solar System's comet disk. *Science*, **307**, 71–75.

63 Lellouch, E., Sicardy, B., de Bergh, C., Käufl, H.-U., Kassi, S., and Campargue, A. (2009) Pluto's lower atmosphere structure and methane abundance from high-resolution spectroscopy and stellar occultations. *Astrochem. Astrophys.*, **459**, L17–L21.

64 (a) Trujillo, C.A., Brown, M.E., Barkume, K.M., Schaller, E.L., and Rabinowitz, D.L. (2007) The surface of 2003 EL_{61} in the near-infrared. *Astrophys. J.*, **655**, 1172–1178.; (b) Barucci, M.A., Cruikshank, D.P., Dotto, E., Merlin, F., Poulet, F., Dalle Ore, C., Fornasier, S., and de Bergh, C. (2005) Is Sedna another Triton? *Astron. Astrophys.*, **439**, L1–L4.

65 (a) Silvera, I.F., and Deemyad, S. (2009) Pathways to metallic hydrogen. *Low. Temp. Phys.*, **35**, 318- 325.; (b) Strobel, T.A., Somayazulu, M., and Hemley, R.J. (2009) Novel pressure-induced interactions in silane-hydrogen. *Phys. Rev. Lett.*, **103**, 065701-1–065701-4.

66 Encrenaz, T. (2005) The outer planets and their moons. *Space Sci. Rev.*, **116**, 99–120.

67 Kunde, V.G., Flasar, F.M., Jennings, D.E., Bézard, B., Strobel, D.F., Conrath, B.J., Nixon, C.A., Bjoraker, G.L., Romani, P.N., Achterberg, R.K., Simon-Miller, A.A., Irwin, P., Brasunas, J.C., Pearl, J.C., Smith, M.D., Orton, G.S., Gierasch, P.J., Spilker, L.J., Carlson, R.C., Mamoutkine, A.A., Calcutt, S.B., Read, P.L., Taylor, F.W., Fouchet, T., Parrish, P., Barucci, A., Courtin, R., Coustenis, A., Gautier, D., Lellouch, E., Marten, A., Prangé, R., Biraud, Y., Ferrari, C., Owen, T.C., Abbas, M.M., Samuelson, R.E., Raulin, F., Ade, P., Césarsky, C.J., Grossman, K.U., and Coradini, A. (2004) Jupiter's atmospheric composition from the Cassini thermal infrared spectroscopy experiment. *Science*, **305**, 1582–1586.

68 Barr, A.C., and Canup, R.M. (2010) Origin of the Ganymede-Callisto dichotomy by impacts during the late heavy bombardment. *Nat. Geosci.*, **3**, 164–167.

69 Carlson, R.W., Kargel, J.S., Douté, S., Soderblom, L.A., and Dalton, J. (2007) Io's surface composition, in *Io after Galileo – A New View of Jupiter's Volcanic Moon* (eds O.M.C. Lopes and J.R. Specer), Springer (Berlin), Praxis Publ., Chichester.

70 Steudel, R., and Steudel, Y. (2004) The thermal decomposition of S_2O forming SO_2, S_3, S_4 and S_5O – an *ab initio* MO study. *Eur. J. Inorg. Chem.*, 3513–3521.

71 (a) Loeffler, M.J., Raut, U., Vidal, R.A., Baragiola, R.A., and Carlson, R.W. (2006) Synthesis of hydrogen peroxide in water ice by ion radiation. *Icarus*, **180**, 265–273.; (b) Pan, X., Bass, A.D., Jay-Gerin, J.-P., and Sanche, L. (2004) A mechanism for the production of hydrogen peroxide and the hydroperoxyl radical on icy satellites by low-energy electrons. *Icarus*, **172**, 521–525.

72 Waite, T.H., Jr., Lewis, W.S., Magee, B.A., Lunine, J.I., McKinnon, W.B., Glein, C.R., Mousis, O., Young, D.T., Brockwell, T., Westlake, J., Nguyen, M.-J., Teolis, B.D., Niemann, H.B., McNutt, R.L., Jr., Perry, M., and Ip, W.-H. (2009) Liquid water on Enceladus from observations of ammonia and ^{40}Ar in the plume. *Nature*, **460**, 487–490.

73 Kieffer, S.W., Lu, X., McFarquhar, G., and Wohletz, K.H. (2009) A redetermination of the ice/vapour ratio of Enceladus' plumes: implications for sublimation and the lack of a liquid water reservoir. *Icarus*, **201**, 238–241.

74 Postberg, F., Kempf, S., Schmidt, J., Brilliantov, A., Abel, B., Buck, U., and Srama, R. (2009) Sodium salts in E-ring ice grains from an ocean below the surface of Enceladus. *Nature*, **459**, 1098–1101.

75 Mousis, O., Lunine, J.I., Pasek, M., Cordier, D., Waite, J.H., Mandt, K.E., Lewis, W.S., and Nguye, M.-J. (2009) A primordial origin of the atmospheric methane of Saturn's moon Titan. *Icarus*, **204**, 749–751.

76 Waite, J.H., Jr., Young, D.T., Cravens, T.A., Coates, A.J., Crary, F.J., Magee, B., and Westlake, J. (2007) The process of tholin formation in Titan's upper atmosphere. *Science*, **316**, 870–875.

77 Iess, L., Rappaport, N.J., Jacobson, R.A., Racioppa, P., Stevenson, D.J., Tortora, P., Armstrong, J.W., and Asmar, S.W. (2010) Gravity Field, Shape, and Moment of Inertia of Titan. *Science*, **327**, 1367–1369.

6
Exoplanets

This chapter will deal with exoplanets, and the relationship between a specific category of exoplanets (the super-Jupiters) on the one hand and dwarf stars (M dwarfs and brown dwarfs) on the other hand.

There is no apparent reason why our Solar System should be the only one of its kind in the Universe. Consequently, astronomers have been searching for planets in other star systems, and to date (May 2010), about 390 such systems with 460 companion objects classified as planets have been traced. An updated list can be found at http://exoplanet.eu/catalog.php. New exoplanets are announced with increasing frequency.

The first of these discoveries, in 1995, is usually attributed to an about Jupiter-mass companion, 51 Peg-b in the constellation of Pegasus [1]. 51 Pegasi is a class G2 star at a distance of 14.7 pc, less than half as old as the Sun and hence with a higher metallicity (for the definition of "metallicity" vide infra). Factually, the first discovery has been reported in 1989 for the object HD 114762b[1] [2], one of the two companions of HD 114762, which is an old (11.8×10^9 a) class F9 star at a distance of 39.5 pc in the constellation of Coma Berenices. HD 114762b has 11 Jupiter masses; for objects of this mass, the correct classification–giant planet vs. brown dwarf–cannot always be carried forth straightforwardly.

Brown dwarfs,[2] also defined as L and T class stars (see the Hertzsprung–Russel diagram, Figure 3.2 in Section 3.1.1), all have about the same radius as Jupiter; they are, however, sufficiently more massive. According to the current working definition, the limiting mass for a brown dwarf is defined as the mass where thermonuclear fusion of deuterium can occur; Eqs. (6.1) and (6.2):

$$^2_1H + {}^2_1H \rightarrow {}^3_1H + {}^1_1H \quad (E = 4.03 \text{ MeV}) \quad (6.1a)$$

$$^2_1H + {}^2_1H \rightarrow {}^3_2He + {}^1_0n \quad (E = 3.27 \text{ MeV}) \quad (6.1b)$$

---

1) HD refers to the *Henry Draper* classification of stars; the lower case letter "b" denotes the star's first discovered companion. All successive companions are labeled according to the alphabet.
2) Brown dwarfs are not intrinsically "brown," a color which is produced by mixing (and thus desaturating) black with red, orange, or yellow. Brown dwarfs are, however, emissive bodies. "Infrared dwarfs" have been proposed as an alternative term by K. Brecher, astrophysicist at Boston University.

Chemistry in Space: From Interstellar Matter to the Origin of Life. Dieter Rehder
© 2010 WILEY-VCH Verlag GmbH & Co. KGaA, Weinheim
ISBN: 978-3-527-32689-1

$$^2_1H + ^3_1H \rightarrow ^4_2He + ^1_0n \quad (E = 17.6 \text{ MeV}) \tag{6.2}$$

This limits the *low* end of the mass range to 13 Jupiter masses, m_J, for objects with the metallicity of the Sun. "Metallicity," commonly expressed as the ratio [Fe/H] relative to that of the Sun, where iron (Fe) stands for all "metals" (elements beyond helium), is defined by the logarithmic expression Eq. (6.3). For the Sun, with an actual metal content of 1.6% by mass, [Fe/H] thus is zero. Younger stars have higher (positive), and older stars lower (negative) [Fe/H] ratios. The *upper* mass limit for brown dwarfs is ca. 80 m_J (0.08 $m_\odot$), which in turn is the lower mass limit for common hydrogen fusion: $4^1H \rightarrow\rightarrow {}^4He + 2\beta^+ + 2\nu$ (Section 3.1.3):

$$[\text{Fe/H}] = \log_{10}(n_{Fe}/n_H)_{\text{star}} - \log_{10}(n_{Fe}/n_H)_{\text{Sun}} \tag{6.3}$$
(n = number of atoms per unit volume [number density])

Brown dwarfs, which are lacking the TiO and VO band typical of M dwarfs (see also further down), are presently subdivided into two categories, the L dwarfs and the T dwarfs. The L class subgroup shows characteristic bands due to metal hydrides (such as FeH, CrH, MgH, and CaH), alkali metal atoms (Na, K, Rb, Cs), and H_2O and CO. T dwarfs are devoid of the FeH and CrH bands; they have prominent features originating from K and Na, and strong absorption bands from CH_4, either exclusively or in addition with H_2O and CO [3]. These are characteristics which they share with several exoplanets, for example, HD 189733b (Table 6.1) at

Table 6.1 Properties of selected exoplanets (compared to Jupiter, in italics) and their central stars (data retrieved from http://exoplanet.eu/catalog.php, from Ref. [4a] for HR 8799b, [4b] for GJ 1214b, and [4c] for WASP 12b. The notations m_J and r_J refer to the mass and mean radius of Jupiter, $m_\oplus$ and $r_\oplus$ to the mass and radius of Earth (provided for the super-Earths only), and $m_\odot$ to the mass of the Sun).

Designation and constellation	m_J ($m_\oplus$)	r_J ($r_\oplus$)	Semimajor axis (AU) [Eccentricity]	Orbital period (d)	Spectral type[a]	$m_\odot$	Metallicity[b]
Jupiter	*1[c]*	*1[d]*	*5.20 [0.0484]*	*4.3 × 10³*	*G2*	*1*	*0*
(1) Hot Jupiters							
WASP 12b Auriga	1.41	1.79	0.023 [0.004]	1.09	G0	1.33	0.3
OGLE-TR 113b Carina	1.32	1.09	0.023 [0]	1.43	K	0.79	0.14
CoRoT 1b Monoceros	1.03	1.49	0.025 [0]	1.51	G0	0.95	−0.3
HD 189733b Vulpecula	1.13	1.14	0.03 [0]	2.22	K2	0.8	−0.03
HD 209458b Pegasus	0.69	1.32	0.047 [0.07]	3.52	G0	1.01	0.04
(2) Super-Jupiters							
CoRoT 3b Aquila	21.7	1.01	0.057 [0]	4.26	F3	1.37	−0.02
HD 114762b Coma Berenices	11	?	0.36 [0.34]	83.9	F9	0.84	−0.71

Table 6.1 Continued

Designation and constellation	m_1 ($m_⊕$)	r_1 ($r_⊕$)	Semimajor axis (AU) [Eccentricity]	Orbital period (d)	Spectral type[a]	$m_⊙$	Metalicity[b]
HR 8799d[e] Pegasus	~10	~1.2	~24 [<0.4]	~365 × 10²	A5	1.47	−0.47
(3) Mass < 10 Earths: Super-Earths							
GJ 1214b Ophiuchus	0.018 (6.6)	0.242 (2.68)	0.014 [<0.27]	1.58	M4.5	0.16	–
CoRoT 7b Monoceros	0.015 (4.8)	0.15 (1.7)	0.017 [0]	0.85	K0	0.93	0.03
Gl 876d Aquarius	0.018 (6.6)	?	0.021 [0.139]	1.94	M4	0.32	−0.12
Gl 581d Libra	0.0223 (7.1)	?	0.22 [0.38]	66.8	M3	0.31	−0.33
Gl 581e Libra	0.0061 (1.9)	?	0.03 [0]	3.14	M3	0.31	−0.33
(4) Exoplanets comparable to the giant planets in the Solar System							
HD 142b Phoenix	1	?	0.98 [0.37]	337	G1	1.1	0.04
HD 147513b Scorpius	1	?	1.26 [0.52]	540	G3/G5	0.92	−0.03
HD 47186c Canis major	0.351	?	2.395 [0.04]	1353.6	G5	0.99	0.23
47 Uma-c Ursa major	0.46	?	3.39 [0.22]	2190	G0	1.03	0

a) Cf. Section 3.1.1.
b) As defined by Eq. (6.3).
c) 317.6 Earths, 10^{-3} Suns.
d) 9.94 Earths.
e) See also Figure 6.1.

a distance of 19.3 pc, an exoplanet of the size and mass of Jupiter circling its central star in the constellation of Vulpecula. Detection of these molecular species in a planet's atmosphere greatly predominated by H_2 is achieved by observation in the near infrared (typically 1.5–2.5 μm, 6670–4000 cm^{-1}) when the object occults part of the stellar disk, allowing a fraction of the star's light to traverse through the planet's atmosphere. The fractional ratios f of gaseous water and methane thus determined in the atmosphere of HD 189733b amount to 5×10^{-4} (H_2O) and 5×10^{-5} (CH_4) [3a]. CH_4, CO_2, CO, and H_2O have recently been detected on HD 189733b [3c] and in the dayside atmosphere of another "hot Jupiter," HD 209458b [3d], at a distance of 47 pc in the constellation of Pegasus.

Table 6.1 contains a selection of exoplanets and their properties, categorized according to the following criteria:

1) "Hot Jupiters": Planets of about the size of Jupiter but on orbits very close (<0.03 AU) to the central star;

2) "Super-Jupiters": Exoplanets exceeding Jupiter in mass by about the tenfold and more, and hence objects which resemble brown dwarfs rather than planets;

3) Small planets with masses < 10 Earths (super-Earths);

4) Planets with masses about those of Saturn or Jupiter, with their central star resembling our Sun, and orbital diameters comparable to those in the Solar System.

The major fraction of the exoplanets is about the size and mass of Jupiter. Contrasting Jupiter, however, they orbit their stars at distances typically at 0.02–0.03 AU (Jupiter: 5.20 AU), which is still critically closer than Mercury (0.387 AU), and certainly provides quite inhospitable conditions. It has become common, therefore, to refer to these objects as "hot Jupiters." More generally, semimajor axes of the exoplanets' orbits are between 0.02 and 4 AU, but can go up to 670 AU and thus well beyond the stretch of the Kuiper belt and even the Scattered Disk in our Solar System. As a consequence of the surprisingly close contact between the exoplanet and its central star, orbital periods are very short, around 1 day (Jupiter: 4.3×10^3 days), but again there are also objects with orbital periods in the order of magnitude of Jupiter's and more, up to 3.2×10^5 days. Eccentricities typically are low, but can go up to 0.93, comparable to highly eccentric comets and the dwarf planet Sedna in our system. The eccentricities of the orbits of the planets listed in Table 6.1 are zero or close to zero in many cases. The eccentricity of the super-Earth Gliese 581d (0.38) compares to the eccentricity 0.249 for the orbit of the dwarf planet Pluto and 0.206 for the planet Mercury.

The notations for the stars listed in Table 6.1 follow indexing common in astronomy:

HD   Henry Draper catalog for star designation;
HR   Harvard Revised number (classification of bright stars);
Gl and GJ   Gliese and (more recent version) Gliese+Jahreiss catalog of near-by stars (up to 25 pc);
CoRoT   Convection Rotation and Planetary Transits, a space mission led by the French Space Agency in cooperation with the ESA (European Space Agency);
WASP   Wide Area Search for Planets, the UK's extrasolar planet detection program;
OGLE   Optical Gravitational Lensing Experiment, an astronomical project of the Warsaw University.

A large percentage of the planet-like objects cataloged so far are "hot Jupiters." They are hot, because they are heated by the stellar irradiance of the close-by sun. But a few considerably smaller and hence Earth-like exoplanets, category (3), referred to as "super-Earths," have also been detected, such as CoRoT 7b with ca. 4.8 $m_\oplus$, Gliese 581e with 1.9 $m_\oplus$, and GJ 1214b with 6.6 $m_\oplus$. Gl 581e and GJ 1214b orbit their rather small sun, red dwarfs (spectral type M3 and M4.5) of about 1/3 and 1/6 the mass of our Sun, at a distance still clearly less than Mercury. GJ 1214b [4b] has a mean density, derived from the mass–radius ratio, of $1.9\,\mathrm{g\,cm^{-3}}$, that is, about 50% of its overall mass should be water, which compares to only 0.06% for Earth. The remaining mass supposedly is represented by a silicate mantle and an Fe–Ni core. The planet's proximity to its sun gives rise to an equilibrium surface temperature between 120 and 280 °C, depending on what albedo is assumed. In any case, the planet's hydrogen–helium envelope, should be rather steamy. The exoplanet CoRoT 7b, which circles a sun with characteristics resembling our Sun, has a density of $5.6\,\mathrm{g\,cm^{-3}}$ and thus compares, in this respect, to Earth ($\rho = 5.515\,\mathrm{g\,cm^{-3}}$).

Another companion of Gliese 581, the super-Earth Gliese 581d, orbits its sun with a semimajor axis of 0.22 AU (the orbital eccentricity is 0.38). This is still within Mercury's orbit. However, Gliese 581d is considerably less "powered" by its much cooler M3 sun (ca. 3500 K) than Mercury is by our G2 Sun (5780 K), and this may provide conditions for some primitive forms of life to exist on Gliese 581d [5].

The likely presence of a steamy ocean covering the surface of GJ 1214b has flared up speculations on potential habitats in extrasolar systems. Habitable zones can be defined as regions around the central star, where liquid water can exist for a sufficiently long period of time to enable biological evolution. As an additional parameter for habitability, sufficiently high planetary atmospheric CO_2 concentrations to sustain photosynthesis may be added – if we restrict "habitability" to life as it exists on Earth (Chapter 7). The atmospheric composition of the super-Earths presently is not available. The atmospheres of the hot Jupiters is dominated by H_2, He, H_2O, and $CO/CO_2/CH_4$, plus minor amounts of other constituents (vide infra) [6]. CO_2 can be removed from the atmosphere by a process termed "weathering" on Earth, that is, the formation of carbonates (of Ca and Mg) from silicates, Eq. (6.4a). But CO_2 can also be redelivered by thermal decomposition of carbonates, Eq. (6.4b). This is a likely process to occur on the hot Jupiters, with atmospheric temperatures of 1000 K and more. Since many if not all these giant planets are tidally locked to their parent star, and hence have rotation and orbital motion synchronized, they dispose of a permanent hot dayside and somewhat cooler nightside, with attenuation of temperature differences coming about by violent atmospheric circulation:

$$MSiO_3 + CO_2 \rightarrow MCO_3 + SiO_2 \;(M = Mg, Ca) \tag{6.4a}$$

$$MCO_3 \rightarrow CO_2 + MO\,(+SiO_2 \rightarrow MSiO_3) \tag{6.4b}$$

Also an interesting subgroup of exoplanets, interesting in the context of versatile chemistry on the molecular level and hence the prospect for Life, encompasses the category (4) planets, as well as the super-Jupiter HR 8799d in the planetary system of the star HR 8799 [4a] (category (2) in Table 6.1). This rather hot A5 star (surface temperature 7430 K), also designated V342 Pegasi, has three companions of seven to ten Jupiter masses, that is, below the mass limit for brown dwarfs,

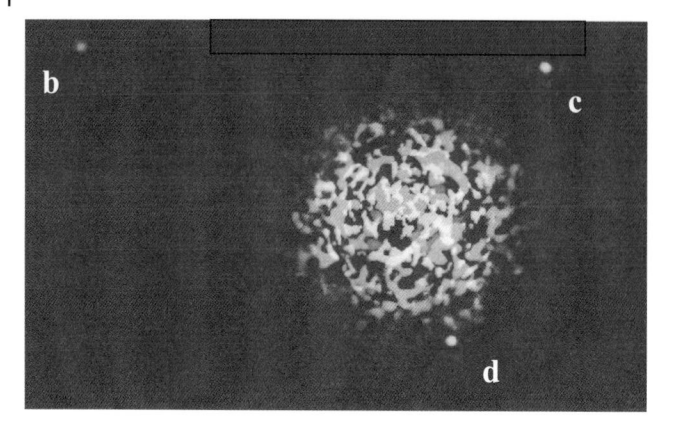

Figure 6.1 Infrared image taken July/ September 2008 from the spectral type A5 star HR 8799 (large blob; in the constellation of Pegasus) and its three giant-Jupiter companions at distances of 24 (d), 38 (c), and 68 AU (b). For data of HR 8799d, see also Table 6.1. © Ref. [4a] Figure 1. Reprinted with permission from AAAS.

orbiting the central star at distances of 24, 38, and 68 AU; Figure 6.1. In addition, there is a dusty debris disk at about 75 AU. These extrasolar planets are in orbits, which compare to those of Uranus (19.2), Neptune (30.1 AU), and the Kuiper belt objects (~35–100 AU), but they receive considerably more energy from their central star than Uranus and Neptune receive from the less hot Sun: HR 8799d, at a distance of 24 AU, is in a similar temperature range as Jupiter at 5.2 AU, a fact which may foster chemistry toward assembling parts for Life. It should be kept in mind, however, that HR 8799, and consequently the complete system, is rather young, viz. ca. 60×10^6 years. Development of Life in the Solar System, which is roughly a hundred times older (4.5×10^9 years) than HR 8799, began approximately 3.6–4 billion years ago. Hence, we can hardly expect Life yet to have developed on these exoplanets orbiting young suns.

Another recent example for a Jupiter-sized exoplanet, discovered by visual spectroscopy in this case [7], is Formalhaut-b. Formalhaut (HD 216956), a very bright star in the constellation of Piscis Austrinus approximately 7.7 pc from Earth, is also surrounded by a dusty disk extending between 133 and 158 AU. The planet Formalhaut-b orbits its central star (spectral class A3, surface temperature 8750 K, age 0.2 billion years) at a distance of 115 AU, which is somewhat beyond the aphelion of the dwarf planet Eris (97.6 AU) in the Solar System, but again in a temperature region corresponding to that of Neptune. The orbital period of this exoplanet is 872 Earth years.

Hot Jupiters have recently been categorized into two subgroups, designated pM and pL [8], where "p" refers to planet, and M and L to the classification scheme of stars. M and L stars are dwarfs and subdwarfs (brown dwarfs; vide supra) at the low luminosity (and temperature) end of the Hertzsprung–Russel diagram. The hotter M stars contain gaseous TiO and VO in their atmospheres, along with TiO_2,

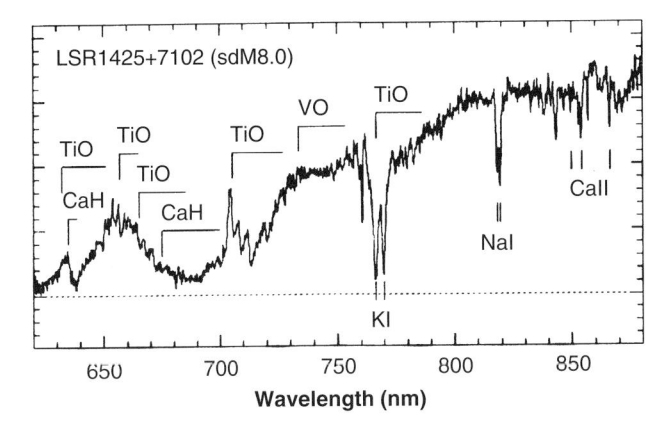

Figure 6.2 Optical to near infrared spectrum of LSR 1425+7102, a spectral class M8 subdwarf (sd). Exoplanets belonging to the pM-type category are expected to have comparable atmospheric composition. KI and NaI represent neutral sodium and potassium atoms, CaII singly charged calcium ions, Ca^+. Reproduced from Ref. [9] by permission of the AAS.

CaO, atomic Na and K, Ca^+ ions, and hydrides such as CaH, MgH, FeH, and CrH. Typical spectral features for an M subdwarf are shown in Figure 6.2. The atmospheres of category pM exoplanets display corresponding patterns. TiO_2 and CaO form by water oxidation of TiO and Ca, respectively, as shown by Eqs. (6.5) and (6.6). Titanium and vanadium oxides, including mixed oxides of composition $(Ti,V)O_2$ and $(V,Fe)O(OH)$ (the Terrestrial mineral of this composition is montroseite) are responsible for the high (optical) opacities of these planets: they absorb the greater part of the incoming light, giving rise to low albedos:

$$TiO + H_2O \rightarrow TiO_2 + H_2 \tag{6.5}$$

$$Ca + H_2O \rightarrow CaO + H_2 \tag{6.6}$$

On the other hand, in pL-type exoplanets, which receive less stellar flux and consequently are somewhat cooler, Ti(O) and VO have condensed out; the main absorbers here are water and carbon monoxide. In case of Solar metallicities, primary condensation products for titanium are titanium nitride TiN (at pressures >30 bar) and calcium titanates such as $Ca_3Ti_2O_7$ and $Ca_4Ti_3O_8$. The final product of this transformation is perowskite $CaTiO_3$ [8]; Eq. (6.7) summarizes the generation of perowskite. Possible condensation products for vanadium are vanadium pentoxide V_2O_5, Eq. (6.8), and rossite, CaV_2O_6, Eq. (6.9):

$$CaO + TiO_2 \rightarrow CaTiO_3 \tag{6.7}$$

$$VO + H_2O \rightarrow VO_2 + H_2, \; 2VO_2 + H_2O \rightarrow V_2O_5 + H_2 \tag{6.8}$$

$$CaO + V_2O_5 \rightarrow CaV_2O_6 \tag{6.9}$$

The pL-planets typically have stratospheric effective daytime temperatures of 1000 K, with the nightside temperature dropping by ca. 250 K. The resulting temperature on the nightside hence comes close to the temperature in the Venusian troposphere (740 K). Dayside temperatures of pM-planets can go up to 2500 K, dropping by around 500 K on the nightside. For pL-planets, the major atmospheric carbon-, oxygen-, and nitrogen-bearing species on the 1 bar (10^5 Pa) level are CH_4, CO, CO_2, N_2, and NH_3; minor species include CH_3, C_2H_4, C_2H_6, CH_2O (formaldehyde), CH_3OH (methanol), HCN, HNC, CH_3NH_2 (methylamine), and HNCO (isocyanic acid). At the 1 kbar (10^2 MPa) level, H_2O becomes the main constituent.

After H, C, N, and O, sulfur and phosphorus are the next most abundant volatiles, and give rise to interesting gas-phase chemistry. This chemistry has been modeled for late type-M dwarfs and L subdwarfs (brown dwarfs) [10] and should also be relevant for hot Jupiters and super-Jupiters. The dominant sulfur-bearing gas in the approximate temperature range 2300–1300 K is H_2S, which is replaced, at high temperatures and low pressures by HS, which further dissociates to form monoatomic sulfur, Eq. (6.10). In the presence of SiO, silicium sulfide can form, Eq. (6.11), provided that silicon is not removed from the atmosphere. The condensation of magnesium silicates (Mg_2SiO_4, $MgSiO_3$) effectively removes both Si and SiO. Sulfur is removed at temperatures of about 1000 K by condensation into Na_2S (Eq. (6.12a)), NH_4SH (Eq. (6.12b)), MnS, and ZnS. Ammonium hydrogensulfide NH_4SH is also a prominent constituent in the troposphere of "out" Jupiter (Section 5.5.1):

$$H_2S \rightarrow HS + \frac{1}{2}H_2; HS \rightarrow S + \frac{1}{2}H_2 \tag{6.10}$$

$$SiO + H_2S \rightarrow SiS + H_2O \tag{6.11}$$

$$2Na + H_2S \rightarrow Na_2S + H_2 \tag{6.12a}$$

$$NH_3 + H_2S \rightarrow NH_4SH \text{ (solid)} \tag{6.12b}$$

The dominant phosphorus species at $T < 1000$ K and pressures below 1 bar is tetraphosphorus hexoxide P_4O_6. As the temperature increases, P_4O_6 is replaced by phosphane PH_3 at pressures > 1 bar, and by diphosphorus P_2 at pressures < 1 bar, Eqs. (6.13) and (6.14). In L-dwarf atmospheres, P_2 and PH_3 have comparable abundances; they can interconvert, as formulated in Eq. (6.15). At temperatures around 2000 K and pressures approaching 10–100 bar (1–10 MPa), hence at deeper atmospheric levels, disintegration of phosphane to PH_2 and hydrogen occurs, Eq. (6.16), and PH_2 becomes the predominant species. As the pressure decreases, that is, as PH_2 is transported into higher atmospheric levels by convection, elemental phosphorus forms either in molecular form (P_2) at $T \approx 1500$ K, or in the form of atomic phosphorus at $T > 2000$ K, Eq. (6.17). Further, PS and PO can form. Example reactions are provided by Eqs. (6.18) and (6.19):

$$P_4O_6 + 12H_2 \rightarrow 4PH_3 + 6H_2O \tag{6.13}$$

$$P_4O_6 + 6H_2 \rightarrow 2P_2 + 6H_2O \tag{6.14}$$

$$2PH_3 \rightleftarrows P_2 + 3H_2 \tag{6.15}$$

$$PH_3 \to PH_2 + \frac{1}{2}H_2 \tag{6.16}$$

$$nPH_2 \to P_n + 1/n\,H_2 \; (n = 1, 2) \tag{6.17}$$

$$P_2 + 2H_2S \to 2PS + 2H_2 \tag{6.18}$$

$$PH_2 + H_2O \to PO + 2H_2 \tag{6.19}$$

The chemistry of exoplanets has still to be revealed directly. So far, extrapolation from the atmospheric chemistry in dwarf and subdwarf stars, low-mass brown dwarfs in particular, and the well-explored chemistry in the atmospheres of Jupiter and Saturn provide reliable scenarios for what can be expected to go on in the atmospheres of hot Jupiters and of super-Jupiters. To reveal the chemistry in extrasolar planets converging toward the size and the mass of Earth, and to an energy flux received by the central star comparable to the energy flux by our Sun, will be a more tedious task – as long as we do not have a clue as to what the proper nature of these planets is. Objects such as Gl 581d, GJ 1214b, and CoRoT 7b (Table 6.1) might be promising candidates for planetary conditions approaching those of our home planet. "Promising conditions" include the presence of water, or $H_2O + CO_2/CH_4$: microbial life at the interface of liquid H_2O/liquid CO_2–CH_4 has been shown to exist in deep sea troughs on Earth; Section 7.3.

Summary

Exoplanets are companions of stars which exhibit properties reminiscent of the planets in the Solar System. They can be categorized into four subgroups: (i) hot Jupiters, which are planets with masses about that of Jupiter, but orbit their sun at the very close distance of typically 0.03 AU; (ii) super-Jupiters, which have masses clearly above Jupiter; (iii) super-Earths with masses corresponding to 2–7 Earths, orbiting comparatively cold M dwarfs; and (iv) about Jupiter-sized exoplanets with a semimajor axis of their orbit comparable to planets in the Solar System, and orbiting stars similar to our Sun. Of particular interest in the context of possible extraterrestrial life are the super-Earths. These have mean densities similar to Earth (CoRoT 7b), suggesting a matchable built-up, or similar to the ice-planets (GJ 1214b), suggesting the presence of major amounts of water.

The atmospheric composition of the super-Jupiters is related to that of M dwarfs and, even more, of brown dwarfs (L and T subdwarfs). Correspondingly, one distinguishes between pM- and pL-type exoplanets. The VO and TiO absorptions typical of M dwarfs are lacking in the colder brown dwarfs. L subdwarfs are characterized by strong absorption features of metal hydrides MH (M = Ca, Mg, Fe, Cr), alkali metal atoms, H_2O, and CO. T subdwarfs are devoid of FeH and CrH, but exhibit strong bands stemming from Na, K, and CH_4.

References

1 Mayor, M., and Queloz, D. (1995) A Jupiter-mass companion to a solar-type star. *Nature*, **378**, 355–359.

2 Latham, D.W., Mazeh, T., Stefanik, R.P., Mayor, M., and Burki, G. (1989) The unseen companion of HD 114762: a probable brown dwarf. *Nature*, **339**, 38–40.

3 (a) Swain, M.R., Vasisht, G., and Tinetti, G. (2008) The presence of methane in the atmosphere of an extrasolar planet. *Nature*, **452**, 329–331.; (b) Grillmair, C.J., Borrows, A., Charbonneau, D., Armus, L., Stauffer, J., Meadows, V., van Cleve, J., von Braun, K., and Levine, D. (2008) Strong water absorption in the dayside emission spectrum of the planet HD 189733b. *Nature*, **456**, 767–769.; (c) Swain, M.R., Deroo, P., Griffith, C.A., Tinetti, G., Thatte, A., Vasisht, G., Chen, P., Bouwman, J., Crossfield, I.J., Angerhausen, D., Afonso, C., and Henning, T. (2010) A ground-based near-infrared emission spectrum of the exoplanet HD 189733b. *Nature*, **463**, 637–639.; (d) Swain, M.R., Tinetti, G., Vasisht, G., Deroo, P., Griffith, C., Bouwman, J., Chen, P., Yung, Y., Burrows, A., Brown, L.R., Matthews, J., Rowe, J.F., Kuschnig, R., and Angerhausen, D. (2009) Water, methane and carbon dioxide present in the dayside spectrum of the exoplanet HD 209458b. *Astrophys. J.*, **704**, 1616–1621.

4 (a) Marois, C., Macintosh, B., Barman, T., Zuckerman, B., Song, I., Patience, J., Lafrenière, D., and Doyon, R. (2008) Direct imaging of multiple planets orbiting the star HR 8799. *Science*, **322**, 1348–1352.; (b) Charbonneau, D., Berta, Z.K., Irwin, J., Burke, C.J., Nutzman, P., Buchhave, L.A., Lovis, C., Bonfils, X., Latham, D.W., Udry, S., Murray-Clay, R.A., Holman, M.J., Falco, E.E., Winn, J.N., Queloz, D., Pepe, F., Mayor, M., Delfosse, X., and Forveille, T. (2009) A super-Earth transiting a nearby low-mass star. *Nature*, **462**, 891–894.; (c) Li, S.-L., Miller, N., Lin, D.N.C., and Fortney, J.J. (2010) WASP-12b as a prolate, inflated and disrupting planet from tidal dissipation. *Nature*, **463**, 1054–1056.

5 von Bloh, W., Bounama, C., Cuntz, M., and Franck, S. (2007) The habitability of super-Earths in Gliese 581. *Astron. Astrophys.*, **476**, 1365–1371.

6 Deming, D., and Seager, S. (2009) Light and shadow from distant worlds. *Nature*, **462**, 301–306.

7 Kalas, P., Graham, J.R., Chiang, E., Fitzgerald, M.P., Clampin, M., Kite, E.S., Stapelfeldt, K., Marois, C., and Krist, J. (2008) Optical images of an extrasolar panet 25 light-years from Earth. *Science*, **322**, 1345–1348.

8 Fortney, J.J., Lodders, K., Marley, M.S., and Freedman, R.S. (2008) A unified theory for the atmospheres of the hot and very hot Jupiters: Two classes of irradiated atmospheres. *Astrophys. J.*, **678**, 1419–1435.

9 Lépine, S., Shara, M.S., and Rich, M. (2003) Discovery of an ultracool subdwarf: LSR 1425+7102, the first star with spectral type sdM8.0. *Astrophys. J.*, **585**, L69–L72.

10 Visscher, C., Lodders, K., and Fegley, B., Jr. (2006) Atmospheric chemistry in giant planets, brown dwarfs, and low-mass dwarf stars. II. Sulfur and phosphorus. *Astrophys. J.*, **648**, 1181–1195.

7
The Origin of Life

7.1
What is Life?

The oldest terrestrial "rock," a micrometer-sized zircon ($ZrSiO_4$) found in an Australian sediment, has been dated back to 4.4 billion years (Ga), which compares with 4.56 Ga for that epoch in time when our planet Earth formed; cf. the timeline, Scheme 7.1. The oldest authentic microfossils are found in stromatolites. These are conical, dome-shaped formations of sedimentary rock, essentially originating from the action of photosynthetic cyanobacteria, also known as (taxonomically incorrect) blue-green algae. Contemporary stromatolite communities exist in various places, including extreme habitats such as salty, nitrate-rich lakes in the high Andes. The dating of the stromatolite *fossils*, 3.45 Ga, allows for a time span of just about 1 billion years for the development of the first simple life forms – at the best, since the so-called heavy bombardment with its sterilizing effect ended only 3.8 Ga ago, leaving just 350 million years for life to come into existence. The heavy bombardment[1] goes back to impacting asteroids and comets after their dislocation from the asteroid and Kuiper belts when the giant planets migrated into their final orbital positions. This cataclysm came about in the time span between 4.1 and 3.8 Ga.

But did life actually develop on Earth? The starting material, simple to complex organic molecules and inorganic templates, catalysts, and electron transfer agents, had been available either by primordial synthesis on the early Earth (these scenarios are addressed in Section 7.2) or carried to Earth by meteorites and interplanetary dust, as outlined in Sections 5.3.2 and 5.3.3. Alternatively, primitive life forms might have been shipped to Earth from extraterrestrial sources. Our neighbor planet Mars and Saturn's large moon Titan, the only moon in the Solar System with an appreciable atmosphere, are potential candidates. The transport of sperms of life to Earth from extraterrestrial sources, known as *panspermia*, is a favorite scheme in science fiction, but should not fall in disgrace as an alternative simply for this reason. Panspermia also encompasses the possibility of propagation of life from Earth to other celestial bodies in our Solar System – and way back to Earth.

---

1) Also termed Lunar cataclysm because possibly responsible for many of the Lunar impact craters.

Chemistry in Space: From Interstellar Matter to the Origin of Life. Dieter Rehder
© 2010 WILEY-VCH Verlag GmbH & Co. KGaA, Weinheim
ISBN: 978-3-527-32689-1

7 The Origin of Life

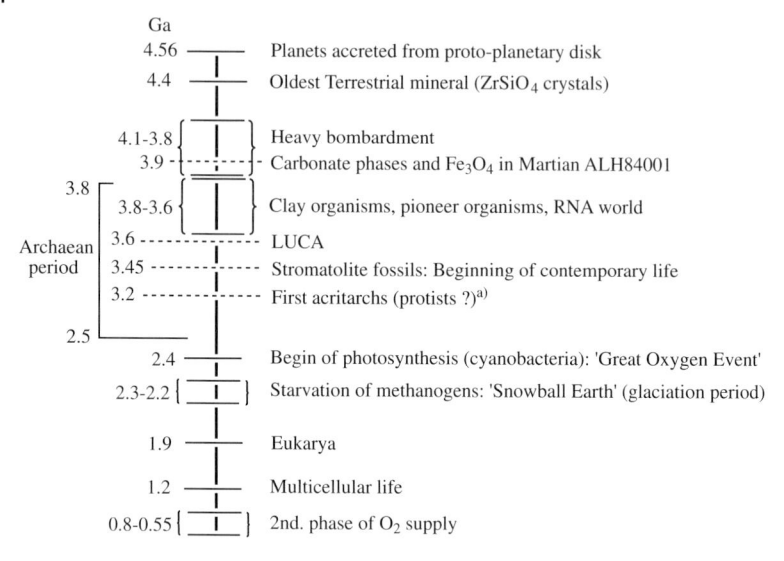

Ga	Event
4.56	Planets accreted from proto-planetary disk
4.4	Oldest Terrestrial mineral (ZrSiO$_4$ crystals)
4.1-3.8	Heavy bombardment
3.9	Carbonate phases and Fe$_3$O$_4$ in Martian ALH84001
3.8-3.6	Clay organisms, pioneer organisms, RNA world
3.6	LUCA
3.45	Stromatolite fossils: Beginning of contemporary life
3.2	First acritarchs (protists ?)[a]
2.4	Begin of photosynthesis (cyanobacteria): 'Great Oxygen Event'
2.3-2.2	Starvation of methanogens: 'Snowball Earth' (glaciation period)
1.9	Eukarya
1.2	Multicellular life
0.8-0.55	2nd. phase of O$_2$ supply

(Archaean period: 3.8 – 2.5)

Scheme 7.1 Cut-out of the timeline (not to scale) for developments on Earth. [a]See: Buick, R. (2010) *Nature*, **463**, 885–886.

Meteorites of Martian origin have been identified on our home planet in the dozens. Likewise, catastrophic touchdowns of asteroidal fragments on Earth can have catapulted terrestrial surface rocks carrying encapsulated life sperms into the cosmos, rocks that were finally captured by our companion planets and their moons to spread life out there.

Certainly, this is speculation to some extent. In any case, life *does* exist on Earth, and we have to address the question of its origin, its limitations, and alternatives. But before going into the details of how life may have come into existence and developed on our planet, "life" should be defined. This is not an easy thing to do; moreover, this may be considered a philosophical question. But even if we disregard the philosophical dimension, the answer is far from being simple, and accordingly, there is no general consensus among natural scientists as it comes to the definition of "life." For a concise and critical account on this issue cf., for example, Ref. [1]. For the time being, let us coin – and scrutinize – the following three paradigms:

1) **Metabolism** Maintenance of metabolic pathways in a chemical system, that is, the synthesis, degradation, and variation of chemical species along oxidative, reductive, and nonredox paths. Metabolism ascertains the supply with essential chemical species, the breakdown of complex molecules into smaller ones for energy release (catabolism), as well as the removal of potentially "useless" (and sometimes toxic) species. Metabolic alteration of an organism's inventory of chemical species commonly is coupled to the production and the consumption of energy. The net balance always is the consumption of energy. Hence, an external energy source is needed. Examples for energy sources are "heat," light, chemical energy released in the course of chemical reactions, and electric energy.

2) **Reproduction** The ability of self-replication, coupled to information transfer from the "parent" organism to its "offspring." In order to enable this information transfer, an information pool (in the form of RNA or DNA; see below) is necessary, which is inherited ("heritage information"), essentially unaltered, by the replicas of the parent.

3) **Evolution** The ability to adapt to changing environmental conditions and to survive in competition with other life forms. This is achieved by implementing defaults in information transfer on replication: Subtle changes in the information library, viz. an imperfect copy of the DNA (or RNA) in the process of replication, and its transfer to the offspring, can eventually turn out to be advantageous for survival. Evolution is thus closely related to imperfect self-replication.

All of these three criteria are viable for a collective of individuals, not necessarily for a single individual. And they have to apply concertedly, that is, metabolism, reproduction, and evolution as such, isolated from each other, are by no means indicators of life: (i) Metabolism also occurs in inanimate systems. Examples are oscillating reactions (such as the famous Belousov Zhabotinsky reaction[2]) or the "blue bottle experiment," the reaction between glucose and oxygen catalyzed by methylene blue: In an alkaline medium, glucose is oxidized to gluconate by methylene blue, which in turn is reduced to its colorless leuco form; the leuco form is then reoxidized by aerial oxygen to methylene-blue when air is allowed to reenter the system; and so forth. (ii) Self-replication is also common in inanimate systems. An example is oversaturated solutions of salts. If an oversaturated solution of sodium chloride is inoculated by a sodium chloride crystal, numerous identical crystals (identical in morphology, not necessarily in size) are spontaneously produced. Another example is the initiation, by a condensation nucleus, of the formation of droplets of rain or crystals of snow flakes, depending on temperature, out of oversaturated clouds of water vapor. (iii) Generally, inanimate systems do not evolve with time: A nanosized diamond crystal from a meteorite about the age of the Solar System (Figure 5.21 in Section 5.3.3) does exhibit exactly the same structural pattern as a diamond produced in an industrial enterprise, and a millimeter-sized quartz crystal from an extraterrestrial source is identical to quartz crystals from the Cambrian period[3] on Earth save, perhaps, for the isotopic ratios of the elements constituting the compound. Phenomena such as metamorphosis of rock by high pressure and temperature, or by weathering, the conversion of carbon-bearing organic matter into coal or crude oil, or the conversion of a metastable modification (such as monoclinic S_8) to a thermodynamically stable modification (orthorhombic S_8) are unidirectional processes slowly advancing as time passes, and may be compared to changes accompanied with evolutionary processes.

---

2) Based on the oscillation between the color orange for Br_2 (formed through the oxidation of bromate by malonic acid, catalyzed by cerium ions) and colorless (removal of Br_2 through the formation of bromomalonic acid).

3) The Cambrian period lasted from 542 to 488 million years ago.

With the definition of Life provided here – the ensemble of metabolism, reproduction and evolution, critical restrictions arise. Viruses, for example, which do not have their own metabolism, are thus not subsumed under life forms despite of the fact that they can self-replicate (with the help of a host) and evolve (mutate). Viruses consist of a genetic information pool in the form of a single- or double-stranded RNA (an example is the HIV virus) or DNA (such as the herpes virus) plus a protein coat (the capsid), and sometimes a fat envelope in addition. They are lacking, however, the membrane typical of all known cellular organisms on Earth: archaea, bacteria, and eukarya, the latter with a differentiated nucleus, the former (archaea and bacteria: prokarya)[4] without a cellular nucleus. Given the fact that viruses do not dispose of a cell membrane, and some viruses employ the more primitive RNA as information carrier (in pro- and eukarya, DNA is used exclusively), it is perhaps tempting to consider viruses as primitive progenitors of prokarya. Very likely, however, viruses are phylogenetically younger than bacteria and archaea. L. Likewise, a prion (proteinaceous infections particle), a polypeptide which can replicate without the involvement of nucleic acid material via conversion of normal proteins into infections ones, and which may well be a relic of an early stage of peptide evolution, cannot be considered a life form by itself.

What then has been the first life form on Earth, from which archaea and bacteria and later also eukarya developed (Scheme 7.2)? Such a preorganism, or *Last Uniform Common Ancestor* (LUCA), must have had available most, but not all, of the ingredients typical of cellular life as we know it today. These constituents of current life are (1) DNA as an information library, allowing for (2) 20 amino acids to be organized into peptides and proteins, plus (3) a pool of enzymes enabling amino acids to be transferred by specific RNA called transfer RNA (tRNA).

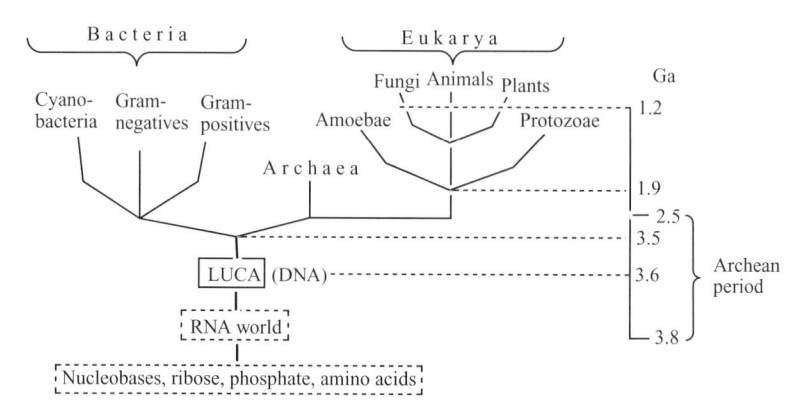

Scheme 7.2 Simplified phylogenetic tree of cellular organisms (time line not to scale), starting with last uniform common ancestor (LUCA) about 3.6 billion years ago. The "RNA world," constituting self-replicating RNA and "ribozymes" (RNA-based enzymes which catalyze the linkages to, and between, amino acids), is a potential prelife form [2]; see text. For the basic building blocks of RNA and DNA, cf. Scheme 7.4.

---

4) For eukarya and prokarya, the terms eukaryotes and prokaryotes are also in use.

These enzymes, termed tRNA synthetases, catalyze the linkage of an activated amino acid (an aminoacyl phosphate) to the tRNA which then delivers this amino acid to the peptide or protein, to which it is supposed to be attached. The DNA is organized in such a way that 64 (partially redundant) combinations of three nucleobases, so-called triplet codons, encode the 20 amino acids plus 3 "stops." The information encoded in the DNA in the form of heredity units (genes) is initially translated into a messenger RNA (mRNA), which transports the information to the ribosomes, the intracellular factories for protein synthesis. The synthesis of the mRNA is catalyzed by an RNA polymerase, supported or, rather, directed by transcription factors. In many cases, these transcription factors are "zinc fingers," proteins with loops stabilized through coordination to zinc ions, which dock into the correct sites of the DNA to "tell" the mRNA polymerase which part of the information in the DNA is to be copied and expressed in a complementary mRNA. The overall process of information transfer is sketched in Scheme 7.3.

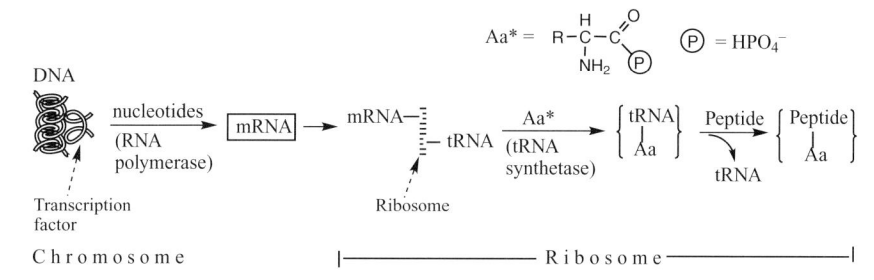

Scheme 7.3 Information *transcription* (DNA → mRNA) as well as *translation* into peptide/protein synthesis via tRNA. Aa = amino acid residue, Aa* = activated amino acid. For details of the primary structure of DNA and RNA, see Scheme 7.4.

Along with RNA processing and the synthesis of specific proteins, the gene pool also provides the work plan for the cellular energetics, the metabolic and catabolic pathways, the shape, envelope, and inner structure of the cell. This multifunctionality sets a lower limit to the minimal amount of information-bearing material, thus a minimal gene set, and hence a minimal size of the living individual. The putative "nanobacteria" in the Mars meteorite Alan Hills 84001, described in more detail in Section 5.2.4.3 (see also Chapter 1), are believed to be too small by an order of magnitude to fulfill this criterion; see, however, also Section 7.5.

RNA and DNA are made up of nucleotides. Nucleotides are ensembles of three fragments: a nucleobases, a sugar (ribose in the case of RNA and 2-desoxiribose in the case of DNA), and phosphate. In Scheme 7.4, fragments of RNA and DNA are illustrated, along with the nucleobases employed in these information pools. Three of the bases – adenine, guanine, and cytosine – are used by both DNA and RNA. The fourth base in RNA is uracil that is replaced by thymine in DNA. Thymine differs from uracil in that it contains a methyl group; thymine is thus more lipophilic. Apart from this minor discrepancy, RNA and DNA are, at first sight, almost identical. There is, however, a strikingly important difference: the

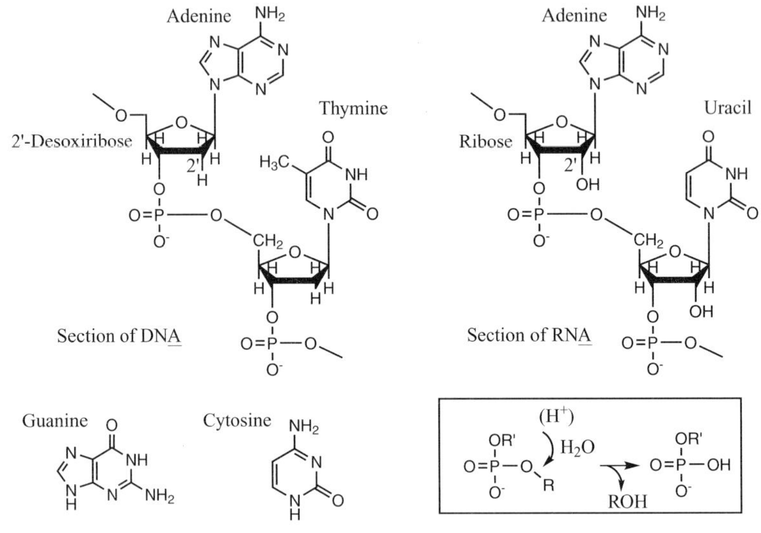

Scheme 7.4 Sections (encompassing two base pairs) of DNA and RNA, and two additional bases (guanine and cytosine) employed unanimously by DNA and RNA. The proton of the OH group of RNA in the 2′ position potentially catalyzes the hydrolysis of the phosphoester bond as shown in the inset (R and R′ here are the nucleoside fragments).

hydroxyl group in the position 2′ of the sugar constituent in RNA is missing in DNA; hence *deoxy*ribonucleic acid. This sugar-OH group can autocatalytically affect the hydrolytic cleavage of the phosphoester linkage (as depicted in the box in Scheme 7.4), and thus destabilize RNA. The DNA pool hence provides a more resistant information library than the RNA pool, and has probably also been employed by LUCA. This view is supported by taking into account the essentially nonoxic conditions in the primordial atmosphere, facilitating the reductive removal of oxygen in the 2′ position, which is otherwise accomplished in living beings by a rather complex, iron-dependent enzyme, ribonucleotide reductase. Taking all this information together, LUCA supposedly was a cellular organism not unlike today's bacteria and archaea, but more primitive: Since archaea are lacking the tRNA synthetase for the amino acids glutamine and asparagine, LUCA will have had a set of tRNA synthetases somewhat reduced compared to that in bacteria. Further, archaea, and thus also LUCA, cannot synthesize fatty acids (they synthesize isoprene-based analogs instead). Given–again–the essentially reducing conditions on early Earth, only nonoxidative pathways (fermentations) to break down glucose will have been available. Further, it is unlikely that copper-based enzymes, which are essential in all later organisms, had been employed–for the simple reason that, in a reducing atmosphere, Cu^+ is the stable oxidation state; and Cu^I salts are essentially insoluble. A comparable problem arose in a later epoch, when the atmosphere became oxidizing, and thus ferric iron Fe^{3+} became the predominant–and insoluble–source for the indispensable iron. However, at that time, life had developed to an extent where evolutionary adaption allowed survival by adjustment or by retreat into oxygen-free niches.

And what came prior to LUCA? It is believed that LUCA was preceded by an elusive RNA world (Scheme 7.2). The simple reason for this assumption roots in the fact that certain RNAs are able to autocatalyze their self-replication; plus the discovery that RNAs more generally can act as catalysts (though not particularly effective ones) also in the synthesis of proteins. In this context, RNA is referred to as ribozyme (not to be mistaken for ribosomes, the protein factories which, though, also exhibit ribozyme activity). Enzyme activity of RNA is also a central component in the functionality of mitochondria. Mitochondria are cell organelles in eukaryan cells, which are dedicated to energy production, and believed to derive from the acquisition of bacterial passengers (endosymbionts), relic bacteria that became incorporated, modified, and reduced to this specific task by the eukarya.

According to the criteria for Life defined earlier, the "RNA world" is a predecessor for life at the best. Somewhere in-between the RNA world and LUCA, the "spark of life" must have ignited a process-enabling interconnected chemical reactions within a protocellular organizational unit, and hence allowing for a metabolic network in accord with the principle of life. A certain degree of "isolation," accomplished by a "membrane" shielding off the cellular organism against external destructive factors, appears to have been (and, of course, still is) an essential requirement for individual development. But interindividual interaction by partial release of this isolation via messenger molecules providing communication is an additional indispensable requirement because cooperation between individuals (such as the establishment of colonies) largely improves the chance of survival in a basically hostile environment.

Along with DNA/RNA, providing information and information transfer, additional macromolecular materials are required to support life: Proteins, which constitute a main fraction of the (intra)cellular structure; polysaccharides, which can act as storage for energy (e.g., starch and glycogen) or provide structure elements (such as cellulose and chitin); and lipids, which again can be involved either in energy metabolism and/or in the buildup of structures, the latter in the form of building panels for cell membranes.

To this point, the question of quantity and quality of Life has been addressed, and it has been taken for granted that polymers such as RNA and proteins, arising from polycondensation, are a prerequisite to enable life forms. Certainly, there is no doubt that life, as we know it, *does* rely on these polycondensates. Nonetheless, the notion of an RNA world preceding LUCA suffers from a paradox: Liquid water, which appears to be essential to support life (and even if this were not the case, polycondensation does *produce* water), also tends to hydrolyze polymers. An additional and even more serious drawback is the extremely low probability that nucleotides condensing to form a polynucleotide will do so in such a way that the "correct" arrangement to enable functionality related to life is accomplished. Other approaches have been proposed. One of these focuses on so-called clay organisms: Several minerals – clays such as montmorillonite in particular – catalyze the undirected abiotic oligomerization of nucleotides to form oligonucleotides and amino acids to form peptides. This issue will briefly be addressed in Section 7.2, and is resumed in more detail in Section 7.4.3. For a critical discussion, see also Ref. [3]. Another approach is the conception of "pioneer organisms": inorganic–organic hybrids in a water-supported "iron–sulfur world," an intriguing alternative (though not free from other shortcomings) to be implemented in Section 7.4.1 in the context of the formation/extraterrestrial supply of basic molecules necessary for the synthesis of macromolecules (Sections 7.4/7.4.4).

Before addressing scenarios that provide basic molecules essential for the machinery of life, the possibility of life forms that do *not* rely on a carbon-based chemistry on the one hand and on water on the other hand is discussed in Section 7.2. In Section 7.3, terrestrial organisms thriving under extreme condition, so-called extremophiles, are introduced in order to delimit the scope of possible extraterrestrial life comparable to life on Earth.

7.2
Putative Non-Carbon and Nonaqueous Life Forms; the Biological Role of Silicate, Phosphate, and Water

Several hypothetical non-carbon biochemistries have been proposed as alternatives to our exclusively carbon-based world. Carbon does have a couple of properties that make it particularly suited for providing the immense variability of organic molecules.[5] Several of these specific properties are listed here:

1) As a main group 14 element, carbon has available 4 valence electrons to form bonds to other carbons, or to atoms of other elements. Depending on whether all of these bonds are single, double, or triple, a carbon atom can have four neighbors (in a tetrahedral geometry, such as in alkanes), three neighbors (in a trigonal planar geometry, as in alkenes and aromatic compounds), or two neighbors (in a linear arrangement, as in alkynes). Dicarbon C_2, formally with a quadruple bond, is present in the gas phase and has been detected in interstellar clouds. In C_2, one of the bonds is antibonding; hence the overall bond order is just (figure 4.13(a) in Section 4.2.3.2). Another uncommon species are carbenes CR_2, in which two of the electrons do not participate in bonding, forming a lone pair (singlet carbenes $|CR_2$) or remaining "isolated" (triplet carbenes $:CR_2$). Further "exotic" carbon species, detected in interstellar clouds and Titan's atmosphere, are CH_3^+ and CH_5^+.[6]

2) For the formation of double and triple bonds, π-type p-orbitals are employed. Carbon belongs to the second row of the periodic table; carbon atoms are thus rather small, and the resulting π bonding interactions are particularly strong.

---

5) The number of inorganic carbon compounds is restricted. Examples are CS, CO, CO_2, carbonates, HCN, cyanides, cyanates, cyanamides NH_2CN, and carbides such as CaC_2.

6) The number of next neighbors of a carbon atom can also exceed 5. An example is interstitial carbon: here, carbon is embedded in, for example, an octahedron or even dodecahedron of metal atoms/ions.

Such π bonding is an essential prerequisite for the formation of homo- and heteroaromatic compounds, abundant life molecules in interstellar dust and meteoritic material.

3) The electronegativities of carbon and hydrogen are almost the same (carbon is just slightly more electronegative), which minimizes polarization in the C—H bond and thus stabilizes the covalent bonding interaction and the compound on the whole.

4) The intermediate electronegativity of carbon (2.5 on the Pauling scale, which spans the range of 0.8 for Cs to 4.0 for F) also implies that bonds to "heteroatoms" such as O, N, and S, although polar, are still essentially covalent, stabilizing the compounds against disruption into solvated ions in polar solvents such as water.

Which elements can potentially substitute for carbon? There are at least two candidates: (1) silicon, also a group 14 element, in the third period and thus just underneath carbon and (2) phosphorus, also in the third period, but in group 15, that is, with one valence electron more than carbon. The choice of phosphorus roots in what is known as "diagonal relationship" in the periodic table: a certain chemical similarity between two diagonally (upper left to lower right) adjacent elements, a similarity that basically comes about by comparable atomic radii.

Silicon biochemistry is the most commonly proposed alternative. There are, however, several problems with silicon. Since silicon atoms are sufficiently larger than carbon atoms ($r = 117.6$ as compared to 77.2 pm), double and triple bonds are weak and only survive in sterically hindered systems. Aromatic silicon compounds do not exist. Silanes Si_nH_{2n+2}, the silicon analogs to alkanes C_nH_{2n+2}, are restricted in chain length, and chemically very reactive. Stable polymeric silicon compounds do exist, but these are compounds of composition $(-SiR_2-O-)_n$, so-called silicones (R represents any carbon-based residue), that is, they contain Si—O—Si units instead of —Si—Si- chains. Silicones are particularly inert and thus essentially resistant against metabolism; metabolism is a prerequisite for life. They might be employed under extreme conditions, for example, in a sulfuric acid environment (Venus' clouds) and at high temperatures. No such environments on Earth have, however, revealed silicon-based life forms.

Another problem is the particularly high affinity of silicon to oxygen. Wherever silicon compounds analogous to carbon compounds are involved in chemical reactions under oxic conditions, silicon dioxide is formed that, contrasting its carbon analog (a gas at ambient conditions), is polymeric and practically insoluble in all solvents. The same applies to silicates, the carbonate analogs, and basis for a broad variety of minerals. The presence of cations in an aqueous medium can increase the solubility of SiO_2 and silicates; seawater in fact contains about 5 ppm of silicate. Certain aquatic organisms, diatoms in particular, employ this silicium source to manufacture protective shells consisting of polymeric amorphous "silica hydrates" of the approximate composition $SiO_n(OH)_{4-2n}$, imbedded in a protein matrix. Many monocotyledonous (grass-like) plants, horsetail (*Equisetum*), and the stinging hairs of nettles (*Utrica*) employ silicas as a structure support. A minor percentage of silicate is constituent of hydroxyapatite, the inorganic contingent of the bony structures of vertebrates, where it is involved in the early stages of bone formation. These examples suggest that extraterrestrial life forms might employ silicon-based molecules in structure scaffolds. But structure alone is not life.

In conclusion, although inorganic silicon compounds are employed by life forms on our planet, and might attain a more prominent role in stabilizing structures in extraterrestrial life forms, it is very unlikely that organic silicon compounds can even partially substitute carbon-based compounds in metabolic pathways, information storage, and information transfer in life. Silicon is the second to most abundant element on Earth (next to oxygen), and here, life is certainly not based on silicon. In cosmic dimensions, silicon is sufficiently less abundant than carbon; the cosmic abundance of carbon to silicon is roughly 10:1. Of the ca. 160 molecular species detected to date (May 2010) in the interstellar medium and in envelopes of developed stars, about 140 are organic carbon-bearing molecules, and only 11 contain silicon, 6 of which (SiC, cyclo-SiC$_2$, cyclo-SiC$_3$,[7] SiC$_4$, SiCN, and SiNC) are silicon-carbon compounds; cf. also the reaction network Scheme 4.5 in Section 4.2.2. Further, except of silicon carbide SiC and silicon nitride SiN, no relevant species have been found in meteorites.

Even so, silicon may have had a part to play in the *origin* of life on Earth. Terrestrial life forms almost exclusively use homochiral molecules, levorotatory (L) amino acids, and dextrorotatory (D) carbohydrates. In the information pool, the two helical strands of double-stranded DNA are right-handed (the helices are winding up clockwise). These preferences raise the question of the origin of chiral molecules. There are several answers to this question (see also Sections 7.4.2 and 7.4.3). One possible clue to this problem is inorganic silicium compounds such as silica and clay. Silica $(SiO_2)_\infty$, the crystalline form of which occurs in the form of several modifications (quartz, tridymite, christobalite, coesite, and stishovite) is made up of edge-sharing SiO_4 tetrahedra. In quartz, one-dimensional helical strands of these SiO_4 tetrahedra are edge-connected to form a three-dimensional lattice. The helices can be left- or right-handed, providing quartz crystals with chirality. In Figure 7.1a, a single (left-handed, winding up anticlockwise) strand is shown in the side and the top view. Quartz surfaces of either D- or L-chirality may have worked as a template in the synthesis of chiral excesses of organic compounds. In addition to silica, clays such as the phyllosilicate montmorillonite (Figure 7.1b) have been shown to act as templates for the polymerization of amino acids (to form polypeptides) [4a] and nucleotides (to form RNA) [4b]. The surface of clays, when in contact with water, becomes negatively charged, facilitating catalysis of polymerization reactions. For details, see Section 7.4.3.

The second candidate for an animate non-carbon world is phosphorus, the diagonal neighbor of carbon, with a similar electronegativity: 2.1 for P versus 2.5 for C. With a radius of 110.5 pm, phosphorus atoms are slightly smaller than silicon (117.6 pm), but still clearly larger than carbon (77.2 pm). Akin problems

---

7) For the molecular structure, see Scheme 4.1 in Section 4.1.

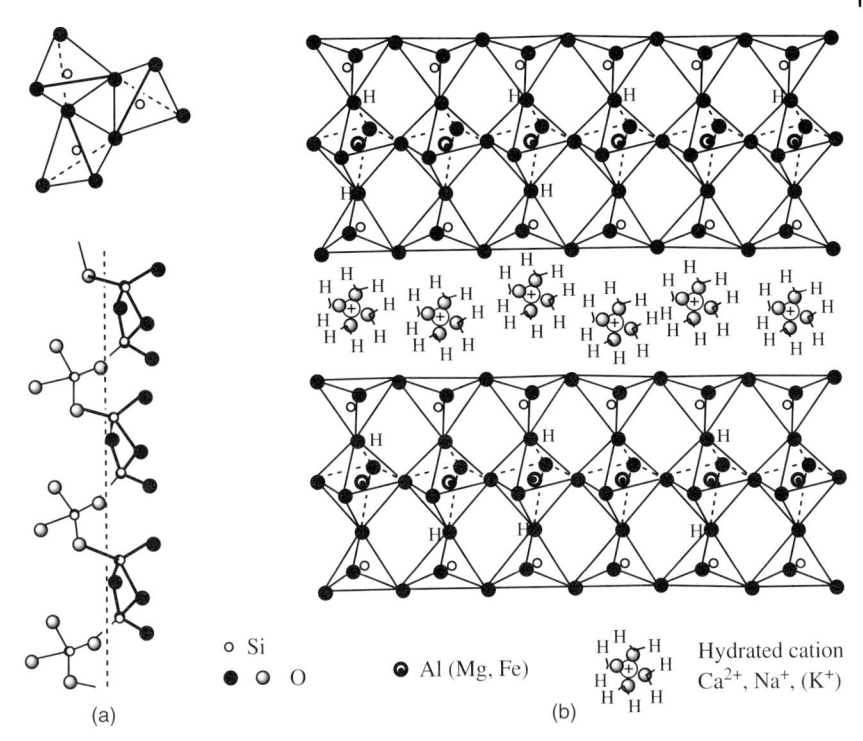

Figure 7.1 (a) Details from the structure of quartz, showing the one-dimensional arrangement of SiO₄ tetrahedra into a left-handed helix (in the top view and the side view). Blackened oxygens are above, and gray oxygens below the plane. In the quartz crystal, these helices are three-dimensionally interconnected. (b) Schematic built-up of the layered structure of the clay montmorillonite $Al_2(OH)_2Si_4O_{10}$ (idealized formula). The layer spacing is 1–2 nm. The hydrated cations in-between the layers are exchangeable, as are the hydroxyl protons of the octahedral aluminum units $\{Al(OH)_xO_{5-x}\}$.

therefore arise as it comes to multiple bond formation: double and triple bonds between P atoms are energetically disfavored. As silicon, phosphorus forms homonuclear chains, but with comparable restrictions as in silicon chemistry: The chain length is limited to a few members at the best, and linear phosphanes P_nH_{2n} or cyclic phosphines P_nH_n are very reactive. Sufficiently more stable, and hence more appropriate, are polyphosphazenes (Scheme 7.5), polymers with the repeating of

Scheme 7.5 Examples for phosphorus-based molecules related to compounds in a carbon world. R is an organic or inorganic substituent.

−N=PR$_2$- units, where R can be an organic or inorganic residue. Also, cyclic systems comparable to benzene, viz. (N=PR$_2$)$_3$, are known and stable. In these triphosphazenes, the π electron system is less effectively delocalized than in benzene due to the differing electronegativities of N (3.1) and P (2.1); they are thus only partially aromatic. In phosphorine (phosphabenzene), a phosphorus analog of pyridine, the aromaticity amounts to about 90% that of benzene.

While, for the utilization of phosphorus or phosphorus–nitrogen as a substitute for carbon-based polymers in living beings, restrictions similar to those formulated for silicon apply, phosphorus, in the form of phosphate, undoubtedly plays a central biochemical role (Scheme 7.6) as a linker for nucleosides, an energizer for biomolecules, a constituent of cell membranes in the form of phospholipids, and a support for bone structures. About 50% of the bony substances of the endoskeletons in vertebrates is, integrated in a collagen protein matrix, hydroxylapatite Ca$_5$(PO$_4$)$_3$OH, also a widespread mineral, plus some fluorapatite Ca$_5$(PO$_4$)$_3$F. Apatite contains isolated phosphate anions. Phosphate can form a variety of biorelevant mono- and diesters. Examples for diesters are the phosphate linkages in RNA and DNA (Scheme 7.4). Two examples for sugar-based monoesters, nicotine-adenine-dinucleotidephosphate NADPH and glucose-6-phosphate, are shown in Scheme 7.6. The dinucleotide NADPH, where the two nucleosides are connected via diphosphate, is a reducing agent, delivering two reduction equivalents in the form of hydride H$^-$ to a substrate. Glucose-6-phosphate is one of the numerous examples for the activation of an organic substrate by phosphorylation, where "activation" adverts to resourcing an organic molecule for metabolism by decreasing the activation barrier for the initiating step in a catabolic path. Phosphorylation of substrates is carried out by adenosinetriphosphate (ATP), in collaboration with magnesium ions Mg^{2+}. ATP is a very generally employed energy pool not only in

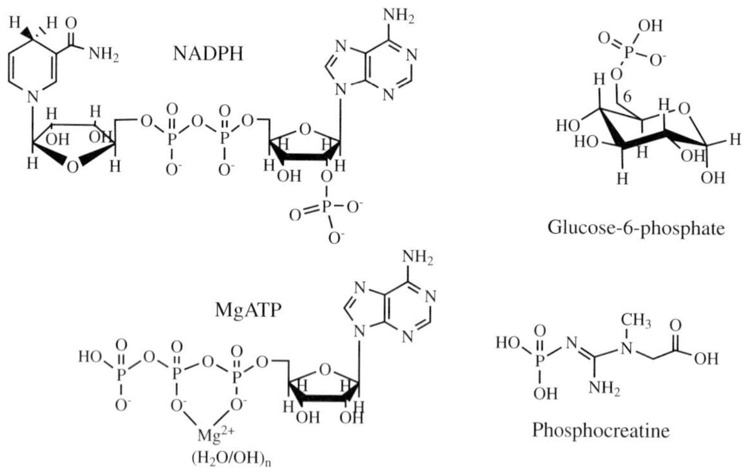

Scheme 7.6 Examples of phosphate-activated molecules utilized as reductants (NADH) or as energy sources in metabolic pathways.

metabolism but also for translocation phenomena, such as the transmembrane transport of cations (Na$^+$, K$^+$, and Ca^{2+}) against a concentration (and electric potential) gradient. High-energy phosphate for rapid energy delivery in skeletal muscles and brain can be provided by creatine phosphate, also known as phosphocreatine (Scheme 7.6), an organic amide of phosphoric acid.

Phosphate esters also play a genuine role as building units in cell membranes of archaea, bacteria, and eukarya. Bacterial and eukaryan cells contain phospholipids in their membranes (Scheme 7.7). The phospholipids are distinct from "normal" lipids inasmuch as one of the terminal glycerol functions is esterified with phosphate. A second position on the phosphate is commonly esterified with choline (trimethylglycine), as shown in Scheme 7.7, or ethanolamine. This arrangement allows for the assembly of a membrane bilayer in which the hydrophilic "head" groups, the glycerol ester moieties, point toward both the hydrophilic external environment and the likewise hydrophilic internal medium, whereas the hydrophobic and thus lipophilic "tails," provided by the long hydrocarbon chains of the lipid acid residues, are directed into the intramembrane space, providing the cell with a protective shield. Gram-negative bacteria additionally have available a coating on the outside of their membrane, consisting of lipopolysaccharides that contain glucosamine-phosphate as an integral constituent, making these bacteria resistant against Gram (crystal violet/safranin) staining. Archaean cell walls are designed somewhat differently: instead of phospholipids, phosphoether lipids are employed to protect the organism from excessive exchange between the hydrophilic intracellular cytosol and extracellular environment. Ether-lipids are distinct from common lipids in that one or two of the ester bonds in common lipids are replaced

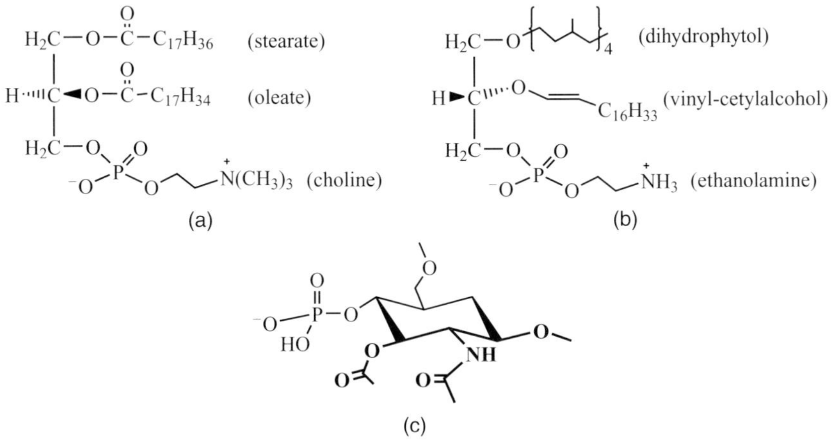

Scheme 7.7 Examples of phospholipids employed as building blocks in cell membranes. (a) A phospholipid used by bacteria and eukarya (with the central carbon of the glycerol backbone in the D-configuration). (b) An ether phospholipid used by archaea (with the central carbon of the glycerol backbone in the L configuration). (c) Glucosamine phosphate is an integral constituent of the coating of the outer membrane of Gram-*negative* bacteria.

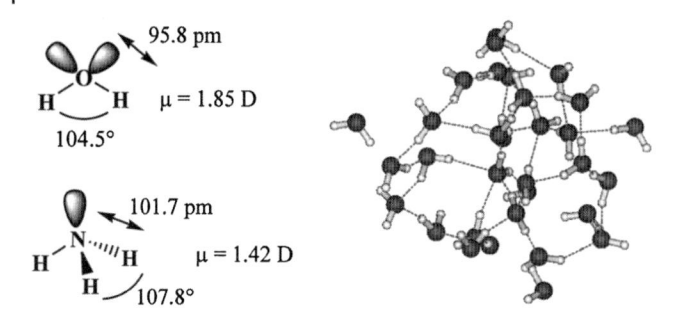

Figure 7.2 Structure parameters of the water and the ammonia molecule, and a snapshot of a 32 molecule cluster in liquid water, as obtained by molecular dynamics calculations on the basis of NMR parameters. Hydrogen-bonding contacts are represented by dashed lines. Water structure: Courtesy by Dr. J. Vaara, Department of Physics, Oulu, Finland.

by ether bonds. As noted, an additional difference between bacterial and archaean cell walls is the deployment of isoprene-based long-chain alcohols as ester or ether by archaea (Scheme 7.7b). For inorganic models of cells, see below.

Life as we know and conceive it is not only intimately connected to carbon-based biomolecules but also to the presence of water. Water does have properties that, taken all together, are not reproduced by any other solvent. These properties are as follows:

1) A rather broad temperature range over which water remains liquid – in a temperature range where (a) organic compounds are not destroyed by heat and (b) the temperature is sufficiently high to enable chemical reactions at "reasonable" reaction rates.

2) A high dipole moment that allows for (a) association of H_2O molecules to short-lived clusters and local "quasicrystalline" areas in the liquid phase (Figure 7.2) providing a high viscosity, surface tension, and a comparatively high boiling point, (b) separation of ionic compounds into cations and anions by hydration (ion–dipole interaction, Eq. (7.1)), and (c) stabilization of the molecular dispersion of hydrophilic molecules such as sugars by hydrogen-bonding interaction in the liquid phase:

$$Na^+Cl^- \xrightarrow{H_2O} [Na(H_2O)_x]^+ + [Cl(H_2O)_y]^- ; x, y \approx 6 \qquad (7.1)$$

3) High heat capacity, useful for temperature "buffering," as a consequence of the dipole–dipole association of water molecules (hydrogen-bonding network in liquid water).

4) Self-dissociation, Eq. (7.2), coupled with its amphoteric nature, that is, its ability to act as a proton donor (to bases) and a proton acceptor (for acids):

$$2H_2O \rightleftarrows H_3O^+ + OH^- \quad [H_3O^+][OH^-] = 10^{-14} \text{ M}^{-2} \qquad (7.2)$$

5) The *decrease* of density as water freezes, providing a floating ice sheet that insulates and thus protects the remaining liquid from freezing solid to the bottom.

Table 7.1 Physical properties of water and solvents that have been proposed as alternatives for supporting life forms.

	Melting point/boiling point (K) at 1 bar (10^5 Pa); temperature range Δ	Heat capacity (J mol^{-1} K^{-1})	Density of liquid (g cm^{-3})	Dipole moment (D)
H_2O	273/373; $\Delta = 100$	75.3 (278 K)	1 (277 K)	1.85
H_2S	191/213; $\Delta = 22$	34.6 (278 K)	0.914 (213 K)	0.97
NH_3	195/240; $\Delta = 45$	80.0 (278 K)	0.823 (298 K)	1.42
$(HF)_n$[a]	190/293; $\Delta = 103$	48.7 (273 K)	0.959 (293 K)	1.91[b]
HCN	260/299; $\Delta = 39$	70.8 (290 K)	0.687 (293 K)	2.98
CH_3OH	175/338; $\Delta = 163$	40.9 (278 K)	0.791 (293 K)	1.70
CH_4	91/112; $\Delta = 21$	53 (93 K)	0.465 (109 K)	0
C_2H_6	89/184; $\Delta = 95$	68.5 (94 K)	0.54 (184 K)	1×10^{-4}
CO_2	195/217 (at 5.2 bar); $\Delta = 22$	109 (273 K)	0.93 (273 K)	0

a) In the liquid state, $n \approx 7$.
b) For $n = 1$.

The temporary local order in liquid water induced by dipole–dipole association provides a closer packing of the water molecules than in ice. In crystalline ice, edge-sharing OH_4 tetrahedra are arranged three-dimensionally in the same spatial design as in the SiO_4 tetrahedra of the SiO_2 modification tridymite (see also Figure 7.1a). The density of water, with a maximum of 1 g cm^{-3} at 4 °C, drops to 0.9999 g cm^{-3} for liquid water at 0 °C and to 0.917 g cm^{-3} for ice at 0 °C. Solutes such as salts, sugar, glycol, and H_2O_2 disrupt the order, resulting in a freezing point depression. Brines of sulfates and perchlorates can have freezing points as low as −70 °C, and a 1:1 (by volume) mixture of water and hydrogen peroxide freezes at −52 °C, which allows "water" to remain liquid in appropriate sites of Mars (Sections 5.2.4.5 and 7.5).

In Table 7.1, properties of water are compared with the respective properties of a selection of other thinkable solvents: hydrogen sulfide, ammonia, hydrogen fluoride, hydrogen cyanide, methanol, methane, ethane, and carbon dioxide. The nonpolar solvents CH_4 and C_2H_6 have been added because methane–ethane lakes have been found on the Saturnian moon Titan and led to speculations on alternative life forms. Carbon dioxide, also devoid of a permanent dipole moment, forms hydrates with water under extreme conditions, and is readdressed in Section 7.3 on terrestrial extremophiles. In the presence of water, both methane and carbon dioxide form clathrate hydrates at low temperature and under high pressure, providing systems where organisms adapted to a nonpolar environment can access some water to sustain life. Clathrates of methane, produced at some point in the past either biologically (by methanogenesis) or geologically (by serpentinization) are discussed as a possible source of temporary local methane plumes on Mars (see Figure 5.14 in Section 5.2.4.4).

Of the six polar solvents, methanol comes closest to water, except that its heat capacity is only about half that of water. Ammonia, on the other hand, has a heat capacity (and thus an inner ordering) that compares to that of water. NH_3 is a well-established alternative solvent (alternative to water) in chemistry, and in fact is the more commonly discussed alternative to support nonterrestrial life forms. A potential problem with NH_3 is the low-temperature range when it is liquid, between -33 to $-78\,°C$ at 1 bar (10^5 Pa), which largely slows down biochemical reactions. However, at pressures more than 10 bar, ammonia is a liquid at room temperature. Ammonia and water easily mix, and hydrates of ammonia, such as $NH_3 \cdot 2H_2O$, $NH_3 \cdot H_2O$, and $NH_3 \cdot ½\ H_2O$ are known. Similar to water, ammonia is amphoteric; the degree of self-dissociation, Eq. (7.3), is less pronounced than in the case of water, though:

$$2NH_3 \rightleftarrows NH_4^+ + NH_2^-\quad [NH_4^+]\cdot[NH_2^-] = 10^{-30}\ M^{-2} \tag{7.3}$$

Taken all this together, ammonia may well be a substitute for water. This view is also supported by the indispensability of ammonia in many biochemical N-metabolic processes. The availability of ammonia is guaranteed by its comparatively high interstellar abundance, amounting to 10^{-7} to 10^{-5} of the hydrogen abundance in interstellar clouds. Ammonia mainly forms by electron capture, Eq. (7.4), with a formation rate of ca. $2 \times 10^{-9}\,cm^{-3}\,s^{-1}$ in clouds of an overall density of $10^5\,cm^{-3}$. The ammonium ion in Eq. (7.4) is generated in a cascade of ion molecule reactions, starting with $N^+ + H_2 \rightarrow NH^+ + H$, as elaborated by Eq. (4.35) in Section 4.2.2. NH_3 is also present in the giant gas planets of the Solar System:

$$NH_4^+ + e^- \rightarrow NH_3 + H \tag{7.4}$$

A further potential water substitute is hydrogen fluoride that associates to oligomers in the liquid state as a consequence of the particularly strong hydrogen bonds (note that fluorine is the most electronegative element). There are, however, several severe restrictions for the use of $(HF)_n$. One is the very low cosmic fluorine abundance: in massive stars, fluorine is "consumed" by nuclear processes (cf. Eq. (3.12), Section 3.1.3). Another restriction arises from the incompatibility of the conjoint presence of HF and silicates: HF reacts with silicates to form fluorides and silica, represented, in a rather generalized form, by Eq. (7.5):

$$Ca_2SiO_4 + 4HF \rightarrow 2CaF_2 + SiO_2 \cdot 2H_2O \tag{7.5}$$

As noted earlier, phospholipids are used as the structural constituents of cell membranes by all terrestrial organisms, providing a protective wall between the hydrophilic aqueous intracellular medium (the cytosol) and the aqueous environment. This is achieved by organization of the lipids in such a way that the hydro*phobic* hydrocarbon *tails* are aligned into a double layer, with the tails directed toward the intramembrane space. Molecules that have available a hydrophilic part (the ester or ether function in the case of lipids) and a hydrophobic part (the hydrocarbon tail) are termed amphipathic. In a hydrophilic medium, such as water, amphipathic molecules spontaneously self-aggregate to energy-minimized spherical micelles (Figure 7.3a), confined by a monolayer with the head groups pointing outward and thus in contact with the surrounding water, and the tails sequestering toward the center of the micelle. Bilayers very much reminiscent of the bacterial cell wall can thus form.

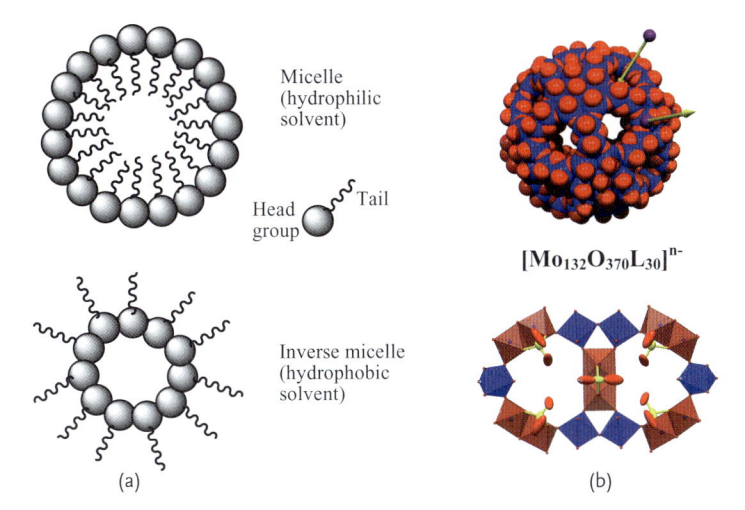

7.2 Putative Non-Carbon and Nonaqueous Life Forms | 229

Figure 7.3 (a) Micelles, inverse micelles and (b) an artificial "mineral cell" based on a polyoxidomolybdate [5]. The arrows in the picture at the top right symbolize cation exchange via the capsule's pores/channels. The detail at the bottom right, for L = SO_4^{2-}, depicts 2 of the 20 Mo_9O_9 pores (inner aperture ca. 120 pm), viewed from the interior. MoO_6 polyhedra are in red and blue, sulfur is yellow, and sulfate oxygen is represented by the red ellipsoids.

The membranes of these so-called micelles, or vesicles, allow, by instantaneously providing very short-lived "openings," for an exchange of chemical species between their interior and the surrounding solvent. *Inverse* micelles (with the hydrophilic head groups pointing inward; Figure 7.3b) formed as amphipathic molecules are dispersed in hydro*phobic* solvents. Cellular life forms present in methane–ethane lakes are expected to be warded off against the hydrophobic medium by "inverse cell membranes" corresponding to the built-up of inverse micelles.

Micelle-like structures can also be provided by merely inorganic frameworks such as naturally occurring zeolites (nanoporous alumosilicates) and artificial polyoxidometalates[8] (POMs). POMs are (usually) highly symmetric anionic macromolecules with oxidometalate polyhedra as building blocks. The larger versions can have a cavity, and pores in their "membrane." As an example, let us consider a polyoxidomolybdate of formula $[Mo_{132}O_{370}L_{30}]^{n-}$, a so-called Keplerate, as shown in Figure 7.3b. The trivial name for this and related polyhedra derives from the relation to Kepler's early model of the cosmos as detailed is his *Mysterium Cosmographicum* (1596; The Cosmographic Mystery). The cluster $\{Mo_{132}O_{370}\}$ contains 12 pentagons (12 pentagons form a dodecahedron; a dodecahedron can be inscribed with Earth's orbit and circumscribed with Jupiter's orbit), linked together by double octahedra (an octahedron inscribes Venus' orbit and circumscribes Mercury's orbit). L is an anionic ligand such as sulfate, acetate, or butyrate. Molybdate is chosen here

---

[8] Poly*oxido*metallates (instead of the more common polyoxometallates) conforms to IUPAC nomenclature.

because molybdenum, in the form of easily water-soluble molybdate MoO_4^{2-}, is the most abundant transition metal present in seawater which, on our planet, is commonly considered the cradle of life. In seawater, the average concentration of molybdate is 100 nm. Molybdenum also is an essential trace element to most if not all life forms. An example is molybdopterine, an enzyme used in methanogenesis (Scheme 5.1 in Section 5.2.4.4).

The hollow POM shown in Figure 7.3b [5] has a diameter of 3 nm and is soluble in water due to its high negative charge, which makes it hydrophilic. The "membrane" consists of a monolayer of pentagonal bipyramids with Mo^{VI} and tetragonal bipyramids (octahedra) with Mo^V. The membrane contains 20 pores at its outer side, extending into channels connecting to the cavity. The ligands L attached to the inner surface (coordinated to the Mo centers) extend into this cavity. The ligands can either be hydrophilic such as sulfate (in that case, the cavity is filled with a highly structured $(H_2O)_{72}$ cluster) or with groups providing a hydrophobic interior such as butyrate ($CH_3CH_2CH_2CO_2^-$) residues. Pores and channels are coated by oxido groups just in the same way as ion channels of living cells that allow for the transport and countertransport of the cations Li^+, Na^+, K^+, Mg^{2+}, and Ca^{2+}. Molybdenum and related transition metals, tungsten and vanadium in particular, are common constituents in rock and thus available on all conceivable worlds. As Mo, V (with 35 nM the second-to-most abundant transition metal in Earth's oceans) and W form easily soluble oxido anions. The relevance of POMs as inorganic mineral cells in facilitating prelife chemical processes certainly remains elusive, but has recently received support in a related context, viz. mineral cells based on iron sulfide, to be addressed in more detail in Section 7.4.1.

7.3
Life Under Extreme Conditions

We have illuminated alternatives to a carbon- and water-based world and concluded that, while silicon and, in particular, phosphorus play an important role in biochemical processes for life as we know it, these elements can hardly substitute for carbon as it comes to the potential of macromolecules responsible for information storage (DNA and RNA), structure (lipids), energy storage (carbohydrate polymers), and proteins. Water as a solvent, and an indispensable medium to support terrestrial life forms, might be replaced by ammonia on planets in extra-solar planetary systems.

In this chapter, scenarios will be described where, under terrestrial conditions, extreme situations are exploited for life to thrive, extreme referring to everything that deviates from our normative surroundings: (i) moderate temperatures (around 15 °C, with the seasonal variations and the ups and downs in tropical and arctic regions), (ii) sufficient supply of water either as the habitat of a life form or as an external source (liquid and vapor), (iii) a moderate pH (around 7, i.e., no excessive alkalinity or acidity), (iv) moderate pressure (something around 1 bar), (v) availability of an electron acceptor such as oxygen (for aerobic organisms) or carbon dioxide (for photosynthetic organisms) to power metabolic processes, (vi) moderate background gamma and X-ray radiation (2 mGy a^{-1}), and (vii) "nontoxic" environments, that is, sufficiently low concentrations of elements that commonly block off metabolic processes, like Cd, Hg, and As. The latter requirement is sometimes circumvented in that specialized microorganisms may resort to Cd[17]) and As, or have developed mechanisms to protect themselves against toxic forms such as ionic mercury Hg_2^{2+} and Hg^{2+}.

Microorganisms are quite versatile in the utilization of available resources of energy, reductants, and carbon to support their metabolic pathways. Scheme 7.8 is an overview of the nomenclature employed to distinguish between the various categories of organisms with respect to how complex organic compounds are synthesized. Most bacteria and archaea living in habitats with extreme conditions are powered by energy from chemical reactions and use inorganic reductants; they are termed chemolithotrophs. Bacteria and archaea exhibit a broad respiratory flexibility in that they can employ non-oxygen sources as terminal electron acceptors in anaerobic environments, including high-valent metal ions such as Fe^{3+}, Mn^{4+}, U^{6+}, and oxidometalates and -pseudometalates, for example, vanadate(V), molybdate(VI), tungstate(VI), arsenate(V), and selenate(VI).

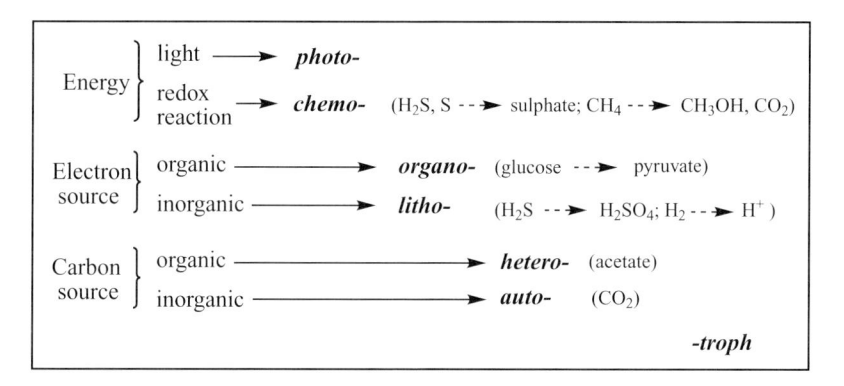

```
Energy      light     ─→  photo-
            redox     ─→  chemo-   (H₂S, S --→ sulphate; CH₄ --→ CH₃OH, CO₂)
            reaction

Electron    organic   ─→  organo-  (glucose --→ pyruvate)
source      inorganic ─→  litho-   (H₂S --→ H₂SO₄; H₂ --→ H⁺)

Carbon      organic   ─→  hetero-  (acetate)
source      inorganic ─→  auto-    (CO₂)
                                                         -troph
```

Scheme 7.8 Classification of organisms according to the sources they resort to in order to support their metabolic pathways. Examples for basic reactions are given in parentheses. A *chemolithoauto*trophic bacterium, for example, is a bacterium that uses chemical energy (energy delivered by a chemical reaction) and an inorganic ("litho") electron source for the reduction of carbon dioxide. Phototrophs use light as an energy source, autotrophs CO_2 as carbon source, and heterotrophs resort to carbon sources other than CO_2.

In Section 7.2, the possibility of life in a nonpolar environment containing niches of water in the form of clathrates has briefly been addressed. Life forms might thrive where, along with water clathrates (water as either the host or the guest), liquid hydrocarbon lakes exist, such as on Titan, or where carbon dioxide is available, for example, in the southern polar area of Mars. As noted in Chapter 6, CO_2 and H_2O supposedly are also present on exoplanets, where these two components can provide environmental conditions, the "correct" pressure and temperature provided, similar to those on our planet.

The discovery of a lake of liquid carbon dioxide (plus ca. 14% of methane) in the Okinawa Trough between Japan and Taiwan, at a depth of about 1400 m [6], has been a bodacious event in this context. At the temperature (around 8 °C) and pressure (140 bar, 14 MPa) pertinent to the bottom of the trough, liquid CO_2 is less dense than water,[9] affording that the lake is maintained in place by a sedimentary pavement (elemental sulfur, quartz, and montmorillonite in this specific case) and a subpavement cap, consisting here of a CO_2 clathrate of the approximate composition $CO_2 \subset (H_2O)_6$. Notably, microbial cells at a concentration of ca. $10^5 \, cm^{-3}$ are present in the liquid CO_2 phase underneath the cap, and a cell count higher by two orders of magnitude at the liquid CO_2–solid clathrate interface.

In this interfacial region, microbial activity has been found, dominated by methane-metabolizing archaea, sulfate-reducing bacteria, and sulfur- and methane-oxidizing chemolithoautotrophs. Sulfate *reduction* occurs stepwise; the overall reaction, leading to hydrogen sulfide, Eq. (7.6a), consumes eight electrons. These electrons can be delivered by various reductants. Examples are hydrogen, Eq. (7.6b), acetate, Eq. (7.6c), and methane, Eq. (7.7). The redox interaction between sulfate and methane, an oxidation with respect to methane and a reduction with respect to sulfate, is one of the reactions carried out by methane-oxidizing archaea (see also below). Sulfur-*oxidizing* microorganisms reverse, the reaction represented by Eq. (7.6a), that is, starting from hydrogen sulfide or elemental sulfur, the sulfur source is oxidized to sulfate (via thiosulfate), Eq. (7.8). Under anaerobic conditions, as in the case of microbial activity in the CO_2 lakes, electron acceptors can be H^+ (to form H_2, as shown in Eq. (7.8)), nitrate (NO_3^-, to end up as N_2), and other oxidants:

$$SO_4^{2-} + 8e^- + 9H^+ \rightarrow HS^- + 4H_2O \quad (7.6a)$$

$$SO_4^{2-} + 4H_2 + H^+ \rightarrow HS^- + 4H_2O \quad (7.6b)$$

$$SO_4^{2-} + CH_3CO_2^- \rightarrow HS^- + 2HCO_3^- \quad (7.6c)$$

$$CH_4 + SO_4^{2-} \rightarrow HS^- + HCO_3^- + H_2O \quad (7.7)$$

$$HS^- + H^+ \rightarrow 1/6 \, S_6 + H_2$$
$$\xrightarrow[2e^-]{} 1/2 \, S_2O_3^{2-} \xrightarrow[4e^-]{} SO_4^{2-} \quad (7.8)$$

Sulfide oxidation is among the energetically most favorable reactions in chemosynthetic processes and therefore utilized by many different lineages of chemolithoautotrophic prokarya, including acidophiles and haloalkaliphiles. The latter are bacteria that thrive in brines such as provided by soda lakes, tolerating salt ($NaHCO_3/Na_2CO_3$ + NaCl) concentrations up to 5 M, and pH values up to 11 [7]. The carbonate/hydrogencarbonate system is an effective buffer (the optimal buffering capacity of the system is close to pH 10), providing ideal conditions for a natural habitat. The high salt concentrations afford adaption of the microorganism to cope with the high environmental osmotic pressure. One strategy is the "in-salt" strategy, that is, setting up a high intracellular ionic strength. The main cation used in this strategy is K^+, which due to its larger size (as compared to Na^+) and

---

9) At a depth between 3000 and 3800 m, liquid CO_2 is denser than water.

thus lower polarizing effect is less hydrated and therefore more mobile than Na$^+$. The second strategy employs water-soluble organic zwitterionic molecules (betains) as osmolytes, in a cytosol that otherwise displays a normal intracellular salt concentration [8]. In Scheme 7.9, some of the more widely employed bacterial organic osmolytes are collated. Scheme 7.9 also contains a yellow pigment, natronochrome, isolated from the cell walls of species belonging to the genus *Thioalkalivibrio*. The pigment, a polyene, may protect the organism against damage by (oxygen) radicals produced as the cells are exposed to UV. Carotenoids, which are present in the membrane of haloalkaliphilic archaea, may attain a similar function, an interesting aspect as it comes to the question of survival of microorganisms on Mars, where the exposure to UV is much more extreme and destructive than on Earth, and has so far prevented, together with oxidants in Mars' surface layers, observation of any evidence hinting toward microbial surface life on Mars.

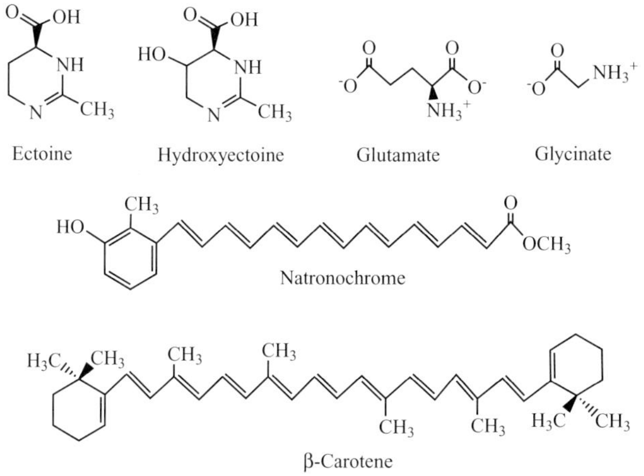

Scheme 7.9 Organic osmoregulators used by haloalkaliphilic bacteria (upper row), and the pigments natronochrome (from bacteria) and β-carotene (from archaea).

Representatives of the *Thioalkalivibrio* strain are particularly versatile oxidizers of sulfur species, including hydrogen sulfide HS$^-$, polysulfide S_n^{2-} (n = 3–8, polysulfides are only stable at higher pH), elemental sulfur, tetrathionate $S_4O_6^{2-}$, sulfite SO_3^{2-}, thiocyanate SCN$^-$, and even carbon disulfide CS$_2$ [7a]. Along with the autotrophic bacteria (*Thioalkalivibrio* is an autotroph), heterotrophs using acetate (such as the genus *Halomonas*) are also to be found among the sulfide oxidizers [7b]. The oxidation of polysulfide, Eq. (7.9) for n = 6, proceeds via a cell-bound intermediate (RS$_n^-$). Electron acceptor is either nitrate, Eq. (7.10), or nitrite, Eq. (7.11a). Instead of nitrite, dinitrogenoxide N$_2$O ("laughing gas") can jump in, Eq. (7.11b). In any case, these reactions classify the respective bacteria as denitrifiers:

$$1/6\,S_6^{2-} \rightarrow 1/n\,RS_n^- \xrightarrow[2e^-]{fast} 1/6\,S_6 \xrightarrow[6e^-]{H_2O,\ slow} SO_4^{2-} \tag{7.9}$$

$$NO_3^- + 2H^+ + 2e^- \rightarrow NO_2^- + H_2O \tag{7.10}$$

$$NO_2^- + 4H^+ + 3e^- \rightarrow \frac{1}{2}N_2 + 2H_2O \tag{7.11a}$$

$$N_2O + 2H^+ + 2e^- \rightarrow N_2 + H_2O \tag{7.11b}$$

Thiocyanate CNS^- is first converted to cyanate CNO^- and H_2S, Eq. (7.12). The daughter products of cyanate are ammonia and carbon dioxide, whereas H_2S is further processed to sulfate:

$$CNS^- + H_2O \rightarrow CNO^- + H_2S \quad \begin{array}{l} \rightarrow NH_3 \text{ and } CO_2 \\ \rightarrow SO_4^{2-} \end{array} \tag{7.12}$$

Sulfide oxidation, along with the oxidation of ferrous to ferric iron, is also carried out by acidophiles, organisms that prosper in environments with an optimum pH below 3 (down to pH 1 for some acidophilic archaea). These can be natural environments, for example, geothermal vents particularly rich in iron sulfides such as pyrite FeS_2 (ferrous disulfide), and man-made environments, for example, acidic mine drainages from sulfidic ores and wastewater processing plants. The association of acidophiles with sulfide deposits is of potential evolutionary importance because metabolic processes might have originated on surfaces of sulfide minerals (cf. Section 7.4.1).

Since, in acidic media, Fe^{3+} is easily soluble, a high tolerance of the bacterial or archaean cell for high levels of Fe^{3+} is afforded, together with pH homeostasis, that is, the maintenance of an about neutral cytosolic pH and hence a pH gradient of several pH units across the cellular membrane. There are several mechanisms to maintain pH homeostasis [9], among these (i) proton efflux by active proton pumping, (ii) influx of potassium ions to maintain an inside positive membrane potential to inhibit H^+ entry, and (iii) cytoplasmatic buffering, for example, by proton-mediated decarboxylation of glutamate, as shown in Eq. (7.13):

$$^+H_3N-\underset{\underset{\underset{CO_2^-}{|}}{\underset{(CH_2)_2}{|}}}{\overset{H}{\underset{|}{C}}}-CO_2^- + H^+ \longrightarrow {}^+H_3N-\underset{\underset{\underset{CH_3}{|}}{\underset{CH_2}{|}}}{\overset{H}{\underset{|}{C}}}-CO_2^- + CO_2 \tag{7.13}$$

An interesting recent discovery is a microbial assemblage in a subglacial pool of marine brine in Antarctica with a chloride ion concentration of 1.375 M [10] (seawater: 0.55 M). This icy brine pool (the temperature of which is −5 °C), which has remained isolated for at least 1.5 million years, is anoxic, about neutral, and highly ferrous: 97% of the total iron, $c(Fe^{II/III}) = 3.45$ mM, is present in the form of Fe^{2+}. The electrons for the reduction of Fe^{3+}, which acts as the terminal electron acceptor for the microbes, are provided by the oxidation of sulfite to sulfate with the aid of sulfite-oxidizing bacteria. Sulfite in turn is produced from sulfate by sulfate-reducing bacteria. Some of the sulfite is further reduced to disulfide, part of which becomes immobilized through the formation of pyrite, but the main amount is recycled. The sulfite–sulfate shuttle thus catalyzes the reduction of ferric to ferrous iron. The overall reaction scheme is depicted by Eq. (7.14). Metabolism is slow – the doubling time for the organisms has been estimated to be ca. 300 years – as a consequence of the low temperature, the restricted availability of organic carbon and inorganic nitrogen, and the limited supply of S^{-II}, which is the sulfur oxidation state in most organic sulfur compounds constituting the biomass of the microorganisms. For bacteria living under normal conditions, the doubling rate is several minutes:

$$Fe^{3+} + \text{\textcircled{e}} \rightarrow Fe^{2+}$$
$$(S_2^{2-} \leftarrow \leftarrow)\ 1/2SO_3^{2-} \rightarrow 1/2SO_4^{2-} + \text{\textcircled{e}} \quad (7.14)$$
$$e^- \text{ (e.g., from acetate oxidation)}$$

Bacteria and archaea can also cope with the extreme situations, coupled with very slow metabolic rates and thus extremely low reproduction rates (of several hundred to thousand years), deep within the Earth's crust, provided that there are pockets of fossil water present, or water has been oozing into niches during the millennia. The very slow metabolism, sometimes accompanied by dramatic shrinkage of the organisms due to long periods of starvation, place these individuals at the very limit of what can be accounted for as "life." These "intraterrestrial" pressure-loving organisms, termed baryophiles or piezophiles, have been retrieved in sedimentary and basaltic rock down to depths of 10 km, corresponding to pressures of up to 800 bar (80 MPa). Since the temperatures go high as one goes down in the crust, they are also thermophiles, where 120 °C possibly is the upper limit to allow life to persist. While sedimentary rock usually contains remnants of organic material stemming from the period of their formation, living conditions are particularly harsh in magmatic rock (such as basalt and granite) where there is hardly any organic carbon available. These organisms therefore rely on carbonate as carbon source and hydrogen as energy source; they are thus lithoautotrophs, which have been living in these habitats isolated from the rest of the world for up to a hundred million years. The hydrogen is formed, for example, from ferrous minerals and water, Eq. (7.15a), and/or by radiolysis of water. Equation (7.15b) is an example for the latter process. The α particles are delivered by radioactive decay of, for example, ^{238}U:

$$2Fe^{2+} + 2H_2O \rightarrow 2FeO(OH) + H_2 \quad (7.15a)$$

$$H_2O \xrightarrow{\alpha} OH + H;\ 2H \rightarrow H_2 \quad (7.15b)$$

While, in extreme environments, microbial *communities* are commonly present, single-species ecosystems, or systems grossly dominated by a single species, are sometimes characteristic of habitats at high depths [11]. Having been isolated from an oxic environment for millions of years, these species lack a defense system to protect them against oxygen. Electron donors are typically hydrogen and formate, Eq. (7.16); electron acceptor is sulfate, Eq. (7.6):

7 The Origin of Life

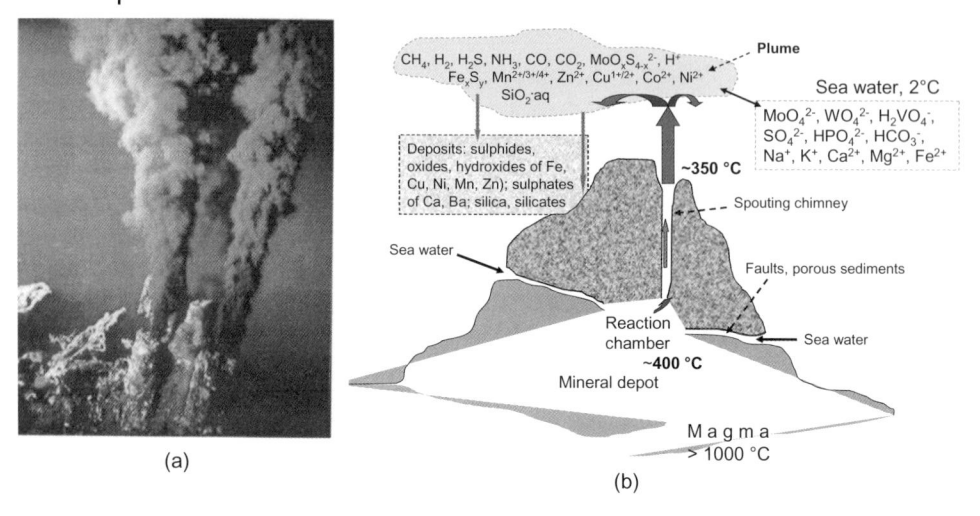

Figure 7.4 (a) Deep-sea black smoker (at 2600 m, volcanic ridge off Costa Rica); source: www.dgukenvis.nic.in/A%20black%20smoker. (b) Schematic representation of a sea floor hydrothermal vent, including a selection of chemical species involved in the formation of plumes and deposits. Black smoke is formed through precipitation of iron, nickel, and copper sulfides, white smoke by zinc sulfide, calcium sulfate, and hydrated silica.

$$HCO_2^- \rightarrow CO_2 + 2e^- + H^+ \tag{7.16}$$

In the deep ocean, hydrothermal vents nourished by close-to-seafloor-surface volcanism discharge a cocktail of chemicals in which gaseous components such as highly reduced sulfur species, viz. H_2S and solid components derived thereof, viz. metal sulfides (predominantly of iron, nickel, manganese, and copper) prevail. The metal sulfides, when exhaled with the acidic extremely hot water, form a black smoke of particulate sulfides on contact with the cold, slightly alkaline (pH ca.7.5) deep-seawater in the surroundings. The vents are commonly termed "black smokers" (Figure 7.4) for this reason. Alternatively, plumes can be ejected that contain silica along with zinc sulfide and sulfates of barium and calcium; these are referred to as "white smokers." Black smokers in particular are locations that represent unusual biota, sheltering a plethora of hyperthermophile (>80 °C) and thermophile (50–80 °C) microorganisms.

LUCA, the common ancestor of bacteria on the one hand, and archaea and eukarya, on the other hand (see Scheme 7.2 in Section 7.1), have possibly been mesophilic (adapting to temperatures <50 °C) [12]. LUCA developed during an epoch of Earth's history, 3.8–3.6 Ga ago, shortly after the heavy bombardment had come to an end. At that time, the young Earth experienced shorter days and the Sun shone somewhat dimmer, being about one-third less luminous than today. When the conditions on early Earth changed to provide a warmer climate ca. 3.5 Ga ago, an improvement of thermotolerance became an essential factor for survival, and consequently thermo- to hyperthermophilic bacteria and archaea/eukarya derived from LUCA. As the Earth-spanning ocean later cooled down from an estimated 70 °C[10] to a temperature of 10 °C at present, readaption to mesophilic situations took place, and thermotolerance of the majority of the organisms decreased. Numerous niches to accommodate thermophiles remained, however, associated with hot springs, geysers, and volcanic exhalations in volcanic surface and subsurface areas as well as in seafloor spots, allowing bacteria and archaea which succeeded to retreat into these niches to survive and evolve. The biota associated with black deep-sea smokers are particularly narrative as it comes to early life forms on our planet, and the possibility of alternative, though related, life forms on other objects in the Solar System (Venus and Mars, Titan, Enceladus and the Galilean moons, Europa in particular) and on exoplanets.

The seawater that enters through faults and pores in the base of the black (and white) smokers becomes heated to temperatures of 400 °C and more, and reacts with the minerals present in the reaction zone. The reaction products, gases and particulate matter – are ejected outside through chimneys to form a plume, or smoke, the constituents of which react with the anions and cations present in the cold (about 2 °C) seawater. Insoluble sulfides, oxides, and silicates settle down to form multicolored sediments. The formation of hydrogen, Eq. (7.17), ammonia, Eq. (7.18), and methylthiol (synonyms are methylsulfide and mercaptomethane), Eq. (7.19), are three examples for reactions involving ferrous sulfide (see also Section 7.4.1). The reactions are exergonic. Both H_2 and NH_3 are important in sustaining the metabolism of microorganisms: H_2 is a source for reduction equivalents, for example,, in the hydrogenation steps of methanogenesis (see below) and NH_3 as a generally accessible nitrogen source for the synthesis of organic nitrogen compounds:

$$FeS + H_2S \rightarrow FeS_2 + H_2 \tag{7.17}$$

$$3FeS + N_2 + 3H_2S \rightarrow 3FeS_2 + 2NH_3 \tag{7.18}$$

$$3FeS + CO_2 + 4H_2S \rightarrow 3FeS_2 + 2H_2O + CH_3SH \tag{7.19}$$

Along with H_2 and NH_3, the gas mix that is exhaled by the vents contains CH_4, CO_2, and H_2S, providing a sound basis for thriving life for both, thermophilic methanogens and methanotrophs.[11] Methano*genic* bacteria and archaea reduce carbon dioxide and other carbon sources, such as formate, to methane in a

---

[10] This view that thermophilic conditions prevailed on early Earth has recently been challenged. According to ^{18}O and ^2H analyses in 3.42 Ga old rocks from South Africa, the ocean temperature supposedly was below ca. 40 °C. See: Hren, M.T., Tice, M.M., Chamberlain, C.P. (2009) Oxygen and hydrogen isotope evidence for a temperate climate 3.42 billion years ago. *Nature*, **462**, 205–208: Blake, R.E., Chang, S.J., and Lepland, A. (2010) Phosphate oxygen isotope evidence for a temperature and biologically active Archean ocean. *Nature*, **464**, 1029–1033.

[11] Methanogenic and methanotrophic microorganisms are not restricted to habitats associated with hydrothermal vents. Hot subterranean springs, but mildly thermophilic and mesophilic environments as well, such as clippings of organic waste, cellulose degradation in the stomach of ruminants, rice fields and so forth also provide methane.

multistep reaction cascade (Scheme 7.10a), delineated in detail in the context of occasional methane plumes observed in Mars' atmosphere (Scheme 5.1 in Section 5.2.4.4). Enzymes catalyzing the various steps depend on (i) molybdenum or, in hyperthermophiles, tungsten (Scheme 7.10b) for initiating two-electron reduction, (ii) iron and nickel in a sulfide-dominated coordination environment for reductive hydrogenation of the intermediate products,[12] and (iii) in the final step, nickel in the center of a porphyrinogenic ligand system. The choice of tungsten instead of molybdenum by thermophiles may root in the availability of these two metals: the mixed oxidosulfido anions $MO_xS_{4-x}^{2-}$ are more readily available for $M = W$. In addition, a higher reduction potential of the tungsto- as compared to the molybdo-enzyme can come in. The reduction of CO_2 to CH_4 is powered mainly by the oxidation of dihydrogen to protons ($H_2 \rightarrow 2H^+ + 2e^-$) and formate to CO_2 Eq. (7.16).

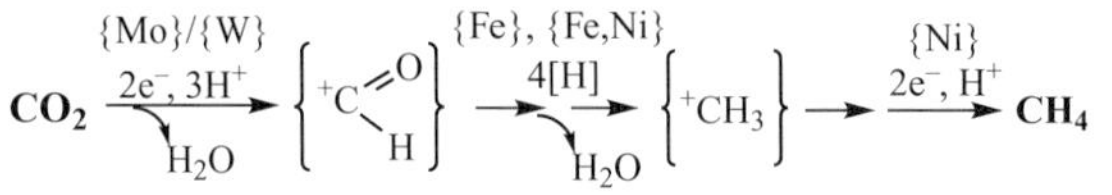

{Mo}/{W} are alternative cofactors (Mo/W-pterins) in CO_2 reductases

{Fe} and {Fe,Ni} represents iron- and iron-nickel-dependent hydrogenases

{Ni} is the cofactor F_{430} in coenzyme-M reductase

(a)

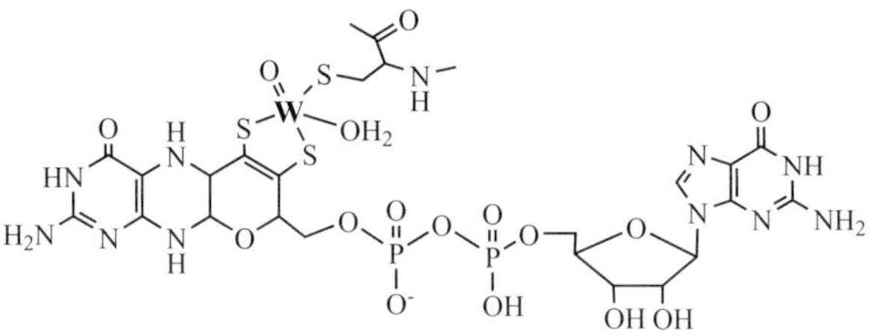

(b)

Scheme 7.10 (a) The reaction cascade for the reduction of CO_2 to CH_4 (for details see Scheme 5.1). (b) The tungsten enzyme responsible for the first reduction step ($CO_2 \rightarrow \{^+CHO\}$) in methanogenesis by the thermophilic archaeon *Methanobacter wolfei*. The two moieties linked to diphosphate are tungstopterin and guanosine.

The reverse reaction, the oxidative conversion of methane to carbon dioxide or hydrogencarbonate, Eq. (7.20), is carried forth by methano*trophic* microorganisms. Here, the initiating step, the oxidation of methane to methanol, is catalyzed by a methane monooxygenase, which depends on iron or copper. For the methanotrophs, CH_4 usually is the sole energy and carbon source; the carbon for incorporation into biomass is set aside from the formate intermediate in the oxidation path. The reduction equivalents set free in this oxidation of CH_4 to CO_2 can be used to

12) In addition, cobalt, in the form of cobalamine, is needed for methyl transfer between organic substrates.

reduce sulfate or Fe^{3+} (present as ferrihydrite $FeO(OH) \cdot \frac{1}{2}H_2O$) and $Mn^{3+/4+}$ (in the form of birnessite $(K,Na,Ca)_x(Mn^{III},Mn^{IV})_2O_4 \cdot 1.5H_2O$, "$MnO_2$"), Eqs. (7.21)–(7.23), or nitrate and nitrite, Eqs. (7.10) and (7.11a), or by (intermittently) delivering. oxygen from nitrate via the decay of nitrous oxide to N_2 and O_2 [13a]. Energetically, the oxidation by ferrihydrite and birnessite is more favorable than by sulfate; the conversion rates of methane are, however, more favorable in the case of sulfate as the oxidant [13b]:

$$CH_4 \xrightarrow[2e^-, 2H^+]{\{Fe\}/\{Cu\} \; H_2O} CH_3OH \xrightarrow[4e^-, 5H^+]{H_2O} HCO_2^- \xrightarrow[2e^-, H^+]{} CO_2 \quad (7.20)$$

$$\downarrow \text{biomass}$$

{Fe} and {Cu} are soluble and particulate monooxygenases, respectively

$$CH_4 + SO_4^{2-} \rightarrow HCO_3^- + HS^- + H_2O \quad (7.21)$$

$$CH_4 + 8FeO(OH) + 15H^+ \rightarrow HCO_3^- + 8Fe^{2+} + 13H_2O \quad (7.22)$$

$$CH_4 + 4MnO_2 + 7H^+ \rightarrow HCO_3^- + 4Mn^{2+} + 5H_2O \quad (7.23)$$

It has been estimated that anaerobic methane oxidation by methanotrophs consumes up to 20% of the methane release by methanogens and abiotic processes. The net result hence is an excess of methane. The greater part of this methane is presently locked off in the form of methane clathrates in deeper and cooler offshore areas of the oceans.

The activity of methanogens started about 3.5 Ga ago and led to a dramatic increase of methane in the atmosphere that, in the forgoing ca. 0.3–0.4 Ga prior to initiation of life, was dominated by N_2, CO_2, and H_2O. This methane release was accompanied by an increase of the temperature and a corresponding adaption of the organisms to rise in global temperature.[10] Only about 1 Ga later, or 2.4 Ga back from now, photosynthesis by cyanobacteria became the dominant bacterial process, rapidly changing the atmospheric composition to a dominance of oxygen (along with N_2), known as the "Great Oxygen Event." The net reaction is formulated in Eq. (7.24); the oxygen is delivered by the CO_2 (indicated by *):

$$n(CO_2^* + H_2O) \rightarrow C_nH_{2n}O_n + nO_2^* \; (\text{e.g., glucose for } n = 6) \quad (7.24)$$

This change from an anaerobic methane to an aerobic oxygen atmosphere[13] may have been supported by a concomitant depletion of nickel in the oceans [14]. Methanogens rely on nickel present in iron–nickel hydrogenases that catalyze the hydrogenation of the formyl intermediate in methanogenesis (Scheme 7.10). Nickel is further employed in the last enzymatic step of methane formation, where S-methyl-ethanesulfonate (= methyl-coenzyme-M) is oxidatively linked via

---

13) Increase of atmospheric oxygen did not occur continuously. A second stage, after a temporary decline of oxygen supply beginning with the Great Oxygen Event 2.4 Ga ago, took place about 800–550 million years ago, ultimately leading to the present atmospheric and oceanic oxygen levels.

a disulfide bond to a second organic sulfide HSR (coenzyme-B) with concomitant reductive release of methane, a reaction that is catalyzed by coenzyme-M reductase, a nickel-dependent porphinogenic enzyme, {Ni}; Eq. (7.25). Depletion of nickel from the oceans as a consequence of reduced nickel influx may have starved methanogens, thus reducing methane production. This gradual removal of methane from the atmosphere also led to a global cooling 2.3–2.2 Ga ago, resulting in a glaciation period of the planet known as "Snowball Earth":

$$H_3C-S-SO_3^- + HSR \xrightarrow{\{Ni\}} CH_4 + RS-S-SO_3^- \quad (7.25)$$

Methyl-CoM   CoB

Summary Sections 7.1–7.3

The principle "life" encompasses (i) self-sustained metabolism, that is, a complex network of chemical reactions toward the synthesis and catabolic alteration of matter with a net energy consumption, (ii) reproduction by self-replication, including complete information transfer from the parent to the replica, and (iii) evolution (interlinked with imperfect self-replication) in the frame of adaption to changing environmental conditions and of competition for the available habitat. Information is stored in a genetic pool, represented by DNA and, in the "RNA world" that came prior to the first authentic life forms, in the form of hydrolytically less-stable RNA, with ribose instead of deoxiribose in the nucleotide polymer. A (proto)cellular organizational unit representing the principle of life requires, along with the information pool, basic modules for organizing and maintaining its internal structure and metabolic pathways (polypeptides/proteins, carbohydrates, and lipids), and a certain degree of protective isolation by means of a membrane, which also allows for an exchange with the surroundings and communication with co-individuals. From the RNA world, which may have developed shortly after the end of the cataclysmic heavy bombardment 3.8 Ga ago, the "last uniform common ancestor," LUCA, evolved ca. 0.2 Ga later. The first fossils representing microbial life shaped in a fashion reminiscent of and comparable to today's lives date back 3.45 Ga.

Potential candidates for non-carbon life forms are silicon and phosphorus. However, both the Si–Si and the P–P bonds are very reactive, discouraging the formation of chain-like structures common with hydrocarbons. Also, π bonding is essentially excluded. In the case of phosphorus, phosphazenes $\{-N=PNR_2-\}_n$ might come in. A further restriction for a silicon world arises from the fact that, in the presence of oxygen, silicon is largely immobilized by formation of SiO_2 and silicates. On the other hand, Si and P are supportive elements in a carbon world: Hydrates of silica are employed in structure scaffolds by many organisms, and silica and clays (such as montmorillonite) may have acted as templates in the formation of polypeptides and polynucleotides ("clay organisms") and in chiral induction. Phosphate is widely used as a linker of and an activator in organic molecules and molecular assemblies, in phospho- and phosphoether lipids and, in the form of apatite, to support bone structures.

The replacement of water as a medium to support life is also hardly conceivable: Water has exceptional properties that are not concurrently available with other solvents. These properties encompass its amphoteric nature, the moderate (and broad) temperature range where water is liquid, its large dipole moment allowing for internal structuring, and solvation of ions and molecules by hydration and hydrogen bonding. Possible alternatives are water–methane clathrates $CH_4 \subset (H_2O)_6$, H_2O–CO_2 interfaces, and NH_3. Ammonia is an abundant molecule in dense interstellar clouds, where its fractional abundance amounts to 10^{-7}–10^{-5}.

Terrestrial microbial life adapted to extreme conditions may provide clues to possible life forms on other planets. These extremophiles are often living on the oxidation of lower oxidation state sulfur compounds to sulfate. Microbial life has been found to thrive in subsurface deep-sea carbon dioxide lakes at the interface of $CO_2 \subset (H_2O)_6$ clathrates. More common extremophiles comprise (i) haloalkaliphiles (in $NaCl/NaHCO_3$ lakes with salt concentrations up to 5 M and a pH up to 11), (ii) acidophiles (pH < 3, plus high Fe^{3+} tolerance), (iii) baryophiles (pressures up to 80 MPa in depths down to 10 km and temperatures T approaching 120 °C), and (iv) thermophiles (T = 50–80 °C) and hyperthermophiles (T = 80 to ca. 120 °C). Baryophiles and microorganisms retrieved from subglacial brine pools with a temperature of −5 °C have reproduction rates of several hundred to thousand years, mainly as a consequence of a vanishing supply of nutrients. Deep-sea hydrothermal vents, so-called black smokers (exhaling metal sulfides) and white smokers (producing plumes of silica, carbonates, and ZnS), are thriving with archaean and bacterial populations of thermophiles, among these methanogens (generation of CH_4) and methanotrophs (consumption of CH_4).

7.4
Scenarios for the Primordial Supply of Basic Life Molecules

The first organic (i.e., carbon-based) molecules needed as modules for the synthesis of more complex organic matter allowing for an organization into "life" may have been carried to Earth by meteorites/comets and interplanetary/interstellar dust, and/or provided by synthesis in the frame of primordial chemical processes on Earth. The extraterrestrial input has been described in Sections 5.3 and 5.4, and is resumed in Section 7.4.4. For the terrestrial genesis of life molecules, there are two main scenarios, based on the Miller–Urey experiments [15] on the one hand and the Wächtershäuser iron–sulfur world [16] on the other hand. Starting point for the latter is a primordial atmospheric composition dominated by CO_2, N_2, H_2O, plus traces of H_2 and O_2, with the energy for enabling chemical reactions and bond formation coming from the oxidation of sulfide S^{2-} (in the form of iron and nickel sulfides, which also provide catalytically active surfaces) to disulfide S_2^{2-}, for example, in the form of pyrite ("fools gold") FeS_2. Miller and Urey had based their experiments on a primordial gas mix consisting of CH_4, NH_3, H_2O, and N_2, and their presumed energy source was essentially electric discharge (lightening). According to present awareness of the atmosphere of the early Earth, the Wächtershäuser supposition "fits better." The main differences between the Miller–Urey and the Wächtershäuser approach are (i) the carbon source: inorganic (CO_2) versus organic (CH_4) and (ii) a reducing Miller–Urey atmosphere versus the Wächtershäuser conjecture that there was some oxygen present, about 0.1%, based on the plausible assumption that lightening, cosmic rays, Solar wind, and radioactive decay provide energy sources for the splitting of the O–H and C–O bond and thus produced some oxygen. The iron–sulfur world is dealt with in Section 7.4.1 and the Miller–Urey syntheses in Section 7.4.2. In Section 7.4.3, a scenario referred to as "clay organisms" shall be addressed. Clays are phyllosilicates that "breathe" by taking up and releasing water, subject to the water vapor pressure, and catalyze condensation reactions leading to peptides and oligonucleotides.

7.4.1
The Iron–Sulfur World ("Pioneer Organisms")

The key reaction in Wächtershäuser's iron–sulfur world is the formation of the S-methyl ester of thioacetic acid ("activated acetic acid") from methylthiol CH_3SH and either CO or CO_2. The reaction between methyl thiol and CO, (Eq. (7.26)), is catalyzed by iron–nickel sulfide, which occurs naturally in the form of the mineral pentlandite $(Fe,Ni)_9S_8$. In the presence of H_2O, acetic acid is obtained, and in the presence of H_2O and aniline, the anilide of acetic acid forms and hence a compound with an amide functionality just as in peptides. The reaction yielding acetic acid is related to the Monsanto process, in which acetic acid is generated from CO and methanol in the presence of iodide and a rhodium-based catalyst:

$$2CH_3SH + CO \xrightarrow{\{FeNiS\}} CH_3-C(=O)SCH_3 + H_2S \quad (7.26)$$

$$\xrightarrow{H_2O} CH_3-CO_2H + CH_3SH$$

$$\xrightarrow{H_2O,\ C_6H_5NH_2} CH_3-C(=O)NH-C_6H_5$$

The reaction between methyl thiol and CO_2 to form thioacetic acid methyl ester, Eq. (7.27), involves a redox process: the mean oxidation state in carbon for the starting products ($2CH_3SH + CO_2$) in Eq. (7.27) is 0, for the reaction product $-2/3$ per carbon. The necessary reduction equivalents (two electrons) are provided by the sulfur of the catalyst and reaction partner FeS and one of the thiol sulfurs. The net energy supply is provided by the two-electron oxidation of sulfide to disulfide, represented in Eq. (7.28), a variant of Eq. (7.17) in Section 7.3, and accounting for the fact that in an aqueous medium hydrogensulfide (HS^-) instead of sulfide (S^{2-}) is by far the dominant anion. The ferrous minerals involved in these reactions in the prebiotic broth are (i) troilite FeS or the slightly iron-deficient pyrrhotite $Fe_{1-x}S$, $x = 0$–0.2 as reductants (sulfur in the oxidation state –II) and (ii) pyrite or marcasite

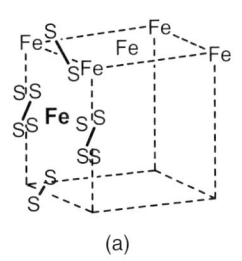

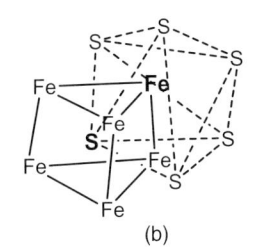

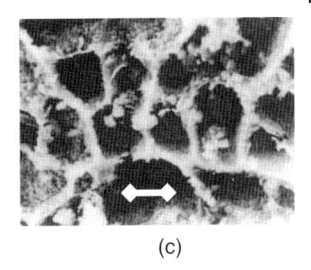

7.4 Scenarios for the Primordial Supply of Basic Life Molecules

Figure 7.5 (a) Section of the structure of pyrite FeS_2 (NaCl structure), showing the octahedral coordination of Fe^{2+} by six S_2^{2-}. (b) Section of the structure of troilite FeS (NiAs structure), highlighting the octahedral (dashed lines) FeS_6 and the trigonal-prismatic (full lines) SFe_6 units. (c) Electron micrograph of the honeycomb structure of artificial FeS (taken from Ref. [17], Figure 2d; permission granted by The Royal Society of Chemistry). The white bar corresponds to a length of 20 μm. Similar microstructures have been inferred for iron sulfide deposits at black smokers.

(both of formula FeS_2) as oxidation products (sulfur in the oxidation state –I). The electrochemical potential (Eq. (7.28)) is –620 mV, corresponding to a free Gibbs energy of –118 kJ mol^{-1}:

$$2CH_3SH + CO_2 + FeS \longrightarrow CH_3-C\underset{SCH_3}{\overset{O}{\diagup\!\!\!\diagdown}} + H_2O + FeS_2 \qquad (7.27)$$

$$FeS + HS^- \rightarrow FeS_2 + H^+ + 2e^-; \Delta E = -620 \text{ mV}, \Delta G = -118 \text{ kJ mol}^{-1} \qquad (7.28)$$

The structures of the two iron minerals pyrite and troilite are shown in Figures 7.5a and b. Pyrite FeS_2 crystallizes in the sodium chloride structure, where S_2^{2-} dumbbells (octahedrally surrounded by six Fe^{2+} neighbors) occupy the positions of Cl$^-$ in the NaCl lattice, and the Fe^{2+} ions (octahedrally surrounded by six S_2^{2-}) occupy the Na$^+$ positions. Troilite FeS has the nickel arsenide structure: S^{2-} is in the center of a trigonal prism formed by six Fe^{2+}, whereas the Fe^{2+} ions are centers in S^{2-} octahedra. In pyrrhotite $Fe_{1-x}S$, voids coming about by Fe^{2+} being under-represented with respect to S^{2-} give rise to surface defects that make this mineral a particularly efficient catalyst. In the surroundings of hydrothermal vents on the ocean floor, such as the black smokers introduced in Section 7.3, iron sulfides initially are deposited in their energy-rich amorphous form, which transforms to the thermodynamically stable crystalline sulfides with a microscopic texture reminiscent of a honeycomb. These "iron sulfide cells," only a few tens of micrometers across [17] (Figure 7.5c), provide a large and reactive surface area as well as "protective" pockets, protective for molecules synthesized on the reactive surface. The concurrence of (i) inorganic sulfidic surfaces/surface pockets acting as catalysts and suppliers of energy for chemical reactions, (ii) basic inventories of organic molecules adsorbed, absorbed, or coordinated to these pocket sites, and (iii) water plus water-dissolved "nutrients," associated with the inorganic substructures has been dubbed "pioneer organisms" [16]. And in fact, basically at least, these pioneer organisms fulfill the three criteria for life as deduced in Section 7.1:

1) **Metabolism** – chemical reactions – occurs;
2) **Reproduction** for some of the organic compounds synthesized, there is an autocatalytic feedback, leading to their replication;
3) **Evolution** there will be errors, and thus variations, in autocatalytic reproduction, akin to evolution.

Since synthetic possibilities in the frame of the inorganic network are restricted, there is some intrinsic determination; the iron–sulfur world therefore does not suffer from the extremely high degree of contingency in the RNA world with its low degree of probability that a directed ("suitable") combination of subunits in the polymer is secured. The iron–sulfur world does, however, not explain the directed polymerization of nucleotides and amino acids either.

The idea of allotting a central role to iron (and nickel) sulfides in the evolution of molecules essential for life is compelling also in the context of those magnetotactic bacteria that have tiny crystals of greigite Fe_3S_4 embedded in cell organelles (magnetosomes) and, even more so, in view of the numerous enzymes relying on iron–sulfur clusters containing Fe^{2+} and Fe^{3+}, most of which are involved in electron transfer processes. These iron–sulfur clusters are "primitive" in the sense that they form by self-assembly from Fe^{2+} and S^{2-} in the presence of an electron acceptor, Eq. (7.29) [18a]. Both, Fe^{2+} and S^{2-}, have been copiously available on early Earth. Trace amounts of oxygen may have functioned as electron acceptor, {EA} in Eq. (7.29). Self-assembly resulting in cubane clusters extends to cyanothiomolybdates (Scheme 7.11) from primordial Mo–S phases, cyanide, and sulfide, a process which may be connected upstream to the evolution of Mo-based enzymes [18b] (which contain the thiolate function of cysteinate coordinated to molybdenum):

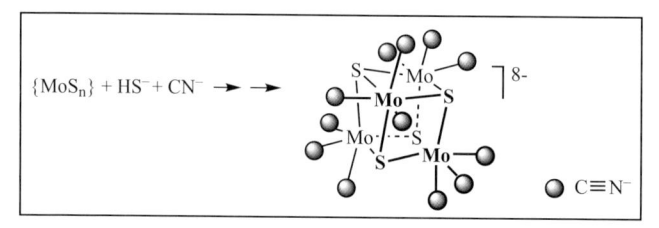

Scheme 7.11 Self-assembly of an anionic cubane-type (D_{2d} symmetry) molybdenum(III)-sulfur cluster, a possible precursor system for molybdopterin-based enzymes.

$$4Fe^{2+} + 4S^{2-} + 4HS^- + \{EA\} \rightarrow [Fe_2^{II}Fe_2^{III}S_4(SH)_4]^{2-} + \{EA^{2-}\} \quad (7.29)$$

A selection of biologically active iron–sulfur clusters is provided in Scheme 7.12, together with an enzyme containing four iron plus two nickel centers in a sulfide-dominated coordination environment. This enzyme, acetyl-coenzyme-A (acetyl-CoA) synthase, catalyzes the transfer of a methyl group (delivered by methyl-cobalamine $Cb\text{-}CH_3$) plus carbon monoxide to coenzyme-A = HS-CoA, a reaction in which a carbon–carbon bond is formed, Eq. (7.30). Mechanistically,

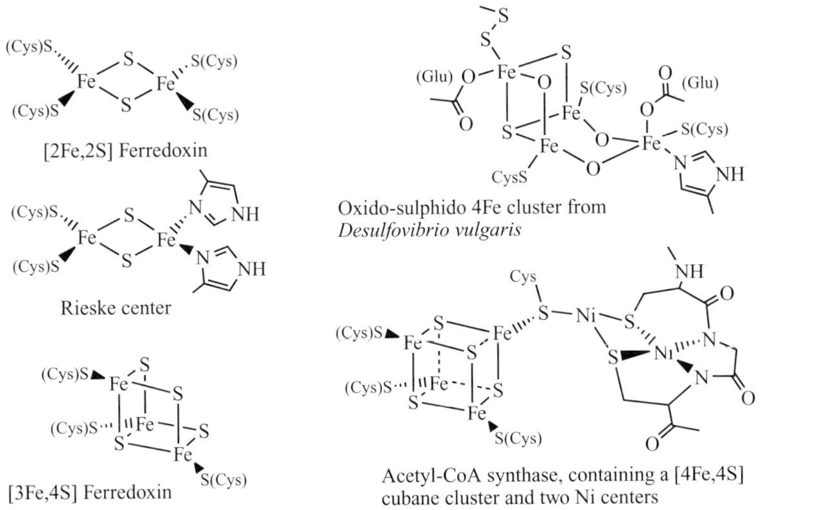

Scheme 7.12 Basic structures of iron–sulfur clusters (overall charges omitted) employed as electron transfer centers in enzymes, and the {Ni₂Fe₄} cofactor of acetyl-CoA synthase that catalyzes the C–C linkage between CO and CH₃, Eq. (7.30). For the formation of a [4Fe,4S] cubane cluster, which is an integral part in acetyl-CoA synthase, see Eq. (7.29).

these reactions are "migratory insertions": the methyl group, activated by coordination to a metal center, migrates and is attached to the carbon of CO, which in turn is activated by coordination to an adjacent metal center. In the case of acetyl-CoA synthase, these two metal centers are Ni. The resulting acetyl-CoA, $CH_3C(O)$S-CoA, is an ester of thioacetic acid – just as the methyl ester $CH_3C(O)SCH_3$ formed according to Eqs. (7.26) and (7.27). Acetyl-coenzyme-A is a highly energetic and thus highly reactive molecule, as much as $CH_3C(O)SCH_3$:

$$CO + HS\text{-}CoA + Cb\text{-}CH_3 + H_2O \longrightarrow \underset{H_3C}{\overset{O}{\underset{\|}{C}}}\text{-}S\text{-}CoA + Cb\text{-}H_2O \qquad (7.30)$$

Methylthiol CH_3SH (the starting molecule in Eqs. (7.26) and (7.27)) is present in gaseous volcanic exhalations. Alternatively, several reaction paths leading to the formation of this thio analog of methanol are conceivable, such as the direct reaction between carbon monoxide and sulfane H_2S under reducing conditions, Eq. (7.31a), and the iron sulfide-mediated reaction between carbon dioxide and H_2S, Eq. (7.31b). In Eq. (7.31b), the reduction equivalents are again (as in Eq. (7.27)) provided by the intrinsic oxidation of sulfide to disulfide. More generally, reduction equivalents can also be provided by H_2 ($\equiv 2H^+ + 2e^-$) or by the catalytic conversion of CO to CO_2, Eq. (7.32):

$$CO + H_2S + 2H_2 \rightarrow CH_3SH + H_2O \qquad (7.31a)$$

$$CO_2 + 3FeS + 4H_2S \rightarrow CH_3SH + 3FeS_2 + 2H_2O \qquad (7.31b)$$

$$CO + H_2O \rightarrow CO_2 + 2H^+ + 2e^-, \Delta H = -41 \text{ kJ mol}^{-1} \tag{7.32}$$

The reduction potential provided in the reaction (7.28) also suffices for nitrogen fixation, that is, the conversion of the particularly inert dinitrogen into ammonia, Eq. (7.33a) [19] or ammonium ion (Eq. (7.33b)), which is the nitrogen form accessible to organisms. This conversion is accomplished *biotically* by nitrogen-fixing organisms such as the free living bacterium *Azotobacter*, the symbiotic companion bacterium *Rhizobium* of legumes, and the cyanobacterium *Anabaena*. Nitrogenase, the enzyme system promoting this reduction at ambient temperature and pressure, contains a complex iron–sulfur cofactor of core composition $\{Fe_7MS_9\}$, where M is molybdenum, vanadium, or iron:

$$N_2 + 3FeS + 3H_2S \rightarrow 2NH_3 + 3FeS_2 \tag{7.33a}$$

$$N_2 + 3FeS + 3H_2S + 2H^+ \rightarrow 2NH_4^+ + 3FeS_2 \tag{7.33b}$$

Acetyl-CoA channels the acetyl fragment into a variety of metabolic paths, and activated acetic acid $CH_3C(O)SCH_3$ can be perceived to have accomplished a similar task in the pseudocellular shelters provided by the honeycomb pockets of iron sulfide or iron–nickel sulfide minerals. This is exemplified by Eq. (7.34) for the generation of activated malonic acid in the case of prebiotic (7.34a) and physiological conditions (7.34b), respectively. The physiological C–C coupling reaction between acetyl-S*CoA* and hydrogencarbonate to form malonyl-S*CoA*, Eq. (7.34b), is powered by the concomitant hydrolysis of ATP (which ends up as adenosinediphosphate ADP and inorganic phosphate HPO_4^{2-}). In the case of the presumed prebiotic reaction (7.34a), the energy may again be provided by chemical energy accompanying oxidation reactions such as $2S^{2-} \rightarrow S_2^{2-} + 2e^-$:

$$H_3C-C(O)S-CH_3 + HCO_3^- \rightarrow {}^-O\text{-}C(O)\text{-}CH_2\text{-}C(O)S\text{-}CH_3 + H_2O \tag{7.34a}$$

$$H_3C-C(O)SCoA + HCO_3^- \xrightarrow[-HPO_4^{2-}]{+ATP} {}^-O\text{-}C(O)\text{-}CH_2\text{-}C(O)SCoA + H_2O \tag{7.34b}$$

The repeated linkage of acetyl fragments activated in the form of acetyl-S*CoA*, with concomitant intermittent reduction by NADPH (for the structure of NADPH see Scheme 7.6), is a central metabolic pathways by which cellular organisms produce the fatty acids necessary to synthesize, inter alia, the phospholipids they need to built up cell walls. As noted in Section 7.1, archaea do not employ common fatty acids but analogs based on oligoisoprenes (terpenes) as building blocks for their membrane structures (Scheme 7.7), and this implicates that LUCA also relied on terpenoid lipids. Isoprene and its oligomers are likewise biosynthesized by linking *CoA*-activated acetyl fragments, coupled with hydrolysis, reduction, and decarboxylation, as sketched in Eq. (7.35) for the formation of dimethylallyl alcohol. For the pioneer organisms, the analogous $CH_3C(O)SCH_3$ might have done the job:

$$2 \text{ Acetyl-CoA (Ac-SCoA)} \rightarrow \text{Acetoacetyl-CoA} \xrightarrow{\text{Ac-SCoA}, H_2O} \text{HMG-CoA} \xrightarrow{\substack{\text{Reduction,} \\ \text{decarboxylation,} \\ \text{activation by diphosphate}}} \text{(isopentenyl)} \xrightleftharpoons{\text{Isomerization}} \text{Dimethylallyl alcohol}$$

(7.35)

7.4.2
The Miller–Urey and Related Experiments

Basic molecules necessary for the assembly of macromolecular life molecules are provided in the frame of the classical experiments carried out by Miller and Urey [15b, c] and related experiments by a variety of other groups during the last six decades. Miller and Urey based their investigations on the assumption that the primordial atmosphere was a reducing one, consisting of H_2, N_2, CH_4, and NH_3, plus minor amounts of CO_2 and CO [15c]. For their experiments, a gas mix of H_2, CH_4, H_2O, and NH_3 was commonly employed, and an electric spark discharge as the energy source. The composition of their experimental atmosphere is not consistent with the *overall* primordial atmosphere according to the present stand of geochemical knowledge; reducing conditions as assumed by Miller and Urey did, however, likely exist *locally*. In any case, the pioneering experiments by Miller and Urey remain a valuable source for deciphering and understanding how amino acids, sugars, and nucleobases can form as an essentially inorganic gas mix is subjected to a burst of energy. In some of their experiments, ammonia was replaced by dinitrogen. The results were grossly the same, although relative yields of the organic products changed.

Miller and Urey assumed a couple of basic reactions to take place in the atmosphere of the young Earth to back up their choice of gas mix. As an example, the lack of CO_2 in any substantial amounts was believed to be due to gradual depletion of atmospheric CO_2 through the reaction with silicates, shown for calcium metasilicate in Eq. (7.36), and the rapid conversion of CO_2 to methane by hydrogen, Eq. (7.37). The oceans have been – and still are – another (temporary) sink for CO_2:

$$CaSiO_3 + CO_2 \rightarrow CaCO_3 + SiO_2 \tag{7.36}$$

$$CO_2 + 4H_2 \rightarrow CH_4 + 2H_2O \tag{7.37}$$

The gases of the secondary atmosphere of our planet about 4 Ga ago were mainly supplied by volcanic exhalations, after the light gases, hydrogen and helium, in the primary atmosphere had escaped into space as a consequence of Earth's insufficiently strong gravitation. The main components probably had been CO_2, N_2, and

H_2O, along with small amounts of CO, CH_4, H_2, O_2, and SO_2, and possibly some C_nH_{2n+2}, H_2S, and PH_3. The Miller–Urey atmospheric constituents NH_3 and CH_4 are photolabile, and the main part will thus have been destroyed by the Sun's UV radiation. More recent experiments conducted in a nonreducing (devoid of H_2) atmosphere composed of CO_2, N_2, and H_2O, subjected to spark discharges between copper electrodes, have demonstrated that amino acids such as glycine and α-alanine can also be formed under nonreducing conditions [20a]. The intermittent formation of oxygen (from CO_2 and H_2O), and of Cu^{2+}, can be an important factor in promoting the synthesis of the amino acids. Cu^{2+} ions can act as promoters for the formation of peptide bonds by condensation of amino acids; see below. Other experiments, for example, with interstellar ice analogs consisting of nonreducing components such as $H_2O/CH_3OH/NH_3/HCN$ and $H_2O/CH_3OH/NH_3/CO_2/CO$ and conducted with UV as the energy source at temperatures between 12 and 15 K and a pressure of 10^{-5} bar (1 Pa) [20b, c] have revealed that the formation of amino acids is a very common and apparently ubiquitous event as soon as carbon-, oxygen-, hydrogen-, and nitrogen-bearing molecules are available along with an energy source, irrespective of the specific composition of the gas mix and the environmental parameters such as the state of aggregation (gas, liquid, or ice), temperature, and pressure.

The synthesis of amino acids (and hydroxy acids) in these simulated prebiotic atmospheres mainly follows the Strecker syntheses, which has been described in some detail, together with alternative pathways, in the context of the occurrence of amino acids in carbonaceous chondrites in Section 5.3.2 (Eqs. (5.55)–(5.57)). Briefly, the synthesis of glycine proceeds by condensation of formaldehyde H_2CO, hydrogen cyanide HCN, and ammonia NH_3 via aminoacetonitrile, Eq. (7.38). Other α-amino acids are formed correspondingly. Hydroxy acids such as glycolic acid are formed by condensation of H_2CO, HCN, and H_2O, (Eq. (7.39)). Formaldehyde and hydrogen cyanide, which are not originally present in the primordial soup in any appreciable amounts, arise from oxygen atoms ($H_2O \rightarrow O + 2H$) and methane, Eq. (7.40a), and from NH_3 and CH_4, Eq. (7.40b), respectively. The latter reaction, which is a strongly endothermic process and thus again affords an efficient energy source, is reminiscent of the Degussa BMA (*Blausäure-Methan-Ammoniak*; Blausäure is HCN) process. In reaction (7.40b), hydrogen is produced as a by-product, that is, the experimental conditions gradually convert the reaction medium into a reducing one, even if the process is started at redox neutral conditions:

$$H_2CO + HCN + NH_3 \rightarrow H_2N-CH_2-C\equiv N + H_2O \tag{7.38a}$$

$$H_2N-CH_2-C\equiv N + 2H_2O \rightarrow H_2N-CH_2-CO_2H + NH_3 \tag{7.38b}$$

$$H_2CO + HCN \rightarrow HO-CH_2-C\equiv N \tag{7.39a}$$

$$HO-CH_2-C\equiv N + 2H_2O \rightarrow HO-CH_2-CO_2H + NH_3 \tag{7.39b}$$

$$CH_4 + 2O \rightarrow CH_2O + H_2O \tag{7.40a}$$

$$CH_4 + NH_3 \rightarrow HCN + 3H_2 \tag{7.40b}$$

Alternatively, a radical-only path to assemble amino acids may take place, as proposed on the basis of theoretical models for photolytic processes on ice-coated dust grains in interstellar clouds [21]. Here, the initiating reactions are the creation of the radicals OH (from water), CH_3 (from methane), and NH_2 (from ammonia). The generation of radicals is followed by the reaction sequence provided by Eq. (7.41) for the formation of glycine:

$$CO + OH \rightarrow CO_2H$$

$$COOH + CH_3 \rightarrow CH_3CO_2H$$

$$CH_3CO_2H + NH_2 \rightarrow H_2N-CH_2-CO_2H + H \quad (7.41)$$

The next step, the formation of peptides by amino acid condensation (water removal), cf. Eq. (7.42) for the dipeptide $(Gly)_2$, is unfavorable thermodynamically and kinetically. Proposals to circumvent this problem encompass (i) catalysis of the formation of the peptide bond by clay minerals, (ii) dehydration in salt solutions of sufficiently high salt concentrations (e.g., NaCl, $c > 3M$), termed "salt-induced peptide formation," and (iii) activation of the amino acids by intermittent coordination to metal ions, in particular Cu^{2+}, in combination with high salt concentrations [4, 22]:

$$H_3N^+-CH_2-CO_2^- + H_3N^+-CH_2-CO_2^- \rightarrow H_3N^+-CH_2-C(O)-NH-CH_2-CO_2^- + H_2O \quad (7.42)$$

While the role of clay minerals in the formation of oligomers of amino acids and nucleotides is debated in Section 7.4.3, the salt- and Cu^{2+}-induced linkage of amino acids to dipeptides is briefly being dealt with at this point: Geochemical data indicate that Cu^{2+} (which, contrasting Cu^+, is easily soluble in water) has been available on early Earth, documented, in the case of Cu^{2+}, by minerals such as malachite $Cu_2CO_3(OH)_2$ and azurite $Cu_3(CO_3)_2(OH)_2$. In a sodium chloride solution containing Cu^{2+}, the main species present is $CuCl^+$ in the form of the pentaqua complex cation $[CuCl(H_2O)_5]^+$. This species can coordinate, by replacing the water ligands, up to two amino acids such as glycine, to form a complex of composition $[CuCl(H_2O)_2gly_2]^+$ (Figure 7.6) in which the two aqua ligands are in *trans* positions, one of the glycines is coordinated in the bidentate mode through the carboxylate oxygen and the amino group, and the other in the monodentate mode via the carboxylic acid oxygen. The coordination to the Cu^{2+} ion diminishes the kinetic barrier for the subsequent condensation to form the dipeptide. The water of condensation is picked up by the sodium ions that, in the brine, are unsaturated with respect to hydration. Homoleptic as well as heteroleptic dipeptides of a variety of amino acids, including lysine, glutamate, proline, histidine, and methionine, have thus been obtained. Additional interesting aspects in this scenario are (i) the preferential choice of α-amino acids (which form thermodynamically favored chelate-*five* rings on coordination to the copper center) and (ii) chiral selection: a preference for L-alanine and L-valine compared with the D-antipodes was observed. The reason for this preference goes back to disparities in ground state energy for the L- and D-forms due to parity violation of weak nuclear forces, a disparity that

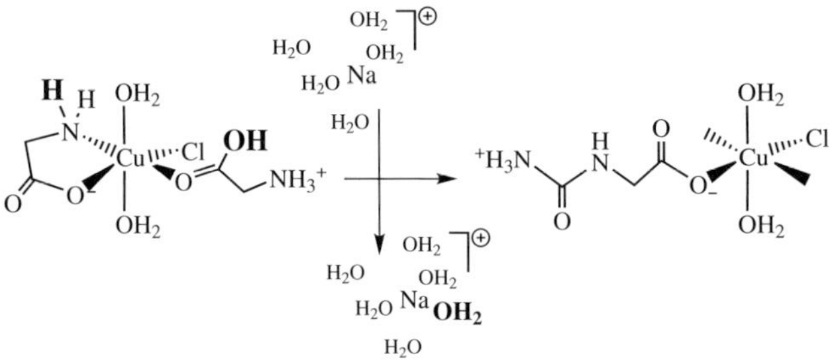

Figure 7.6 Dipeptide formation catalyzed by Cu^{2+} in ca. 5 M aqueous NaCl. The equatorial coordination sphere $(NO)_{gly}O_{gly}Cl$ in $[CuCl(H_2O)_2gly_2]^+$ is somewhat distorted toward tetrahedral coordination, providing central (at the Cu^{2+}) chirality. For the net reaction, see Eq. (7.42). The water liberated (in bold) in this reaction is taken up into the hydration sphere of the partially dehydrated Na^+.

is amplified, in favor of the L-enantiomer, by coordination of the amino acid to the copper center. In addition, there is some distortion of the equatorial coordination plane toward tetrahedral geometry. This distortion creates chirality at the Cu^{2+} center, helping to amplify chiral disparity in the resulting dipeptide. In the case of valine, where this effect is particularly pronounced, the dipeptide ratio $(L-Val)_2/(D-Val)_2$ is almost 8 [4a].

Let us now consider the formation of carbohydrates, ribose in particular, from simple building blocks formed in the primordial broth. Ribose, an integral constituent of RNAs, is formed by condensation reactions according to the formose (or Butlerov) reaction, starting from formaldehyde, and catalyzed in alkaline media by hydroxides such as $Ca(OH)_2$. The overall reaction scheme is depicted in Scheme 7.13. The reaction starts with the formation of glycolaldehyde from two molecules of formaldehyde. Isomerization of glycolaldehyde to its enediol form allows for the aldol addition, catalyzed by OH^-, of a third formaldehyde molecule. Glyceraldehyde thus formed isomerizes to dihydroxyacetone, which reacts with glycolaldehyde to give ribulose, which isomerizes to ribose. In Scheme 7.13, the β-D-pyranose form of ribose is displayed; the β isomer is the variant employed in RNA nucleotides. It has been suggested that ribose can be stabilized and stored by binding to free valences present at the surface of borate minerals (inset in Scheme 7.13) such as kernite $Na_2[B_4O_6(OH)_2] \cdot 3H_2O$ [23a], a polymeric borate consisting of interconnected (via oxo linkers) $\{BO_4^-\}$ and $\{BO_2(OH)\}$ moieties. Similarly, silicate promotes the formation and stabilization of sugars [25b].

Having synthesized ribose as one of the constituents of nucleotides, the formation of the nucleobases shall be considered next: formation of the purines adenine (A) and guanine (G) on the one hand, and the pyrimidines thymine (T), cytosine (C), and uracil (U) on the other hand; for formulae, see Scheme 7.4 in Section 7.1. The base pairs used in RNA are A+C and G+U; in DNA, these are the base pairs A+T and G+C. Adenine is formally a pentamer of hydrogen cyanide, $(HCN)_5$, and in fact can be generated from HCN by oligomerization if an energy source such

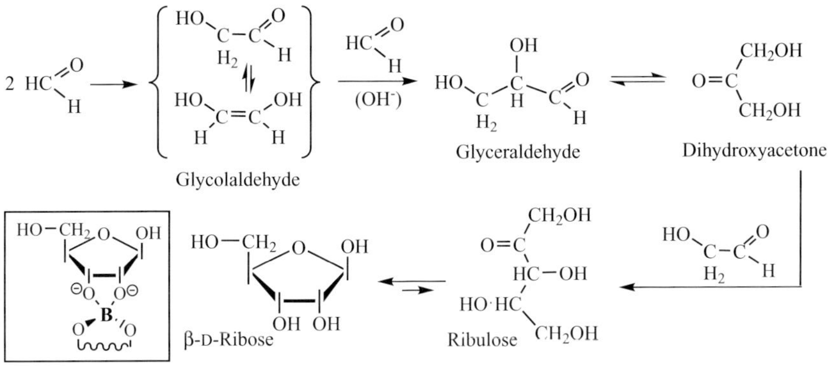

Scheme 7.13 The formation of ribose in the formose reaction, also known as Butlerov reaction. The inset (lower left) shows the proposed stabilization by binding to borate mineral surfaces.

as lightning discharge or ultraviolet radiation is available. The various intermediates in this oligomerization are provided in Scheme 7.14. Replacing the third HCN in this reaction sequence by formaldehyde H_2CO provides amino-formyl-acetonitrile instead of amino-malodinitrile, ending up in guanine after incorporation of two additional HCN molecules (4HCN + H_2CO → guanine).

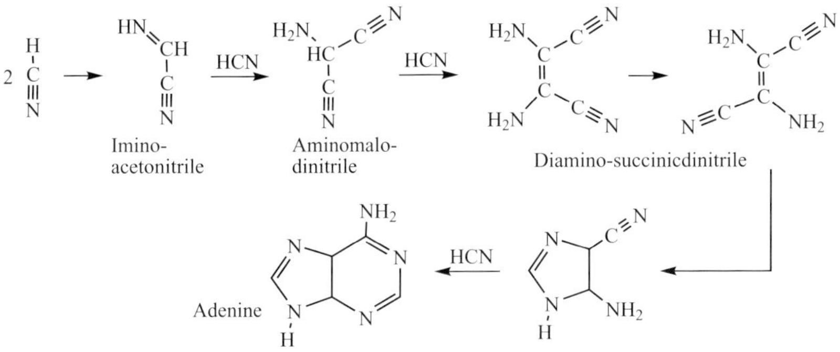

Scheme 7.14 The oligomerization of five molecules of hydrogen cyanide to form adenine.

Interestingly, also in the context of the Wächtershäuser scenario (Section 7.4.1), adenine is also formed from formamide under mild conditions in the presence of iron sulfide or iron–copper sulfide catalysts, Eq. (7.43) [24]. Efficient catalysts are pyrite FeS_2 and Fe–Cu minerals such as chalcopyrite FeCuS. Formamide HC(O)NH_2 is, along with hydrogen cyanate HNCO, the simplest molecule containing all of the four basic elements necessary for the formation of biomolecules. Formamide has been detected in the interstellar medium and in meteorites, and likely also formed in the primordial atmosphere of Earth from HCN and H_2O according to Eq. (7.44):

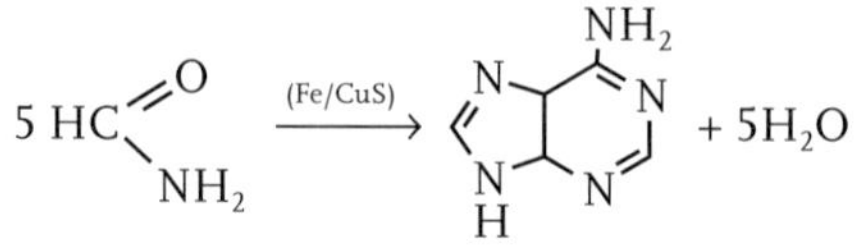

$$5\ HC\overset{O}{\underset{NH_2}{\diagup}} \xrightarrow{(Fe/CuS)} \text{[adenine]} + 5H_2O \qquad (7.43)$$

$$HC\equiv N + H_2O \longrightarrow HC\overset{O}{\underset{NH_2}{\diagup}} \qquad (7.44)$$

The electric discharge in a gas mix containing methane and dinitrogen delivers cyanoacetylene that adds water to form hydroxyvinyl-cyanide. Hydroxyvinyl-cyanide in turn can condense with urea, giving cytosine and, by hydrolysis of cytosine, uracil. Scheme 7.15 summarizes the reactions that yield these two RNA pyrimidines. Under primordial conditions, urea can be obtained (along with HCN and H_2CO) by subjecting a mixture of CO, N_2, and H_2O to an electric discharge. Alternatively, urea may be formed in a procedure comparable with the industrial Bosch-Meier process from ammonia and carbon dioxide via ammonium carbamate, Eq. (7.45), or in analogy to Wöhler's urea synthesis from ammonia and hydrogen cyanate, Eq. (7.46). Finally, hydrolysis of cyanamide also yields urea; cyanamide is generated by, for example,, UV irradiation of HCN + H_2O, Eq. (7.47). For the interstellar production of cyanamide, cf. Scheme 4.4. in Section 4.2.2.

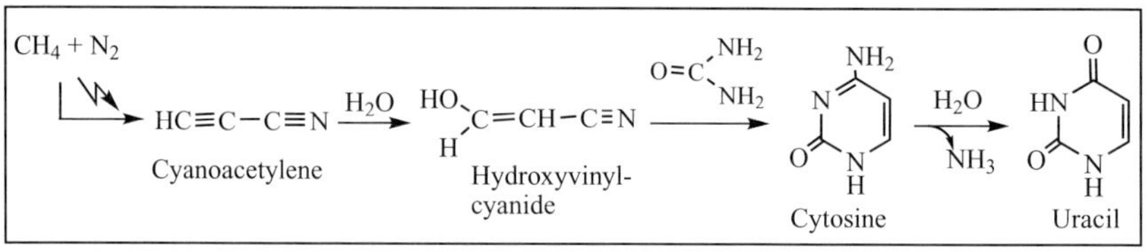

Scheme 7.15 Formation of the pyrimidone bases cytosine and uracil.

$$2NH_3 + CO_2 \longrightarrow O=C\overset{NH_2}{\underset{O^- NH_4^+}{\diagup}} \xrightarrow{H_2O} O=C\overset{NH_2}{\underset{NH_2}{\diagup}} + H_2O \qquad (7.45)$$

$$NH_3 + \overset{H-O}{\underset{C\equiv N}{\diagup}} \longrightarrow NH_4^+\ {}^-O-C\equiv N \longrightarrow O=C\overset{NH_2}{\underset{NH_2}{\diagup}} \qquad (7.46)$$

$$3\ HC\equiv N + 2H_2O \xrightarrow{UV} H_2N-CH_2-CO_2H + \left.\begin{array}{c} H_2N-C\equiv N \\ \text{Cyanamide} \\ \updownarrow \\ HN=C=NH \\ \text{Carbodiimide} \end{array}\right\} \xrightarrow{H_2O} O=C\overset{NH_2}{\underset{NH_2}{\diagup}} \qquad (7.47)$$

Glycine

With ribose and the pyrimidine/purine bases available, nucleosides should be formed. The direct conjunction of ribose and purines is, however, rather ineffective, and direct nucleoside formation from ribose and pyrimidines does in fact not occur at all. Mineral surfaces may help, acting as catalysts to promote the condensation between ribose and the RNA bases in a way similar to the condensation of amino acids to peptides and of nucleotides to oligonucleotides (see Section 7.5). A prebiotically plausible way to *pyrimidine* ribonucleotides, circumventing this problem and thus providing an alternative efficient reaction path, emerges from the common precursor molecule 2-aminooxazole [25a] (Scheme 7.16). This key compound forms by condensation of glycolaldehyde with cyanamide at neutral pH in the presence of phosphate as buffer and catalyst. Further condensation with glyceraldehyde results in the formation of arabinose-aminooxazoline, which picks up canoacetylene to furnish an intermediate (anhydro-arabinosenucleoside) that, in the presence of diphosphate, urea, and ammonium ions, undergoes phosphorylation and rearrangement to the 2′,3′-phosphate of cytidine. Photochemically assisted hydrolysis provides the corresponding uracil derivative uridine.

The beautiful findings displayed by the reaction sequence in Scheme 7.16 have contributed much to the relief of the advocates of an RNA world preceding LUCA. In the overall reaction scheme, phosphate takes over several central functions in that it is effective (i) as a catalyst, (ii) as a buffer, (iii) in the annihilation of "undesired" by-products, and (iv) in stabilizing cytidine by formation of a cyclic phosphodiester in the 2′,3′-positions, very much as borate stabilizes ribose (Scheme 7.13). Under primordial conditions, surface defects in phosphorus minerals may

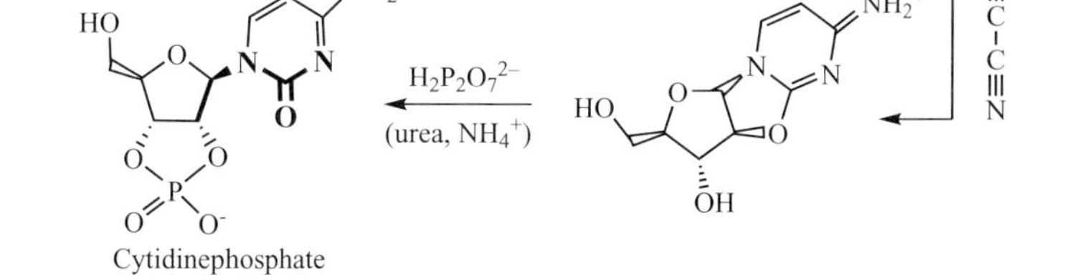

Scheme 7.16 The assembly of cytidine via 2-aminooxazol (in the box) from simple precursor molecules present in the prebiotic broth. Those of the cytidine moieties that originate from 2-aminooxazol are emphasized in bold. Cytidinephosphate is partly hydrolyzed to uridine phosphate, initiated by UV irradiation (see Scheme 7.15 for the conversion of cytosine to uracil). For the formation of the various precursor molecules, cf. Eq. (7.47) (cyanamide), Scheme 7.13 (glycolaldehyde and glyceraldehyde), and Scheme 7.15 (cyanoacetylene).

have provided the thermodynamic basis for stabilizing and storing nucleosides through binding to surface phosphate groups, but "mobile" phosphate in wet environments may also have come in. In any case, phosphate apparently is an essential coreactant, catalyst, and energizer in nucleoside formation and stabilization, and phosphate is, of course, an indispensable constituent – the third building unit along with ribose and the nucleobases – in RNA (as well as in DNA) nucleotides.

However, the availability of *mobile* phosphate in the primordial soup is tied to another problem: its accessibility. In Earth's crust, phosphorus – present in the form of phosphates – is about half as abundant as carbon, and slightly more abundant than sulfur. But phosphate is not easily mobilized from phosphate minerals, which are typically tertiary orthophosphates (i.e., they contain the PO_4^{3-} anion) and thus practically insoluble under ambient pH conditions. Typical phosphate minerals are wavellite $Al_3(PO_4)_2(OH)_3 \cdot 5H_2O$, lazulite $(Mg,Fe^{II})Al_2(PO_4)(OH)_2$, and apatite $Ca_5(PO_4)_3(OH/F)$. The latter, with fluorine contents in most cases <0.1%, is also, as noted, the inorganic constituent of the vertebrates' bones and teeth. There are four major possibilities by which mobilization of phosphate from insoluble deposits can occur:

1) Solubility increases as the pH decreases. The pK_2 for phosphoric acid (defined by the equilibrium $HPO_4^{2-} + H_3O^+ \rightleftarrows H_2PO_4^- + H_2O$) is 7.21. At a pH of 7, around 62% of the phosphate is thus present in the form of $H_2PO_4^-$ (the remaining 38% is HPO_4^{2-}). As from pH 4.5 downward, primary phosphate (dihydrogenphosphate $H_2PO_4^-$) is the exclusive species, and all salts of primary phosphate, irrespective oft the countercation, are water soluble. Consequently, acidic environments will allow for the mobilization of phosphate. Many bacteria and fungi take advantage of this by excreting organic acids such as citric, lactic, gluconic, 2-ketogluconic, oxalic, tartric, fumaric, succinic, and acetic acid, to have access to phosphate under phosphate-stress situations. Sufficiently acidic environments, provided by either inorganic acids (H_2SO_4) or organic acids formed locally under primordial conditions, are likely to have been available. Acidic media provide, however, a severe disadvantage in condensation reactions such as peptide and phosphoester bonds: protons catalyze the backward reaction.

2) Multifunctional organic acids (including amino acids) and other organic compounds, for example, sugars, can efficiently chelate cations such as Ca^{2+}, $Fe^{2+/3+}$, and Al^{3+} and thus set free phosphate by "masking" the cationic part of the phosphate mineral under rather moderate acidic conditions.

3) Oligophosphates, in particular diphosphate (mainly $HP_2O_7^{3-}$ at pH 7) and triphosphate (mainly $H_2P_3O_{10}^{3-}$ at pH 7), are much easier soluble than monophosphates in neutral, slightly acidic, and even slightly alkaline aqueous media. The pentasodium salt of triphosphate, for example, is employed to soften hard water: its calcium salts are soluble. Di- and triphosphates are formed from monophosphates by condensation. Cyanamide and its valence isomer carbodiamide (see Eq. (7.47)) effectively promote the condensation of phosphate to diphosphate, Eq. (7.48).

4) Condensation affected by cyanamide also keeps phosphate in a soluble form by its attachment to (esterification with) organic water-soluble molecules. An example is 6-glucosephosphate, Eq. (7.49); for this activated sugar, see also Scheme 7.6. Cyanamide is converted to urea in these reactions.

$$2H_2PO_4^- + H_2N-C\equiv N \rightarrow H_2P_2O_7^{2-} + (H_2N)_2C=O \tag{7.48}$$

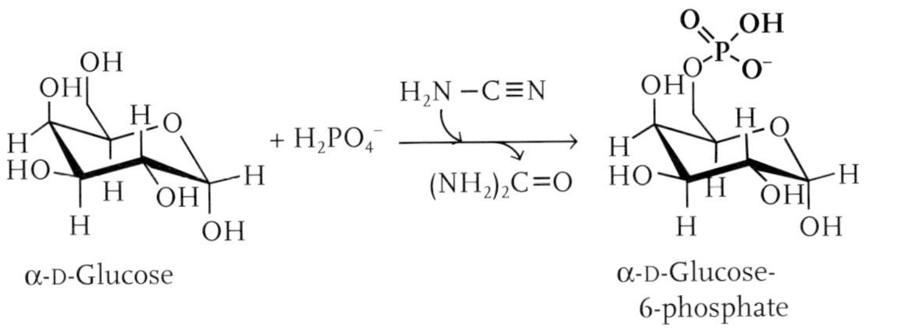

α-D-Glucose + H$_2$PO$_4^-$ $\xrightarrow[(NH_2)_2C=O]{H_2N-C\equiv N}$ α-D-Glucose-6-phosphate (7.49)

Meteoritic schreibersite (Fe, Ni)$_3$P may also be employed as a source for diphosphate, and thus the formation of organophosphates {O$_3$P–O–C} and phosphonates ({O$_3$P–C}, found in the Murchison meteorite): Aqueous corrosion of schreibersite produces H$_2$P$_2$O$_7^{2-}$, likely via phosphite radicals xPO$_3^{2-}$, along with H$_2$ as reduction equivalents [25b].

Equation (7.49) is representative for the phosphorylation of biologically important molecules. Commonly, this phosphorylation goes along with an increase in intrinsic energy and thus an increase in reactivity of the molecule in metabolic pathways. This includes the last step of nucleotide formation, namely, the introduction of phosphate into the 5′ position of the ribose in the nucleoside. The increase in intrinsic energy implicates a decrease of stability. It has therefore been hypothesized that a more robust RNA analog may have preceded RNA. Such analogs may have been provided by peptide nucleic acids (PNAs) [26a], in which a neutral pseudopeptide backbone replaces the ribose phosphodiester backbone. Such a peptide scaffold can be assembled from diamino acids, for example, 2,3-diaminopropanoic acid (aminoethylglycine) or 2,4-diaminobutanoic acid, Scheme 7.17, which have been found in meteorites (the Murchison meteorite is an example), and synthesized by UV irradiation of ices based on CO, CO$_2$, CH$_3$OH, NH$_3$, and H$_2$O [26b].

Scheme 7.18 summarizes the formation of the basic building blocks, synthesized from simple precursor molecules in the primordial soup or synthesized in

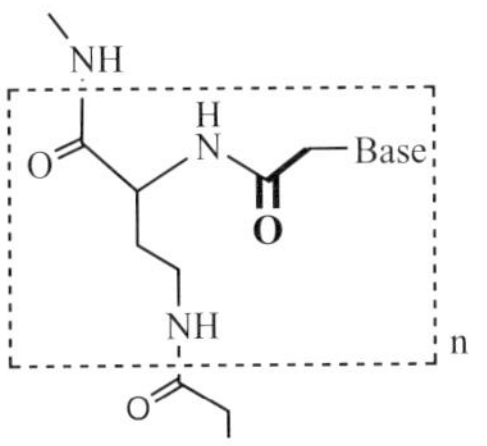

Scheme 7.17 PNA derived from nucleobases derivatized with 2,4-diaminobutanoic acetyl. The monomeric unit is framed; the acetyl moiety highlighted in bold.

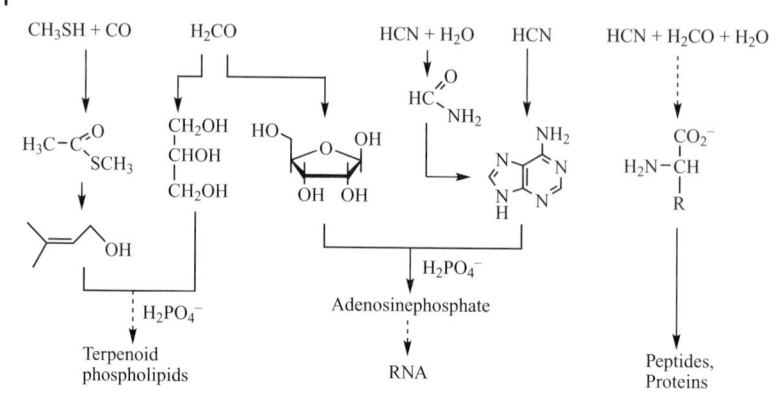

Scheme 7.18 Summary of central reaction paths, starting from simple precursor molecules (top row) and preceding to more complex units and further to oligomers and polymers that serve as building blocks for membranes (terpenoid phospholipids), for information storage and information transfer (RNA), and for intracellular structures (proteins) in primitive pre-LUCA life forms. The formation of a *purine* nucleotide is exemplified here for adenosine; for the formation of *pyrimidine* nucleotides, cf. Scheme 7.16. Dashed arrows indicate that additional molecular fragments are needed for product formation.

interstellar clouds and brought to Earth by extraterrestrial messengers such as meteorites and interplanetary dust.

As far as the primordial synthesis of amino acids is concerned, racemates are obtained, that is, mixtures of equal amounts of the D- and L-enantiomers. Terrestrial organisms almost exclusively use just the L-version of the mirror images. Since there is no apparent reason for this choice, it is the obvious thing that either one of the optical isomers has been used preferentially in peptide formation, as in the salt-induced, Cu^{2+}-assisted condensation (Figure 7.6), or has been available in excess and therefore employed preferentially. What then can have caused the selective formation of one of the antipodes? This question had already been addressed in the context of the L-amino acid enrichment in meteorites (Section 5.3.2) and interstellar dust grains (Section 4.2.5.2), and shall be resumed here in more detail. There are several processes allowing for a very efficient amplification of asymmetry, a slight initial symmetry braking and thus imbalance between L- and D-forms provided, summarized in Scheme 7.19.

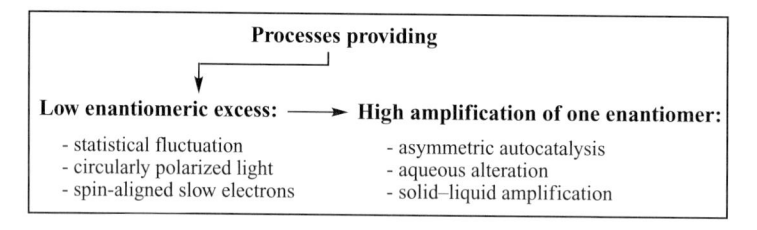

Processes providing

Low enantiomeric excess: ⟶ **High amplification of one enantiomer:**

- statistical fluctuation
- circularly polarized light
- spin-aligned slow electrons

- asymmetric autocatalysis
- aqueous alteration
- solid–liquid amplification

Scheme 7.19 An overview of selected processes leading to enantiomeric enhancement.

The degree of asymmetry is quantified by the percental enantiomeric excess (*ee*) defined in Eq. (7.50). In principle, a very low *ee* can come about by statistical fluctuation. In practice, a low imbalance between the two enantiomers can result from irradiation with circular polarized vacuum ultraviolet (CPVU). Vacuum UV brakes down organic molecules such as amino acids (to form volatiles such as CO_2 and NH_3); CPVU does so selectively because it is preferentially absorbed by either the left- or the right-handed amino acid,[14] leaving one of the enantiomers in excess. An enrichment of L-leucine by 0.9% has thus been achieved by irradiating racemic leucine with left CPVU at 182 nm [27a]. Less-destructive circularly polarized UV, with wavelengths between 270 and 320 nm, selectively eliminates one of the enantiomers of (R,S)-aldehydes, yielding better *ee*s. Aldehydes substantially enriched in one of the isomers, for example, (S)-2-methylbutyraldehyde, may have served as precursors for L-amino acids, L-isoleucine in this specific case [27b], Eq. (7.51). Potential sources for CPVU are reflection nebula in star formation regions, magnetic supernova remnants (neutron stars), magnetic white dwarfs, and light scattering from elongated interstellar dust grains oriented along the intragalactic magnetic field lines. Chirality-dependent bond cleavage has also been observed for R,S-butanol subjected to slow, spin-polarized electrons [28]. Electrons, liberated from atoms or molecules by ionizing radiation, can align their spins in a magnetic field as provided by a magnetic substrate such as magnetite Fe_3O_4:

$$ee = 100 \times (c_L - c_R)/(c_L + c_R)\% \tag{7.50}$$

where c_L and c_R are concentrations of the L- and D-enantiomers, respectively; $c_L > c_R$.

$$\text{(S)-2-Methylbutyraldehyde} + HCN + H_2O \longrightarrow \text{L-Isoleucine} \tag{7.51}$$

All of the chiral 20 α-amino acids used in peptide and protein structures of known living organisms carry a hydrogen atom at the carbon in the α-position and are thus, at ambient temperature, subject to slow racemization. On timescales of a hundred million years, that is, on timescales afforded for the development of prebiotic material into oligomeric and polymeric structures necessary for the self-organization of the first primitive life forms, all minor magnifications of the L-enantiomers will have been equalized. Stabilization and hence outwearing of the required time span may have been achieved by absorption on surfaces of chiral minerals that, like quartz (Figure 7.1a in Section 7.2), occasionally occur in just one of their antipodes. More efficient than CPVU and spin-aligned electrons are chemical reactions that can amplify one enantiomer to dominance: asymmetric autocatalysis, solid–liquid amplification, and aqueous alteration.

14) A phenomenon termed dichroism. One of the enantiomers has a greater extinction coefficient, thus absorbs more photons of a defined energy, and consequently is photolyzed faster.

Asymmetric autocatalysis is the catalytic replication of a chiral compound, that is, a chiral product acts as a catalyst for its own production (suppressing the production of its antipode). A key reaction in this respect, known as the Soai reaction, is the alkylation of 2-methylpyrimidine-5-carbaldehyde with di*iso*propyl zinc to form the chiral *iso*propyl alcohol of 2-methylpyrimidine with an *ee* > 95% when initiated with ca. 2% *ee* L- or D-leucine (Leu), Eq. (7.52), or -valine [29]:

$$\text{Methylpyrimidine-5-carbaldehyde} + \text{Zn}(i\text{Pr})_2 \xrightarrow{\text{}^+\text{H}_3\text{N-CHR-CO}_2^-\ \text{Leu}} \text{Methylpyrimidine-5-}iso\text{propanol} \qquad (7.52)$$

Solid state single chirality by solid–liquid amplification is achieved in a system, where crystals of the L- and D-amino acids (with one of the enantiomers in a slight excess) are in equilibrium with a saturated solution of the racemic amino acid. This reaction affords the presence of a catalyst that mediates solution-phase racemization. The effect is enhanced by mechanical attrition. An example is the evolution of solid, enantio-pure L-aspartic acid, starting from a racemate of solids with a minor excess of the L-isomer [30], Eq. (7.53):

$$\text{solid D-enantiomer} \rightleftarrows \left\{ \text{solution phase} \xrightleftharpoons{\text{(cat)}} \right\} \rightleftarrows \text{solid L-enantiomer}$$

$$\text{cat} = \text{2-hydroxybenzaldehyde} + \text{CH}_3\text{CO}_2\text{H} \qquad (7.53)$$

Meteorites, mainly carbonaceous chondrites (Section 5.3.2), can contain a comparatively large excess of the amino acids in their L-form, topped by 18.5% *ee* of L-isovaline in the Murchison meteorite. Isovaline, without a hydrogen atom on Cα, is resistant toward racemization. The *ee* in meteoritic amino acids is commonly traced back to aqueous alteration of the originally racemic product during development of the asteroidal parent body of the meteorite [31]. Aqueous alterations are modifications to minerals that occur through reactions with water. The extent of aqueous alteration and the enantiomeric excess are correlated, suggesting that aqueous alteration has been influential for the enrichment of L-isomers. A possible mechanism by which symmetry braking can occur is a mutual chiral induction, between the mineral phase and the amino acid adsorbed to it, in the course of aqueous mineral transformation. Plausible candidates for suitable minerals are calcite and clays.

Selective binding of L-amino acids to Lewis sites of surfaces of chiral clayey minerals may induce chiral amplification. In any case, clays do enable condensation reactions between amino acids to form peptides and between nucleotides to form oligonucleotides. In this respect, clayey materials are popularly referred to as "clay organisms." Before returning, in Section 7.4.4, to the origin of basic molecules and the possible input from extraterrestrial sources, the role of "clay organisms" in oligomer formation will be highlighted in Section 7.4.3.

7.4.3
"Clay Organisms"

Typical clays are alumosilicates such as montmorillonite (Figure 7.1b in Section 7.2), forming alternate layers of interconnected (partially protonated) tetrahedral $\{Si^{4+}(O^{2-})_4\}$ and octahedral $\{Al^{3+}(O^{2-})_6\}$ units, and providing the ability to accommodate water and hydrated mono- and divalent cations in-between the layers. This internal surface of clay minerals has available Lewis acid sites (the Al^{3+} cations) for the absorption and thus immobilization of an amino acid via its carboxylate functionality, and sites (the O^{2-} or OH^- groups of the clay scaffold) for interaction, via hydrogen bonding, with the amino functionality of the amino acid, lowering the activation barrier for condensation, and removing the water of reaction (thus forestalling the back-reaction) by keeping it coordinated to Al^{3+} or directing water molecules into the hydration sphere of the interlayer cations. This is sketched in Figure 7.7 for the formation of glycyl–glycine. Oligomerization up to $(Gly)_6$ has been observed, as well as the formation of heterodipeptides, among these glycyl–alanine, and a preference of L-amino acids [32a]. Model calculations for the formation of glycylamide from glycine and ammonia on a feldspar surface have shown that the activation barrier is reduced by $134\,kJ\,mol^{-1}$ with respect to the gas-phase reaction [32b].

Activated alumina Al_2O_3 appears to be even more effective in the formation of the peptide bond, which is possibly due to the fact that hydration of alumina (through the water released by the condensation reactions) is exothermic, and the substrate thus also functions as an energy source. Alumina-catalyzed oligopeptide formation yielded peptides such as $(Gly)_{11}$, $(Gly)_2Ala$, $Ala(Gly)_2$, $(Gly)_2Leu$, $(Gly)_2Pro$, and $(Leu)_2$ (Ala = alanine, Leu = leucine, and Pro = proline). Mechanistically, the

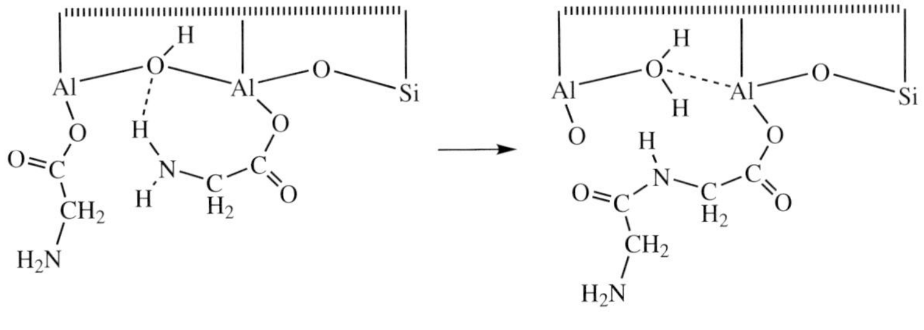

Figure 7.7 Activation of glycine and its condensation to glycyl–glycine on a montmorillonite or related clay surface. For details of the build-up of montmorillonite, see Figure 7.1.

formation of a tripeptide such as Gly–Gly–Ala, Eq. (7.54), appears to proceed via a cyclic intermediate of glycyl–glycine, formed by alumina-assisted condensation [33]. It should be cautioned, however, that, under prebiotic conditions, terminators of chain growth – simple carboxylic acids and amines – were available in higher concentrations than chain-extending amino acids. Note that, while a plethora of amino acids is present in meteoritic materials, no peptides have been found:

$$\text{(7.54)}$$

Clays, in particular conditioned montmorillonites, also efficiently catalyze the condensation of nucleotides to oligonucleotides up to 50-mers [4b] and thus to a size where information storage and catalytic ribozyme-like functions could become feasible. Montmorillonites are formed by weathering of volcanic ashes. They contain, along with the main scaffold cations silicon and aluminum, varying small amounts of other cations, both as substitutes for Al^{3+} and Si^{4+} and in the interlayer space, among these Cu^{2+}, Fe^{3+}, Zn^{2+}, and Mg^{2+}. Partial replacement of Al^{3+} and Si^{4+} by lower valence cations leaves negative charges on the lattice structure. This supports substrate binding that can be important for catalytic activity. When present in the interlayers, these ions act, however, as catalyst poisons. Conditioning of the clay thus includes removal of all of the interlayer cations by acid treatment and partial restoring of the metal cation load with Na^+. For catalytic activity, balanced interlayer amounts of Na^+ and H_3O^+ appears to be essential, a balance that is achieved by adjusting the pH to 6–7. Interestingly, montmorillonites are also present on Mars.

For the oligomerization, *activated* nucleotides have to be employed. Activation is achieved by introducing a phosphoamidate bond through the reaction of a suitable amine with ATP in the presence of Mg^{2+}. An adequate amine is imidazole or, even more efficient, the cationic imidazole derivative 1-methyladeninium. With the latter, 50-mers are produced at 25 °C within less than a day, and with up to 70% specificity in 3′,5′-linkages, which is the ribose-phosphodiester-ribose mode of linkage also relevant for RNA. The reaction starts with the intercalation of the activated nucleotide monomers in-between the clay platelets. The first step in the clay-catalyzed reaction, leading to the formation of an activated dinucleotide, is shown in Scheme 7.20.

The successful oligomerization of adenine and uridine nucleotides by clayey minerals that very likely have been present also on early Earth, viz. the oligomerization up to chain lengths conspicuously close to what supposedly is the lower limit for information storage and ribozyme activity, has substantially attributed to the revival of the notion that "clay organisms" have preceded the RNA world and LUCA. But do these doubtlessly fascinating laboratory experiments reproduce the

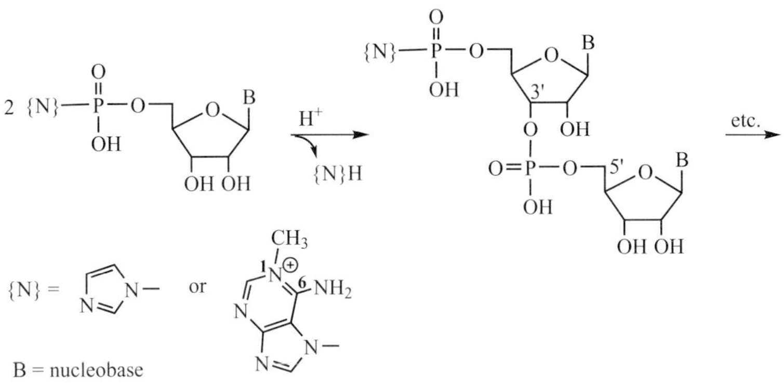

Scheme 7.20 Formation of oligomeric nucleotides from activated monomers, catalyzed by conditioned montmorillonite [4b]. Depending on the activator {N}, the nucleobase B stands either for adenine and/or uracil ({N} = 1-methyladenylium) or for adenine, uracil, guanine, and cytosine ({N} = imidazolyl). Along with the 3′,5′-linkages (shown), 2′,5′-links are also formed.

feasibility on the primordial Earth some 3.8 or so billion years ago? This notion has sensibly been criticized [2, 3], and a briefing on some of the points of criticism is provided here:

- Modified ("conditioned") montmorillonites are employed. The native clays contain transition metal ions that interfere with phosphorylation. In fact, native clays are inactive.

- The "activator" 1-methyladenylium will have been around in vanishing amounts only. The prebiotic formation of adenine from HCN or formamide is already comparatively ineffective. 1-Methyladenylium can be synthesized from adenine and methylamine (which had been readily available) under rigorous conditions only. But even then, the by far dominant methylation product is N^6-methyladenine (adenine methylated at the *exo*-nitrogen in position 6; cf. Scheme 7.10).

- As in the case of the growth of peptide chains of noteworthy lengths, the growth of oligonucleotide chains will have been counteracted by readily available terminators, alcohol-functional molecules in this case, that interfere competitively with phosphorylation processes, leading to phosphoesters of ribose.

- As already pointed out earlier, it is extremely unlikely that a randomly organized 50-mer RNA would act as a RNA replicase or a ribozyme for protein synthesis.

Although principally conceivable, the possibility that "clay organisms" have played a role in initiating the supply of oligomers and polymers used in cellular life forms is thus just minute. A more direct doorway to oligonucleotides – though less efficient as it comes to the size of the oligomer – is the oligomerization, by γ-rays, of adenosylmonophosphate to 8-mers on the surface of basaltic volcanic ashes under acidic conditions typical of postvolcanic activity [34].

7.4.4
Extraterrestrial Input

Chemical processes in interstellar clouds, in gas and dust envelopes of developed stars such as preplanetary and planetary nebulae, and in protoplanetary clouds in the process of star formation, did generate and are continuously generating a suite of fundamental molecules akin to those that had been used as seed molecules in primordial and prebiotic chemistry on Earth described in the preceding sections. The chemistry in interstellar (molecular) clouds and in the icy surface layers of micrometer- and submicrometer-sized dust particles, which are constituents of these clouds, has been exemplified in Chapter 4 and detailed in the network charts introduced there (Schemes 4.3–4.6). Here, a brief summary on selected reaction paths that deliver key molecules to prebiotic chemistry is being provided.

Simple *key molecules* for prebiotic chemistry, employed in experiments mimicking prebiotic conditions by Miller and Urey (Section 7.4.2), Wächtershäuser (Section 7.4.1), and others, are based on five elements (H, C, O, N, and S) constituting the main part of organic matter. These elements are delivered by water H_2O, formaldehyde H_2CO, hydrogen cyanide HCN, ammonia NH_3, and hydrogen sulfide H_2S. The synthesis of these key molecules, exemplified by Eqs. (7.55)–(7.59), usually takes place in the gas phase under the extremely low-pressure and -temperature conditions typical of the common interstellar clouds, restricting reactions to bimolecular events with low or nonexisting kinetic barriers (no activation energy afforded), thermodynamic feasibility (exergonic reactions), and transfer of the impact (and bond formation) energy to a second reaction product or a third reaction partner. In many cases, the second reaction product responsible for the off-transport of energy is a hydrogen atom or a photon $h\nu$. Since hydrogen is by far the most abundant element in the interstellar medium, key reactions often involve H_2 that by itself is formed in a three-body reaction (two hydrogen atoms plus grain surface; Section 4.2.5.1). Reactions that fulfill kinetic and thermodynamic restrictions are ion–molecule (ion–neutral) reactions – the initiating steps in reactions (7.55)–(7.59) – and dissociative recombination: the final electron capture steps in these reactions.

$$CO + H_3^+ \xrightarrow[H_2]{H_2} HCO^+ \xrightarrow[h\nu]{H_2} H_3CO^+ \xrightarrow[H]{e^-} H_2CO \quad (7.55)$$

$$HCO^+ + H_2 \rightarrow CH^+ + H_2O \quad (7.56)$$

$$N^+ \xrightarrow{H_2\,(3x)} \xrightarrow[H]{} NH_3^+ \xrightarrow[Mg^+]{Mg} NH_3 \quad (7.57)$$

$$NH_3 \xrightarrow[H]{C^+} H_2CN^+ \xrightarrow[H]{e^-} HCN + CNH \quad (7.58)$$

7.4 Scenarios for the Primordial Supply of Basic Life Molecules

$$S + H_3^+ \underset{H}{\overset{}{\rightarrow}} H_2S^+ \underset{H}{\overset{H_2}{\rightarrow}} H_3S^+ \underset{H}{\overset{e^-}{\rightarrow}} H_2S \tag{7.59}$$

Ion–neutral reactions followed by electron capture leading to the formation of the somewhat more complex molecules acetonitrile CH_3CN, methylamine CH_3NH_2, and aminocyanide NH_2CN; Eqs. (7.60)–(7.62):

$$HCN + CH_3^+ \underset{h\nu}{\overset{}{\rightarrow}} CH_3CNH^+ \underset{H}{\overset{e^-}{\rightarrow}} CH_3CN \tag{7.60}$$

$$NH_3 + CH_3^+ \underset{H}{\overset{}{\rightarrow}} CH_3NH_2^+ \underset{h\nu}{\overset{e^-}{\rightarrow}} CH_3NH_2 \tag{7.61}$$

$$NH_3^+ + HCN \underset{H}{\overset{}{\rightarrow}} NH_3CN^+ \underset{H}{\overset{e^-}{\rightarrow}} NH_2CN \tag{7.62}$$

Even more complex precursors for biomolecules, for example, acetamide and glycolaldehyde in Scheme 7.21 (for a more complete list of complex molecules see Scheme 4.7 in Section 4.2.4), afford higher temperatures and denser regions for their formation, such as the "Large Molecular Heimat" (Sgr B2) in the constellation of Sagittarius. Three rather exciting molecules that have recently been detected in this rich source, aminoacetonitrile (the direct glycine precursor) [35a], ethylformate, and propylcyanide [35b], are included in Scheme 7.21. For aminoacetonitrile, see also the cover of this book. Once formed in the ice layers of dust grains, the molecules are protected from (radiative) destruction. Such protection definitely is essential for sensitive molecules, amino acids in particular, the generation of which in the frame of grain chemistry is described in detail by Eqs. (4.75)–(4.84). Meteorites, and thus the meteorites' parent bodies, the asteroids, are another extraterrestrial source for amino acids. As argued in Section 5.3.2, Eqs. (5.55)–(5.57), precursor molecules from interstellar sources have again been the primary

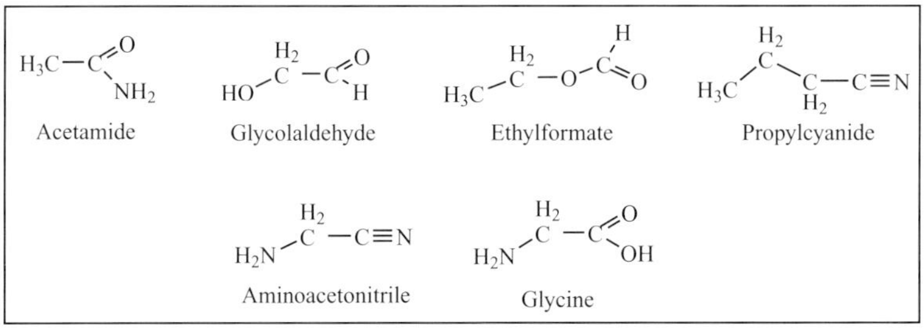

Scheme 7.21 A selection of biologically relevant complex molecules detected in interstellar environments. Aminoacetonitrile, ethylformate, and propylcyanide are three recently detected molecules in the Sagittarius B2 star formation region "Large Molecular Heimat." Glycine, collected from the coma of comet Wild 2 by the spacecraft Stardust, is also included.

supply, followed by processing on the asteroid itself or the premeteoritic fragments torn out of the asteroids. Finally, all of the essential basic molecules are present in comets (Section 5.4). The discovery of glycine (Scheme 7.21) in the dense gas and dust surrounding the nucleus of comet Wild 2, identified in samples collected by the spacecraft Stardust in 2004 [36], rates high in this context. Comets, like meteorites, may have supplied life's ingredients on occasion of impacts long ago.

In any case, the bulk of organic carbon on Earth, and also a substantial part of the water on our planet, has been brought in from interstellar space via meteorites, comets, and interplanetary dust grains, "unharmed" in the case of the latter, because dust is not subjected to heat shocks when entering Earth's atmosphere. Assuming an average carbon content of 2%, the amount of dust-bound carbon falling down on Earth from the interplanetary space amounts to ca. 10^5 kg per year; it should have been higher by several orders of magnitude during late accretionary infall about four billion years ago. Comets, the main constituent of which is water ice, nonetheless can carry a large amount of organic matter to Earth, as already proposed by Oró half a century ago [37]. Equation (7.63) provides a rough estimate of the organic material a comet can carry. Halley's comet, for example, with a radius of 5 km, a density of 0.26 g cm^{-3}, and a fraction of organic matter of 0.19, contains organic material that corresponds to about 10% of the current biomass on Earth. The organic material brought to Earth from extraterrestrial sources, with its ultimate origin coupled with the carbon-rich chemistry of preplanetary nebulae (Section 4.1), may have been the basis for, or contributed to, the development of a versatile organic chemistry on Earth:

$$m_{org} = f_{org} \times m_{comet} = f_{org} \times 4/3 \pi \rho r^3 \qquad (7.63)$$

where m = mass, f_{org} = fraction of organic matter, r = radius, and ρ = density.

A substantial amount of water on our planet was delivered by comets and icy meteorites. However, as revealed by the ^{2}H/^{1}H and ^{18}O/^{16}O isotope ratios, by far not all of the water on Earth had been provided by comets and icy meteorites. Water and OH had been comparatively abundant in the disk surrounding the young Sun. Recent observations show that the column density in protoplanetary disks of near-by objects is $N(H_2O)$ ~10^{17}–10^{18} cm^{-2} (~10^{-3}–10^{-2} mol m^{-2}). Water is efficiently produced by gas-phase reaction according to Eq. (7.64) at temperatures exceeding 300 K. The backward reaction is induced by far UV of 110–180 nm. Absorption of far UV at the disk surface protects water (and organics) within the disk from destruction. The persistence of H_2O in the protoplanetary disk is thus a consequence of self-shielding [38]:

$$O + H_2 \rightarrow OH + H \qquad (7.64a)$$

$$OH + H_2 \rightarrow H_2O + H \qquad (7.64b)$$

About 80 amino acids have been found in meteorites, only 8 of which match with the 20 (or 21 if selenocysteine is included) amino acids essential for life: glycine, α-alanine, aspartic acid (Asp), glutamic acid (Glu), valine, leucine, isoleucine, and proline.[15] Of interest in this context is that archaea – and thus also LUCA – have employed more acidic amino acids (Glu and Asp) in their protein structures than bacteria and eukarya. Along with the eight essential amino acids, the three nucleobases adenine, guanine, and uracil are present in trace amounts in meteorites such as Murchison, Orgeuil, and Murray. While prebiotic delivery of molecules from exogenous sources can thus explain the early availability of some of the amino acids and nucleobases on Earth, other building blocks for life, the remaining amino acids and nucleobases, sugars (in particular ribose),[16] and isoprenoids are less easily traced to extraterrestrial origin. This may reflect insufficiently sensible detection methods, and thus a problem that will be overcome with time. On the other hand, the missing part of the inventory for life as we know it should hint toward at least a participation of Earth-bound prebiotic synthons.

7.5
Extraterrestrial Life?

Although the formation of structural peptides and information-carrying oligonucleotides is, in principle, conceivable along the routes outlined above by resorting to the supplies of simple and less simple precursor molecules (of whatever origin) available on the primordial Earth, the probability that, in such surroundings, the organic macromolecules became infected by the spark of life is an extremely improbable event, clearly minimizing the chance that a similar "incident" had occurred elsewhere. If there is life somewhere else in our Solar System (and even some place in our Milky Way galaxy), its origin through panspermia is in fact more likely than its *de novo* invention. In this chapter, it is thus not intended to provide "proofs" for the existence of extraterrestrial life. Rather, conditions and chemical reactions will be identified, which are being discussed as settings for developments analogous to Earth, and thus as imaginable scenarios for the presence of extraterrestrial life.

Other celestial bodies in the Solar System have been subjected to the same input of cometary, meteoritic, and interplanetary dust sources as Earth, and hence will have received the same suite of organics that, in principle, is needed for the construction of biomolecules. Further, from a chemistry point of view, there is no apparent reason why, on other planets comparable to Earth (Mars and Venus) and on moons with features reminiscent of Earth (Titan, Encephalus, and Europa), the prevailing conditions should not have allowed for the synthesis of, say, the amino acid α-alanine, the nucleobase adenine, and the sugar ribose. Hydrogencyanide is a major component in the inventory of interstellar clouds as well as in condensed matter of whatever provenience, and easily oligo-/polymerizes and becomes

---

15) Sarcosine, aminoisobutyric acid, and β-alanine are also present; these amino acids have restricted biological occurance, for example, in the bacterial cell wall.

16) Sugar-related compounds such as sugar alcohols, sugar acids, dihydroxyacetone, and glyceraldehyde are present in meteorites.

further processed in the presence of water and/or ammonia [37b, c], two other well-established and universal constituents. As has been shown earlier (Scheme 7.14), adenine does form by pentamerization of HCN. Other pyrimidine and purine bases, as well as amino acids, have been detected in laboratory experiments where HCN polymers were subjected to thermochemolysis in the presence of the alkylating agent tetramethylammonium hydroxide [39].

Today's Venus, with her surface temperature of 480 °C and no liquid (surface) water, is, judged to our standard, quite inhabitable. But has this been always so? And what about the possibility that habitable niches exist somewhere on Venus today?

Venus has very likely experienced climate changes during her history and may have harbored oceans during earlier, cooler periods for a time span long enough to allow for the development of life from biomolecules carried from space or generated by prebiotic synthesis on Venus. A cooler period has been evoked from a less bright Sun 4 Ga ago, where the intensity of the Sun's radiation was only ca. 70% that of today. The presence of tremolite, a mineral that is formed by the reaction between dolomite, silica, and water, Eq. (7.65), could provide evidence for an ancient ocean. A closer look to Venus in one of the oncoming space missions might reveal the presence of tremolite:

$$5CaMg(CO_3)_2 + 8SiO_2 + H_2O \rightarrow Ca_2Mg_5(OH)_2(Si_4O_{11})_2 + 3CaCO_3 + 7CO_2$$
$$\text{Dolomite} \qquad\qquad\qquad\qquad \text{Tremolite} \qquad\quad \text{Calcite}$$

(7.65)

There are several parallel and ensuing processes that give rise to warming or cooling effects and thus to climate changes: SO_2 and H_2O released by global volcanic activity react, catalyzed by nitrous oxide or dust surfaces, to produce sulfuric acid (Section 5.2.3) that forms thick clouds, effectively warding off the main amount of sunlight, and thus having provided cooler conditions perhaps a billion years ago. Both water and sulfur dioxide (and, of course, CO_2 which is the dominating atmospheric constituent today) are efficient green house gases. Depletion of H_2SO_4 and SO_2 by reaction with surface minerals might have thinned out the clouds, allowing the planet to warm. This warming-up may have been counteracted by diminishing the green house gas SO_2 and by the depletion of water, which is split by lightning and UV into hydrogen (which escapes) and oxygen (which is lost by surface reactions) [40].

The Venusian oceans, whenever they existed, are lost for sure. But bacteria that may have evolved three to four billion years ago (or carried to Venus by Earth- or Mars-borne meteorites) might have adapted to the conditions in the Venusian clouds where, at a height of 50 km, temperatures of 50–70 °C and water droplets allow for the thriving of acidophilic thermophiles. Unexpected abundances of trace gases in Venus' atmosphere have been tentatively interpreted in terms of microbial activity. Thus, carbon monoxide (17 ppm) is scarce, although it should be produced in larger quantities from CO_2 by lightening and UV. The removal of CO from the atmosphere by the action of bacterial carbon monoxide dehydrogenase (CODH), Eq. (7.66), could account for this depletion. The notation [H] in Eq. (7.66) and the subsequent equations stands for either H_2, or $2H^+ + 2e^-$, or $H^+ + H^-$:

$$CO + H_2O \xrightarrow{(CODH)} CO_2 + 2[H] \tag{7.66}$$

The coexistence of H_2S and SO_2 in the Venusian atmosphere, gases that spontaneously comproportionate to sulfur in the presence of water, might go back to the activity of sulfur bacteria, an assumption that gains support by the concomitant presence of COS and CS, both of which are part of the global sulfur cycle. H_2S is produced by sulfate-reducing bacteria also common in hydrothermal vents (cf. Eq. (7.6)), and H_2S can be reoxidized by sulfide-oxidizing bacteria. In both reactions, summarized by Eq. (7.67), SO_2 is formed as an intermediate product. Other possible routes to hydrogen sulfide start from methane sulfonic acid, Eq. (7.68), or from organic sulfur compounds {S} such as cysteine, Eq. (7.69). In the latter case, carbonyl sulfide COS is released as an intermediate. Its further conversion to H_2S and CO_2 is catalyzed by carboanhydrase (CA), a widespread enzyme depending almost exclusively on zinc but, as an exception, can also built cadmium into its active site,[17] an element that otherwise is highly toxic for organisms:

$$H_2SO_4 + 8[H] \rightleftharpoons H_2S + 4H_2O \tag{7.67}$$

(with intermediates H_2SO_3, $SO_2 + H_2O$, SO_2)

$$CH_3SO_3H + 8[H] \rightarrow CH_4 + H_2S + 3H_2O \tag{7.68}$$

$$\{S\} \rightsquigarrow COS \xrightarrow{H_2O \ (CA)} CO_2 + H_2S \tag{7.69}$$

As noted in Section 5.2.3.3, iron sulfide minerals of compositions varying between pyrite FeS_2 and pyrrhotite $Fe_{1-x}S$ (e.g., $x = 0.1$: Fe_7S_8) have probably been present on Venus. These ferrous sulfides can provide the conditions for a pre-RNA world such as the "iron–sulfur world" described in Section 7.4.1, a world that hosts the imaginary Wächterhäuser "pioneer organisms," based on pseudocellular environmental frames relying on iron sulfide and powered by the reduction of sulfide to disulfide. Although such an alternative "metabolism first" scenario fulfils the criteria for "life" coined in Section 7.1 (metabolism, replication, and evolution), it hardly satisfies our (perhaps Victorian) notion of life.

Mars (Section 5.2.4) is *that* celestial body in the Solar System that provides conditions for potential life pretty close to those encountered on Earth. Although receiving only an average of 12% of the intensity of the sunlight *we* have available, the temperature, ranging between ca. -5 and $-85\,°C$ is comparatively moderate. Since Mars is sufficiently smaller and lighter than Earth (the volume is 0.151, the mass 0.107 Earths), it has lost most of its CO_2-dominated atmosphere; the mean surface level pressure is just 6 mbar (600 Pa). Another consequence of its much

---

[17] A Cd-dependent carboanhydrase has been isolated from the marine diatom *Thalissiosira weissflogii*. The exceptional use of the toxic Cd instead of Zn is a further example for organisms classified as "extremophile" (Section 7.3).

smaller mass and greater distance from the Sun is the lack of a magnetic field. Most importantly, Mars does harbor plenty of water, present in the form of atmospheric haze, water ice as constituent of the polar caps, and subsoil liquid water and/or water ice. The surface topology of Mars, with its fluvial features, suggests several wet climate cycles that have contributed in shaping the surface, pointing to the presence of liquid surface water in olden times. This view is supported by minerals such as hematite Fe_2O_3, jarosite $KFe_3^{III}(OH)_6(SO_4)_2$, and goethite $FeO(OH)$, which form by aqueous alteration of ferriferous (iron-bearing) rock. Also, deposits of clays, carbonates, sulfates, and chlorides strongly hint toward water. For a detailed discussion, see Sections 5.2.4.3 and 5.2.4.5.

Liquid subsoil water may still be present on Mars today – in the form of H_2O–H_2O_2 eutectics, saturated aqueous perchlorate, and weathering fluids loaded with a diversity of other salts. Mixtures of water and hydrogen peroxide can remain liquid down to −56.5 °C (61.2% H_2O_2) and further tend to supercool, that is, to form glasses when cooled beyond the freezing point, helping potential microbes living in these habitats to escape mechanical destruction that would occur in the case of crystalline freezing. H_2O_2 is present in the Martian atmosphere. It is formed from the lysis products essentially of water and CO_2, for example, as shown in Eqs. (7.70a) and (7.70b). Superoxide HO_2 by itself, as well as the hydroxyl radical OH, formulated as intermediates in Eq. (7.70), is a powerful oxidizing agent. Energy sources for their formation are UV photons and electrostatic fields, the latter generated by charged dust particles during sand storms and dust devils. But H_2O_2 could also be produced biochemically, along with formaldehyde in the sunlight-driven reaction between CO_2 and H_2O, Eq. (7.71) [41]. H_2O_2 is susceptible to decay into H_2O and O_2. But due to its hygroscopic nature, H_2O_2 is also readily absorbed by water. On first sight, habitats containing H_2O_2 appear to be hostile. And they are, in most cases: H_2O_2, a common disinfectant, is toxic for most microorganisms due to its high oxidizing power – to *most* microorganisms, though. *Acetobacter peroxidans*, which belongs to the family of bacteria that produce (acetic) acid from sugars and alcohols, is an example for a terrestrial organism that *uses* H_2O_2 (in the oxidation of substrates such as ethanol, acetate, and formate) as a terminal electron acceptor to *support* life, Eq. (7.72):

$$H_2O + h\nu \rightarrow HO + H; \; 2HO + M \rightarrow H_2O_2 + M \tag{7.70a}$$

$$CO_2 \rightarrow CO + O; \; O + OH \rightarrow HO_2; \; 2HO_2 \rightarrow H_2O_2 + O_2 \tag{7.70b}$$

$$CO_2 + 3H_2O \rightarrow H_2CO + 2H_2O_2 \tag{7.71}$$

$$H_2O_2 + 2[H] \rightarrow 2H_2O \tag{7.72}$$

The presence of perchlorate in the subsurface soil of Mars has been inferred from measurements carried out by the Phoenix lander (Section 5.2.4.3). Saturated aqueous perchlorate solutions remain liquid down to −70 °C and thus might provide the liquid life-supporting phase for primitive organisms. As H_2O_2, perchlorate ClO_4^- is a particularly efficient oxidizing agent. It is nonetheless used, at low concentrations, by several strains of bacteria (such as the genus *Dechloromonas*) as electron acceptor in respiration. The Phoenix lander did detect nil organics, which perhaps argues against perchlorate brines as life-supporting media, although a secondary process, the liberation of reactive oxygen from perchlorate, may be responsible for the destruction of organics that have eventually been present. Other solutes, salts in particular, can of course depress the melting point of water as well, and we are aware of this through the use of road salt in winter. Weathering fluids loaded with constituents of basalts of chemical composition closely related to basalts at Mars landing sites have been modeled, showing that, down to temperatures of 223 K, a certain percentage of the original water reservoir remains in the liquid state [42]. Extremophiles adapted to high salt (NaCl and $NaHCO_3/Na_2CO_3$) concentrations thrive on Earth (Section 7.3). Hence, from this point of view, nothing argues against corresponding habitats for microbial life on Mars. The low temperatures will slow down metabolic rates considerably, but again, very slow metabolism, with reproduction rates of several hundred years, has been inferred for terrestrial extremophiles living in chilly subglacial brine niches.

Methane has been proposed to be an indicator of microbial activity on Mars. The gas is present in the Martian atmosphere in an average concentration of 10 ppb, along with traces of formaldehyde as an oxidation product of CH_4. Methane can be provided exogenously by meteorites, comets, and interplanetary dust, and internally by volcanism. In addition, internal supply can be achieved abiotically (hydrogeochemically by serpentinization) and biotically; examples are provided by the *net* reactions, Eqs. (7.73) and (7.74); for additional details, cf. Sections 5.2.4.4 and 7.3:

$$6Fe_2SiO_4 + CO_2 + 2H_2O \rightarrow 4Fe_3O_4 + 6SiO_2 + CH_4 \quad (7.73)$$
$$\text{Fayalite} \qquad\qquad\qquad \text{Magnetite}$$

$$CO_2 + 8[H] \rightarrow CH_4 + 2H_2O \quad (7.74)$$

The *temporal* and *local* reoccurrence of methane plumes in the Martian atmosphere with methane concentrations >250 ppb has led to the assumption that such outbursts could indicate methanogenesis by prokarya. On Earth, about 95% of the methane released into the atmosphere is of biological origin. A potential biogenic source for methane on Mars need not be current. If biogenic methane had been produced in the past, it could have been stored in the form of methane clathrates (methane hydrates; Figure 5.14 in Section 5.2.4.4), forming deposits, from which it is released occasionally. The isotopic patterns of methane, the ratios $^{12}CH_4/^{13}CH_4$ and $C(^1H)_4/C(^1H)_3^2H$, could reveal whether Martian methane is of biogenic origin. In *bio*syntheses of organic matter on Earth, the heavier isotopomers are depleted with respect to the lighter ones as a consequence of the kinetic isotope effect, Section 5.3.2. The amount of CH_4 emitted on Mars by sporadic sources (irrespective of biogenic or nonbiogenic origin) has been estimated to 150×10^3 tons per year. Depletion of methane (and potential other organics) occurs rapidly, on timescales of a few hundred years (for atmospheric methane) to a few hours (for methane near the surface). Loss processes [43] include photochemical destruction, oxidation by photochemically formed OH and singlet-O, and – particularly efficient–oxidation by H_2O_2 or other strong oxidants present in surface regolith. Biogenic methane loss through methanotrophs is also conceivable. Methanotrophs are prokarya using CH_4 as carbon source, either aerobically, Eq. (7.75), or anaerobically. The anaerobic oxidation of methane to carbon dioxide is coupled to the reduction of sulfate, Eq. (7.22) in Section 7.3, or nitrate, Eq. (7.76). Hydrogen-peroxide may attain a comparable function as an electron acceptor; Eq. (7.77):

$$CH_4 + O_2 \rightarrow H_2O + H_2CO\ (\rightarrow\rightarrow \text{biomass}) \tag{7.75}$$

$$5CH_4 + 8NO_3^- + 3H^+ \rightarrow 5HCO_3^- + 4N_2 + 9H_2O \tag{7.76}$$

$$CH_4 + 4H_2O_2 \rightarrow HCO_3^- + H^+ + 5H_2O \tag{7.77}$$

The puzzling fact that no organics have been detected in Mars' surface soil does *not* prove that Mars is devoid of microbial life. Even if there were no organics indigenous to Mars, organic material must have been delivered to Mars exogenously in the same way as it has been delivered to Earth. The absence, so far, of any organics just is indicative of a particularly reactive, oxidizing surface.

Organics, in the form of polyaromatic hydrocarbons, are in fact present in Martian meteorites such as in the meteorite ALH84001 picked up in the Alan Hills fields, Antarctica, in 1984. In ALH84001, the polyaromatic hydrocarbons are associated with magnetite (Fe_3O_4) crystallites and carbonate structures, which originally have been believed to represent fossilized remains of nanosized microbes. As discussed in Section 5.2.4.3 (see also Chapter 1), the early enthusiasm accompanying this interpretation has been damped by serious arguments against such an interpretation, damped almost to the point where the notion that these remains are associated with nanobacteria became discarded in the scientific community. Very recent reinvestigation and reinterpretation of these structures [44a] may very well rejuvenate broader interest in the possibility of ancient life on our companion planet. The reinvestigations are centered on nanosized magnetite crystallites as potential *biomarkers*: The association of these magnetite crystals with carbonate disks (Figure 7.8b), their specific size and shape (Figure 7.8c) and, in particular, the lack of impurities, agree with magnetite assembled in the magnetosomes of magnetobacteria in terrestrial habitats (Figure 7.8a), and thus appear to point toward a biogenic origin of Fe_3O_4 in ALH84001. It has been argued that iron-based carbonates, such as siderite $FeCO_3$, when heated to temperatures of 500–700 °C in a CO_2 atmosphere, convert to magnetite, Eq. (7.78). Temperatures of 500 °C and more go along with impact events like the one that blasted ALH84001 off Mars ca. 15 million years ago, and the magnetites found in ALH84001 have thus been considered to originate from such a "baking" process. However, siderite minerals of geochemical origin are hardly devoid of other divalent metal ions such as Mg^{2+}, Ca^{2+}, and Mn^{2+} integrated into the lattice structure. The lack of these "impurities" in magnetite crystallites from ALH84001 is supportive of a biogenic origin:

$$3FeCO_3 \xrightarrow{\Delta} Fe_3O_4 + CO + 2CO_2 \tag{7.78}$$

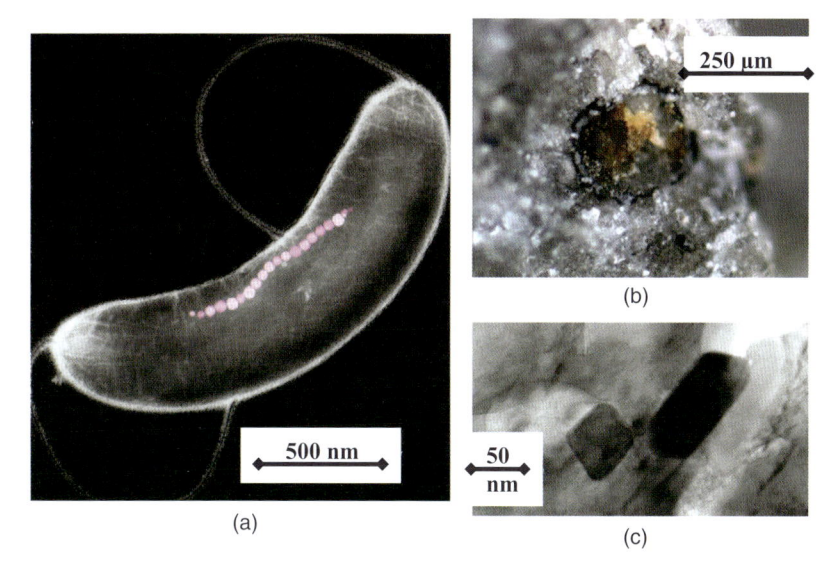

Figure 7.8 (a) Magnetite crystals (highlighted in pink) in a magnetotactic terrestrial bacterium. (b) Carbonate disks like the one shown here are typical features of ALH84001. The black rim is primarily composed of magnetite embedded in (Fe,Mg)CO$_3$; the white bands in the dark rim are magnesite MgCO$_3$. (c) Magnetite crystals from the disk center. Source: 403100main_life_on_mars; by permission by Kathie L. Thomas-Keprta.

An additional argument put forward in disfavor of a biogenic origin of Fe$_3$O$_4$ is the lack of a permanent magnetic field for Mars, and hence the uselessness of magnetic materials for a microorganisms. However, Mars may very well have had an intrinsic magnetic field in his early time. Spacecrafts orbiting Mars have detected weak magnetic fields in several areas, and these may well represent remnants of a once global magnetic field, disappearing when Mars was about 4 Ga old.

The carbonate disks in ALH84001 accommodating the magnetite crystallites have been dated back to 3.9 billion years; igneous crystallization of the matrix rock occured 4.1 Ga ago [41b]. This implies that life on Mars came prior to life on Earth. Terrestrial life is estimated to go back to 3.8–3.5 Ga, suggesting that terrestrial life originates from our neighbor planet – if life forms in the Solar System have developed here at all; see also below.

Life does not only depend on more or less complex organic compounds but also on transition metals, which are often integral constituent of organic molecules, providing in many cases catalytic centers for a variety of metabolic and catabolic processes. Examples for inorganic metal ions in the active centers of enzymes are iron, copper, zinc, nickel, cobalt, molybdenum, and vanadium. The latter two, Mo and V, are the most abundant metals in seawater, where they are present in the form of easily soluble molybdate MoO$_4^{2-}$ (100 nM) and vanadate H$_2$VO$_4^-$ (35 nM), exceeding (in oxic environments) the presence of directly available iron by 1–3 orders of magnitude, depending on the degree of oxygenation. Molybdenum plays an important role as an essential trace element in bacteria and archaea thriving in hydrothermal vents (black and white smokers, Section 7.3) in volcanic deep-sea ocean areas, but also in enzymes essential for humans.[18] Molybdenum has further been addressed in the context of mineral-based cells (Section 7.2) and the self-assembly of thiomolybdenum clusters (Section 7.4.1).

Vanadium [45] has been traced as an essential element (alternative to Mo) in nitrogen fixation by bacteria such as *Azotobacter* and in haloperoxidases (enzymes that catalyze the two-electron oxidation of halide by H_2O_2) of marine macroalgae, terrestrial fungi, and lichens. Even more intriguing are apparently *non*functional vanadium compounds found in living organisms, probably derived from "extinct" enzymes, that is, enzymes that have been employed in an early epoch of the development of life and discarded at a later stage. An example is amavadin in *Amanita* mushrooms such as the fly agaric, a molecular V^{IV} complex with an organic (hydroxyimino-di*iso*propionate) coordination environment. Vanadium can take over catalytic functions both by acting as a Lewis acid and by switching – in the range of physiological electrochemical potentials – between the oxidation states +V, +IV, and +III. It thus has available the catalytic potential of metals primarily operating via the ease of change in oxidation state (iron, molybdenum, and copper), and of metals the catalytic activity of which roots in their Lewis acid activity (zinc). The detection of abundant VO in the atmospheres of brown dwarfs and certain Jupiter-sized exoplanets (Chapter 6), thus is noteworthy because vanadium may have supported, through its potential as a versatile catalyst, primordial processes leading to the manifestation and establishment of life.

Along with the planets Mars and Venus, other Solar System candidates for harboring life are the Jovian moons Ganymed, Callisto, and Europa (Section 5.6.2), the Saturnian moons Enceladus and Titan, and the Neptunian moon Triton (Section 5.6.3). Common to all of these moons is a dominance of water ice in the crust, usually admixed with CO_2, CO, NH_3, CH_4, and HCN, and hence with suppliers of essential life elements in a reactive form, plus H_2O_2 as a potential source for energy supply; vide supra. These molecules are also constituents of the very faint (except of Titan) atmospheres typical of these moons. Water ice with additional ingredients is also the dominant material constituting the Kuiper belt objects (Section 5.5). For the Kuiper belt dwarf planets, beyond Neptune's orbit, temperatures are, however, too low for water ice to become liquefied. This appears not to be the case for the six big moons just mentioned, one of which, Triton, factually is a captured Kuiper belt object. For these moons, extended subsurface salty liquid water oceans presumably exist, and hence conditions that are believed to be essential for life in forms we are aware of. Titan, which carries a thick atmosphere (the ground level pressure is 1.5 that of Earth) and has available a rich stock of organics, plus methane–ethane lakes and (possibly) subsurface water–ammonia oceans, is a case of particular interest in the context of extraterrestrial life. Another

---

18) An example is xanthine oxygenase, a molybdenum-based enzyme involved in the catabolism of organic nitrogen compounds to uric acid.

favorite candidate is Enceladus with its liquid salt water reservoirs close to the surface that eventually undergo eruptive breakthroughs to the surface, driven by volcanism, and accompanied by daunting outbursts into space, Figure 5.29 in Section 5.6.3.

Assuming primitive life on all of these Solar System bodies certainly remains speculative as long as we do not have available more precise data. Mars, as outlined above, makes a difference. Presently, the majority of the scientific community agrees that the conditions on Mars have been – or still are being – in favor of hosting simple life forms akin to terrestrial bacteria and archaea. These organisms may have developed on Mars even before life begun on Earth, and disseminated over our planet by live-carrying rocks knocked out of the Martian surface through impacting meteorites about four billion years ago, rocks that eventually became subjected to the influence of the terrestrial gravitational field, and were finally captured by Earth. The Martian meteorite ALH84001 is an intriguing example.

But even if we assume that terrestrial life had been transferred to Earth from the red planet: Has Mars actually been the cradle of life? Our Solar System is pretty young – just 4.6 billion years old, which compares to 13.7 billion years for the age of the Universe. Life may thus well have developed somewhere else in our galaxy, and bacterial spores may have survived, incorporated, and thus protected from destructive forces in dust grains, blown off from dying star systems into preplanetary disks that merged with interstellar clouds from which newly born suns and their protoplanetary system arose, finally supplying adequate planets (i.e., planets with overall characteristics such as Mars and Earth) with the germ of Life. What are the probabilities for such a scenario?

As outlined previously, one of the major criteria for life is its dependence on information storage and transfer. Today's life forms employ DNA as information bank; early life forms preprokarya, and pre-LUCAs, are likely to have employed RNA. It has been proposed that the minimum number of nucleotides necessary to establish information storage and information transfer (through replication and the catalysis of polypeptide formation) is about 50. In the case of RNA, the four nucleobases employed are uracil, cytosine, adenine, and thymine. To start with, let us assume that these nucleobases have already been transformed into the monomers of RNA, the nucleotides (nucleobase + ribose + phosphate in the correct mode of link-up, that is, with the base in the 1′ and the phosphate in the 5′ positions of the β anomer of ribose). If we allow for 48 random combinations of these nucleotides via phosphodiester linkages (and again in the "correct," i.e., the 3′,5′ connection mode), the number of possible 48-mers is $4! \times 12! \approx 1.5 \times 10^{10}$. A few dozens of these combinations will have had the adequate sequence afforded for a *functional* RNA. This doesn't sound too bad, given the fact that there are ca. 10^{11} stars in our Milky Way galaxy, and assuming that every fiftieth of these stellar systems homes a planet comparable to Earth or Mars. The odds deteriorate, however, if we allow for other linkages than those mentioned. The oligomerization of nucleotides on montmorillonite surfaces, for example, revealed that about half of the 40- to 50-mers are present with 2′,5′ instead of the "correct" 3′,5′ linkages.

Given all this, there is but a minor chance that life as we know it could have developed more than once in our home galaxy – and hence, if not Earth, it is yet at least our Solar System that remains unique.

The low probability for life having developed a second time does not discourage endeavors to locate exoplanets in our galaxy. The search for exoplanets has just begun, and among the ca. 460 exoplanets that have so far been characterized (Chapter 6), less than a handful provide conditions at least faintly resembling those on Earth. At present, the detection of planets with the size and orbital characteristics of Earth is still hampered by insufficiently sensitive detection methods. Examples for exoplanets that are believed to be similar to Earth are Gl 581d and Gl 581e, which orbit their sun Gliese 581 at a distance of 0.22 and 0.03 AU, respectively [46]. Gliese 581 (also denoted GJ 581) is an eight billion year old M3 dwarf (6.26 pc away, in the constellation of Libra) of about one-third the mass of our Sun. The planet Gl 518e has 1.9 Earth-masses and is the smallest exoplanet so far detected. It is supposedly rocky like Earth, but orbits its sun at a distance where life based on organics would be pyrolyzed. In contrast, Gl 481d, with about 7 Earth-masses, probably consist of ice and rock; it orbits in the habitable zone of its sun and may thus have available a water-world.

The overall estimated number of galaxies in the Universe is close to 10^{10}. There is no reason to assume that the chemistry out there should be different from the one we are familiar with. If, in each of these galaxies, there has been a single event enabling the development of life comparable to that in the Milky Way galaxy, 10 billion habitats exist in the Universe. Given the distances – our next neighbor, the Andromeda galaxy, is 2.5×10^6 light-years away – we will hardly ever become aware of such a world. Provided that we do not admit to jumps through hyperspace.

Summary Sections 7.4 and 7.5

For the initiation of Earth-based primordial prelife forms, "pioneer organisms" have been inferred from what is termed "iron–sulfur world" (Wächtershäuser), developing from a primordial atmosphere consisting of CO_2, N_2, and H_2O as the main constituents, plus minor amounts of H_2 and O_2. The basic metabolic step for these pioneer organisms is the exergonic formation of the methyl ester of thioacetic acid (activated acetic acid, an analog of acetyl-CoA) from CO_2 and CH_3SH with iron or iron–nickel sulfide as a catalysts and electron donor (oxidation of FeS to pyrite FeS_2), in a pseudocellular scaffold provided by a honeycomb network of ferrous sulfide. Such an iron–sulfur world is attractive with respect to consecutive reactions enabled by activated acetic acid (C–C bond formation, generation of amide bonds), and the ubiquitous employment of self-organizing iron–sulfur clusters in the enzymatic electron shuttle by all life forms.

On the other hand, in the Miller–Urey experiments, a reducing primordial atmosphere composed of H_2, CH_4, H_2O, and NH_3 (or N_2) was employed, and subjected to electric spark discharges as energy source. These experiments deliver a variety of molecules, including various amino acids along the cyanohydrine pathway or along radical chain reactions. Amino acids also form as a nonreducing atmosphere is subjected to an energy source such as discharge or UV. If copper ions and high concentrations of NaCl ("salt-induced peptide formation") are available, peptide bonds are tied, with a preference of peptides containing L-amino acids. The building blocks of nucleosides, ribose and the nucleobases, accrue from formaldehyde H_2CO (ribose; so-called formose reaction), HCN or formamide (adenine), or HCN + H_2CO (guanine). The pyrimidine bases cytosine and uracil form from urea and cyanoacetylene. The compound 2-aminooxazol can be a key molecule in the generation of the nucleotide cytidinephosphate. The preferential use in life of L-enantiomers (in the case of amino acids) and D-enantiomers (in the case of sugars) may root in a small initial chiral imbalance caused by circularly polarized UV or spin-polarized slow electrons, followed by amplification via asymmetric autocatalysis, solid–liquid amplification, or aqueous alteration.

Basic key molecules for the generation of the more complex matter have been available in space and can thus have been carried to Earth. Examples for basic molecules providing the main elements constituting organic matter are H_2O, H_2CO, HCN, NH_3, and H_2S. Secondary molecules derived thereof are, for example, CH_3CN, CH_3NH_2, and NH_2CN. More complex molecules, such as aminoacetonitrile, propylcyanide, and ethylformate, formed by further processing, for example, in ice layers of interstellar dust grains. Glycine has been found in the coma of comet Wild 2.

Life may have developed on early Venus and retreated into niches when the Venusian oceans disappeared and the surface temperature increased to 460 °C. Water droplets in Venus' atmosphere at an altitude of 50 km, where the temperature is moderate, might accommodate acidophilic chemolithotrophs energized by sulfur redox chemistry. On Mars, water is amply present, at least as subsoil ice, but probably also in subsoil pools containing liquid brines or H_2O–H_2O_2 eutectics. Surface structures on Mars, as well as deposits of sulfates, carbonates, and minerals such as goethite FeO(OH) are very much in support of flows of liquid surface water in earlier times. Subsoil brines may provide a habitat for primitive life, very much as subglacial brine niches on Earth accommodate life. Occasional local and seasonal outbursts of methane on Mars have also been related to microbial life – but can originate from aerothermal activity as well. Finally, crystallites of pure magnetite Fe_3O_4 found in the 4.1 Ga old Martian meteorite ALH84001 may indicate early Martian life; terrestrial magnetobacteria synthesize magnetite in specialized cell organelles.

Life may have been distributed within the Solar System, or even within our Milky Way Galaxy, by panspermia. The likeliness that Life has evolved more than once in our galaxy is close to zero, that is, the odds are against the notion that additional independent events generating life (as we know it) have occurred elsewhere.

References

1. Plaxco, K.W., and Gross, M. (2006) *Astrobiology*, The John Hopkins University Press, Baltimore.
2. (a) Robertson, M.P., and Scott, W.G. (2007) The structural basis of ribozyme catalysed RNA assembly. *Science*, **315**, 1549–1553.; (b) Orgel, L.E. (2004) Prebiotic chemistry and the origin of the RNA world. *Crit. Rev. Biochem. Mol. Biol.*, **39**, 99–123.
3. Shapiro, R. (2006) Small molecule interactions were central to the origin of life. *Quart. Rev. Biol.*, **81**, 105–112.
4. (a) Fitz, D., Reiner, H., and Rode, B.M. (2007) Chemical evolution towards the origin of life. *Pure Appl. Chem.*, **79**, 2101–2117.; (b) Joshi, P.C., Aldersley, M.F., Delano, J.W., and Ferris, J.P. (2009) Mechanism of momtmorillonite catalysis in the formation of RNA oligomers. *J. Am. Chem. Soc.*, **131**, 13369–13374.
5. (a) Merca, A., Haupt, E.T.K., Mitra, T., Bögge, H., Rehder, D., and Müller, A. (2007) Mimicking biological cation-transport based on sphere-surface supramolecular chemistry: simultaneous interaction of porous capsules with molecular plugs and passing cations. *Chem. Eur. J.*, **13**, 7650–7658.; (b) Schäffer, C., Bögge, H., Merca, A., Weinstock, I.A., Rehder, D., Haupt, E.T.K., and Müller, A. (2009) A spherical 24 butyrate aggregate with a hydrophobic cavity in a capsule with flexible pores: confinement effects and uptake-release equilibria at elevated temperatures. *Angew. Chem. Int. Ed.*, **48**, 8051–8056.
6. Inagaki, F., Kuypers, M.M.M., Tsunogai, U., Ishibashi, J.-I., Nakamura, K.-I., Treude, T., Ohkubo, S., Nakaseama, M., Gena, K., Chiba, H., Hirayama, H., Nunoura, T., Takai, K., Jørgensen, B.B., Horikoshi, K., and Boetius, A. (2006) Microbial community in a sediment-hosted CO_2 lake of the southern Okinawa Trough hydrothermal system. *Proc. Natl. Acad. Sci. U. S. A.*, **38**, 14164–14169.; (b) Nealson, K. (2006) Lakes of liquid CO_2 in the deep sea. *Proc. Natl. Acad. Sci. U. S. A.*, **38**, 13903–13904.
7. (a) Sorokin, D.Y., and Kuenen, J.G. (2004) Haloalkaliphilic sulfur-oxidizing bacteria in soda lakes. *FEMS Microbiol. Rev.*, **29**, 685–702.; (b) Shapovalova, A.A., Khijniak, T.V., Tourova, T.P., Muyzer, G., and Sorokin, D.Y. (2008) Heterotrophic denitrification at extremely high salt and pH by haloalkaliphilic gammaproteobacteria from hypersaline soda lakes. *Extremophiles*, **12**, 619–625.
8. Boltyanskaya, Y.V., Detkova, D.N., Shumskii, A.N., Dulov, L.E., and Pusheva, M.A. (2005) Osmoadaption in representatives of haloalkaliphilic bacteria from soda lakes. *Microbiology*, **74**, 640–645.
9. Baker-Austin, C., and Dopson, M. (2007) Life in acid: pH homeostasis in acidophiles. *Trends Microbiol.*, **15**, 165–171.
10. Mikucki, J.A., Pearson, A., Johnston, D.T., Turchyn, A.V., Farquhar, J., Schrag, D.P., Anbar, A.D., Priscu, J.C., and Lee, P.A. (2009) A contemporary microbially maintained subglacial ferrous "ocean". *Science*, **324**, 397–400.
11. (a) Lin, L.-H., Wang, P.-L., Rumble, D., Lippmann-Pipke, J., Boice, E., Pratt, L.M., Sherwood Lollar, B., Brodie, E.L., Hazen, T.C., Andersen, G.L., DeSantis, T.Z., Moser, D.P., Kershaw, D., and Onstott, T.C. (2006) Long-term sustainability of a high-energy, low-diversity crustal biome. *Science*, **314**, 479–482.; (b) Chivian, D., Brodie, E.L., Alm, E.J., Culley, D.E., Dehal, P.S., DeSantis, T.Z., Gihring, T.M., Lapidus, A., Lin, L.-H., Lowry, S.R., Moser, D.P., Richardson, P.M., Southam, G., Wanger, G., Pratt, L.M., Andersen, G.L., Hazen, T.C., Brockman, F.J., Arkin, A.P., and Onstott, T.C. (2008) Environmental genomics reveals a single-species ecosystem deep within Earth. *Science*, **314**, 275–268.
12. Boussau, B., Blanquart, S., Necsulea, A., Lartillot, N., and Gouy, M. (2008) Parallel adaptions to high temperatures

in the archaean eon. *Nature*, **456**, 942–945.

13 (a) Ormland, R.S. (2010) No connection with methane. *Nature*, **464**, 500–501.; (b) Beal, E.J., House, C.H., and Orphan, V.J. (2009) Manganese- and iron-dependent marine methane oxidation. *Nature*, **325**, 184–187.

14 Konhauser, K.O., Pecoits, E., Lalonde, S.V., Papineau, D., Nisbet, E.G., Barley, M.E., Arndt, N.T., Zahnle, K., and Kamber, B.S. (2009) Organic nickel depletion and a methanogen famine before the Great Oxygen Event. *Nature*, **458**, 750–753.

15 (a) Urey, H.C. (1952) On the early chemical history of the Earth and the origin of life. *Geophysics*, **38**, 351–363.; (b) Miller, S.L. (1953) A production of amino acids under possible primitive Earth conditions. *Science*, **117**, 528–529.; (c) Miller, S.L., and Urey, H.C. (1959) Organic compound synthesis on the primitive Earth. *Science*, **130**, 245–251.

16 Wächterhäuser, G. (2007) On the chemistry and evolution of the pioneer organism. *Chem. Biodivers.*, **4**, 584–602.

17 Martin, W., and Russel, M.J. (2003) On the origins of cells: a hypothesis for the evolutionary transitions from abiotic geochemistry to chemoautotrophic prokaryotes, and from prokaryotes to nucleated cells. *Philos. Trans. R. Soc. London B*, **358**, 59–85.

18 (a) Müller, A., and Schladerbeck, N.H. (1986) Einfache aerobe Bildung eines $\{Fe_4S_4\}^{2+}$-Clusterzentrums. *Naturwissenschaften*, **73**, 669–670.; (b) Müller, A., Etzner, W., Bögge, H., and Jostes, R. (1982) $[Mo_4^{III}S_4(CN)_{12}]^{8-}$, a cluster with high negative charge and cubane-like Mo_4S_4-moiety – on the significance of cyanothiomolybdates for prebiotic evolution. *Angew. Chem. Int. Ed. Engl.*, **21**, 795–796.

19 Dörr, M., Käßbohrer, J., Grunert, R., Kreisel, G., Brand, W.A., Werner, R.A., Geilmann, H., Apfel, C., Robel, C., and Weigand, W. (2003) A possible prebiotic formation of ammonia from dinitrogen on iron disulfide surfaces. *Angew. Chem. Int. Ed.*, **42**, 1540–1543.

20 (a) Plankensteiner, K., Reiner, H., Schranz, B., and Rode, B.M. (2004) Simulation of the prebiotic formation of amino acids in a neutral atmosphere by electric spark discharge. *Angew. Chem. Int. Ed.*, **43**, 1886–1888.; (b) Bernstein, M.P., Dworkin, J.P., Sandford, S.A., Cooper, G.W., and Allamandola, L.J. (2002) Racemic amino acids from the ultraviolet photolysis of interstellar ice analogues. *Nature*, **416**, 401–403.; (c) Muñoz Caro, G.M., Meierhenrich, U.J., Schutte, W.A., Barbier, A., Arcones Segovia, A., Rosenbauer, H., Thiemann, W.H.-P., Brack, A., and Greenberg, J.M. (2002) Amino acids from ultraviolet irradiation of interstellar ice analogues. *Nature*, **416**, 403–406.

21 Sorrell, W.H. (2001) Origin of amino acids and organic sugars in interstellar clouds. *Astrophys. J.*, **555**, L129–L132.

22 Li, F., Fitz, D., Fraser, D.G., and Rode, B.M. (2007) Methionine peptide formation under primordial earth conditions. *J. Inorg. Biochem.*, **102**, 1212–1217.

23 (a) Ricardo, A., Carrigan, M.A., Olcott, A.N., and Benner, S.A. (2004) Borate minerals stabilise ribose. *Science*, **303**, 196.; (b) Lambert, J.B., Gurusamy-Thangavelu, S.A., and Ma, K. (2010) The silicate-mediated formose reaction: Bottom-up synthesis of sugar silicates. *Science*, **327**, 984–986.

24 (a) Costanzo, G., Saladino, R., Crestini, C., Ciciriello, F., and Di Mauro, E. (2007) Formamide as the main building block in the origin of nucleic acids. *BMC Evolution. Biol.*, **7**, 1–8.; (b) Saladino, R., Neri, V., Crestini, C., Costanzo, G., Graciotti, M., and Di Mauro, E. (2008) Synthesis and degradation of nucleic acid components by formamide and iron sulfur minerals. *J. Am. Chem. Soc.*, **130**, 15512–15518.

25 (a) Powner, M.W., Gerland, B., and Sutherland, J.D. (2009) Synthesis of activated pyrimidone ribonucleotides in prebiotically plausible conditions. *Nature*, **459**, 239–242.; (b) Pasek, M.A., Dworkin, J.P., and Lauretta, D.S. (2007) A radical pathway for organic phosphorylation during schreibersite corrosion with implications for the

origin of life. *Geochim. Cosmochim. Acta*, **71**, 1721–1736.

26 (a) Nielsen, P.E., and Egholm, M. (1999) An introduction to peptide nucleic acid. *Curr. Issues Mol. Biol.*, **1**, 89–104.; (b) Meierhenrich, U.J., Muñoz Caro, G.M., Bredehöft, J.H., Jessberger, E.K., and Thiemann, H.-P. (2004) Identification of diamino acids in the Murchison meteorite. *Proc. Natl. Acad. Sci. U. S. A.*, **101**, 9182–9186.

27 (a) Meierhenrich, U.J., Nahon, L., Alcaraz, C., Bredehöft, J.H., Hoffmann, S.V., Barbier, B., and Brack, A. (2005) Asymmetric vacuum UV photolysis of the amino acid leucine in the solid state. *Angew. Chem. Int. Ed.*, **44**, 5630–5634.; (b) Pizzarello, S., Huang, Y., and Alexandre, M.R. (2008) Molecular asymmetry in extraterrestrial chemistry: insights from a pristine meteorite. *Proc. Natl. Acad. Sci. U. S. A.*, **105**, 3700–3704.

28 Rosenberg, R.A., Abu Haia, M., and Ryan, P.J. (2008) Chiral-selective chemistry induced by spin-polarized secondary electrons from a magnetic substrate. *Phys. Rev. Lett.*, **101**, 178301–178304.

29 Shibata, T., Yamamoto, J., Matsumoto, N., Yonekubo, A., Osanai, S., and Soai, K. (1998) Amplification of a slight enantiomeric imbalance in molecules based on asymmetric autocatalysis: the first correlation between high enantiomeric enrichment in a chiral molecule and circularly polarized light. *J. Am. Chem. Soc.*, **120**, 12157–12158.

30 Viedma, C., Ortiz, J.E., de Torres, T., Izumi, T., and Blackmond, D.G. (2008) Evolution of solid phase homochirality for a proteinogenic amino acid. *J. Am. Chem. Soc.*, **130**, 15274–15275.

31 Glavin, D.P., and Dworkin, J.P. (2009) Enrichment of the amino acid L-isovaline by aqueous alteration on CI and CM meteorite parent bodies. *Proc. Natl. Acad. Sci. U. S. A.*, **106**, 5487–5492.

32 (a) Bujdák, J., Remko, M., and Rode, B.M. (2005) Selective adsorption and reactivity of dipeptide stereoisomers in clay mineral suspension. *J. Colloid Interface Sci.*, **294**, 304–308.; (b) Rimola, A., Sodupe, M., and Ugliengo, P. (2007) Alumosilicate surfaces as promoters for peptide bond formation: an assessment of Bernal's hypothesis by *ab initio* methods. *J. Am. Chem. Soc.*, **129**, 8333–8344.

33 Bujdák, J., and Rode, B.M. (2003) Peptide formation on the surface of activated alumina: peptide chain elongation. *Catal. Lett.*, **91**, 149–154.

34 Otroshenko, V.A., Vasilyeva, N.V., and Kopilov, A.M. (1985) Abiotic formation of oligonucleotides on basalt surfaces. *Orig. Life Evol. Biosph.*, **15**, 115–120.

35 (a) Belloche, A., Menten, K.M., Comito, C., Müller, H.S.P., Schilke, P., Ott, J., Thorwirth, S., and Hieret, C. (2008) Detection of amino acetonitrile in Sgr B2(N). *Astron. Astrophys.*, **482**, 179–196.; (b) Belloche, A., Garrod, R.T., Müller, H.S.P., Menten, K.M., Comito, C., and Schilke, P. (2009) Increased complexity in interstellar chemistry: detection and chemical modelling of ethyl formate and n-propyl cyanide in Sagittarius B2(N). *Astron. Astrophys.*, **499**, 215–232.

36 Elsila, J.E., Glavin, D.P., and Dworkin, J.P. (2009) Cometary glycine detected in samples returned by Stardust. *Meteorit. Planet. Sci.*, **44**, 1323–1330.

37 (a) Oró, J. (1961) Comets and the formation of biochemical compounds on the primitive Earth. *Nature*, **190**, 389–390.; (b) Oró, J., and Kamat, S.S. (1961) Amino acid synthesis from hydrogen cyanide under possible primitive Earth conditions. *Nature*, **190**, 442–443.; (c) Oró, J. (1961) Mechanism of synthesis of adenine from hydrogen cyanide under possible primitive Earth conditions. *Nature*, **191**, 1193–1194.

38 Bethell, T., and Bergin, E. (2009) Formation and survival of water vapour in the Terrestrial planet-forming region. *Science*, **326**, 1675–1677.

39 Mathews, C.N., and Minard, R.D. (2006) Hydrogen cyanide polymers, comets and the origin of life. *Faraday Discuss.*, **133**, 393–401.

40 Bullock, M.A., and Grinspoon, D.H. (2001) The recent evolution of climate on Venus. *Icarus*, **150**, 19–37.

41 (a) Houtkooper, J.M., and Schulze-Makuch, D. (2007) A possible biogenic

origin for hydrogen peroxide on Mars: the Viking results reinterpreted. *Int. J. Astrobiology*, **6**, 147–152.;
(b) Houtkooper, J.M., and Schulze-Makuch, D. (2009) Possibilities for the detection of hydrogen peroxide-water-based life on Mars by the Phoenix Lander. *Planet. Space Sci.*, **54**, 449–453.

42 Fairén, A.G., Davila, A.F., Gago-Duport, L., Amils, R., and McKay, C.P. (2009) Stability against freezing of aqueous solutions on early Mars. *Nature*, **459**, 401–404.

43 (a) Atreya, S.K., Mahaffy, P.R., and Wong, A.-S. (2007) Methane and related trace species on Mars: origin, loss, implications for life, and habitability. *Planet. Space Sci.*, **55**, 358–369.;
(b) Lefèvre, F., and Forget, F. (2009) Observed variations of methane on Mars unexplained by known atmospheric chemistry and physics. *Nature*, **460**, 720–723.

44 (a) Thomas-Keprta, K.L., Clemett, S.J., McKay, D.S., Gibson, E.K., and Wentworth, S.J. (2009) Origins of the magnetic nanocrystals in Martian meteorite ALH84001. *Geochim. Cosmochim. Acta*, **73**, 6631–6677.;
(b) Lapen, T.J., Righter, M., Brandon, A.D., Debaille, V., Beard, B.L., Shafer, J.T., and Peslier, A.H. (2010) A younger age for ALH 84001 and its geochemical link to shergottite sources in Mars. *Science*, **328**, 347–351.

45 Rehder, D. (2008) *Bioinorganic Vanadium Chemistry*, John Wiley & Sons, Ltd, Chichester.

46 Mayor, M., Bonfils, X., Forveille, T., Delfosse, X., Udry, S., Berteaux, J.-L., Beust, H., Bouchy, F., Lovis, C., Pepe, F., Perrier, C., Queloz, D., and Santos, N.C. (2009) The HARPS search for southern extra-solar planets. *Astron. Astrophys.*, **507**, 487–494.

Index

a

abiotic methanogenesis   134
absorption   81
abundance ratio   4
accretion   25
Acetobacter peroxidans   268
acetonitrile   75, 91, 263
acetyl-coenzyme-A (acetyl-CoA)   245
acetyl-coenzyme-A synthase   244
acetylenes   59
achondrite   103, 150f.
– F-type   151
acid-induced CO_2 mobilization   119
acidophiles   232ff.
adenine   251
adenosinetriphosphate (ATP)   224
adsorption   81
age determination
– Lunar material   14
– meteoritic material   13
alanine   76, 91, 161
– α-alanine   91f.
– β-alanine   91, 161
– L-alanine   249
albedo   179
Alcor   21
ALH84001 meteorite   270ff.
aliphatic hydrocarbon   86
Allende meteorite   152ff.
alpha (α) capture   29
alpha process   28
Alten-type asteroid   103
alumina   259
alumosilicate
– nanoporous   229
Amanita   272
amino acid   57, 158, 248, 264
– α-amino acid   249
– β-amino acid   161

– D-configuration   159
– L-configuration (levorotatory)   159, 222, 259
– prebiotic formation   160
amino-acetonitrile   76, 89f.
amino-formylacetonitrile   251
aminocyanide   263
aminoethylglycine   255
aminonitrile   91, 160
ammonia   57, 228ff., 248
– photolysis   185
ammonium hydrogensulfide   210
Anabaena   246
Antares   46
anthracene   195
antigravitational effect   10
antimatter   9
antineutrino   31ff.
antineutrino capture   33
antineutron   9
antiproton   9
antiquark   8
apatite   254
aphelion   104
Apollo-type asteroid   103
Areology   129
Armalcolite   108
Armor asteroid   103
aromatic hydrocarbon   195
arsane   183
association   42
asteroid   146
– Alten-type   103
– Apollo-type   103
– Armor   103
– classification   146
asymmetric amplification   160
asymmetric autocatalysis   258
asymptotic giant branch (AGB)   22

Chemistry in Space: From Interstellar Matter to the Origin of Life. Dieter Rehder
© 2010 WILEY-VCH Verlag GmbH & Co. KGaA, Weinheim
ISBN: 978-3-527-32689-1

– chemistry   35
– phase   35
– star   35ff., 46, 156
atmospheric window   61
Azotobacter   246, 272
azurite   249

b
bacteria
– Gram-negative   225
– magnetotactic   244
– methanogenic   237ff.
– sulfate-reducing   232, 267
– sulfide-oxidizing   267
Balmer series   63
baryogenesis   8
baryon   8
baryonic matter   10
baryophiles   235ff.
basalt
– olivine-rich   109
Belousov Zhabotinsky reaction   215
benzene   184ff.
betains   233
Betelgeuse   46
Bethe-Weizsäcker cycle   27, 71, 148
Bi–Po–Pb cycle   34
Big Bang   7ff.
biogenic methanogenesis   136
biomarker   270
birnessite   239
birth   42
– star   21
birth line   21
black hole   23ff., 43
black smoker   236
blue bottle experiment   215
borate mineral   250
brightness   21
brown dwarf   23, 203ff.
Buckminster fullerene   155
Butlerov reaction   250f.

c
calcite   119ff., 137f.
calcium titanates   209
calcium- and aluminium-rich inclusion (CAI)   151, 176
Callisto   186ff.
Cameleopardalis   68
α capture   29
carbenes   220
carboanhydrase (CA)   267
carbohydrate
– dextrorotatory (D)   222

carbon   68
– burning   22ff.
– polymorphs   154f.
carbon chain   68
carbon dioxide   86, 136, 227ff., 247
– acid-induced mobilization   119
– clathrates   232
– lake   231
– photodissociation   121, 142
carbon family   54ff.
carbon monoxide   68, 142, 266
– hydrogenation   93
carbon nanotube   154
carbon-based world   220
carbon–nitrogen–oxygen (CNO) cycle   27ff., 43
carbonaceous chondrites   150ff., 248, 258
– carbon-bearing components   153
– CB (Bencubbin)   153
– CH (high in Fe)   153
– CI (Ivuna)   152
– CK (Karoonda)   153
– CM (Mighei)   152
– CO (Ornans)   152
– CR (Renazzo)   153
– CV (Vigarano)   152
carbonates   150
carbonyl sulfide   122, 267
carotenoid   233
cationic methyl   136
α-Centauri   104
ω Centauri   42
centaurs   103, 168
Centaurus   73
Ceres   103, 146ff., 176
chalcopyrite   251
Chandrasekhar limit   23f.
charge transfer reaction   53
charged current interaction   27
Chassignites   131
chemical bond formation   39
chemical sputtering   114
chemolithoautotrophic prokarya   232
chemolithotrophic microbes   139, 231
chirality   222
Chiron   103
chlorine   188
chondrites   151
– carbonaceous, *see* carbonaceous chondrites
chondrules   151
chrysotile   134
circular polarized vacuum ultraviolet (CPVU)   257

classical Kuiper type object   177
clathrates
– CO_2 232
– methane   227
– water-ice   182
clay organism   219, 242, 259ff.
clays   138, 176
cloud   45
cluster   40
cold dark matter (CDM) model
– cosmic evolution   10
color index   20, 179
column abundance   4
column amount   4
column density   4, 69f.
Coma Berenices   203
comet   103ff., 167ff.
– chemistry   171
complex molecule   74
concentration   4
Coronet cluster   18
CoRoT (convection rotation and planetary transits)   206
CoRoT 7b   206ff.
corundum   163
Cosimo's stars (Cosmica Sidera)   1
cosmic evolution
– cold dark matter (CDM) model   10
cosmic microwave background (CMB) radiation   7
cosmo-chronometry   12
– thorium- and uranium-based   13
cosmological clock   13
critical mass   18
cronstedtite   150
crystallinity   179
cubewanos (QB1os)   177
cyan radical   68
cyanamide   255
cyanate   234
cyanide
– chemistry   172
– metal   38
cyano species   172
cyanoacetylene   172, 252
cyanobacteria   239
cyanodecapentayne   49, 94
cyanohydrin   160
cyanopolyynes   38, 59ff.
cyclo-oligomerization   37
cyclopropenone   77
cyclopropylidene   77
cyclotrimerization   37
Cygnus   68
cysteine   267

cytidine   253
cytosine   252

d

dark cloud   45ff.
dark energy   10
dark matter haloes   40
dark molecular cloud   78
dark nebula   94
Deimos   107, 127ff.
density   118
deuterium   60ff.
deuterium fractionation   60, 182
2,4-diaminobutanoic acid   255
2,3-diaminopropanoic acid   255
diamond   154ff.
diaspore   130
diatomic molecule   68
dicarbon   68ff., 220
dicyan   195
diffuse cloud   73
di*iso*propyl zinc   258
dinitrogen   68, 193
diopside   123, 175
dissociative attachment   79
dissociative electron attachment   53, 78
dissociative electron capture   194
dissociative recombination   52, 92
DNA (deoxyribonucleic acid)   216ff.
dolomite   119, 266
Doppler shift radiation   7
dry ice   86
dust grain   45ff.
dust particle   35ff., 80, 100
dwarf   208
dwarf elliptical galaxy   31
dwarf planet   101ff., 176
– characteristics   104
dynamic pressure   118
dynamo effect   111

e

Earth   58, 106
– body and orbital characteristics   102
– Moon   107, 186ff.
eccentricity   127, 206
eccentricity seasons   127
eddy current   183
electron   9
electron attachment   53
– dissociative   53, 78
– radiative   78
electron capture   120, 262f.
– dissociative   194

electron degeneracy pressure   28
electron radiation   93
electron–positron pair   11
elementary particle   9
ellipticals   40f.
emission nebula   94
enantiomeric enhancement   256
Enceladus   189ff.
energy   5
enstatite   150, 175
entry channel   39
Eris   103ff., 176ff.
ethane   93, 172
ethene   184
ethylcyanide   76
ethyne   173
Europa   186ff.
evolution   215
EX Lupi   176
exit channel   39
exoplanet   203ff.
EXor   23, 176
Extended Scattered Disk   103, 176
extraterrestrial input   262
extraterrestrial life   265

f

fayalite   125ff., 193, 269
feldspars   120
fermion   8
ferric hydroxides   134
ferric silicates   134
ferrihydrite   239
ferrosilite   126
fluorapatite   224
formaldehyde   57, 93, 136
Formalhaut-b   208
formamide   251
formose reaction   250f.
formyl cation   136
formylmethanofuran   136
formylmethanofuran dehydrogenase   136
forsterite   138, 163, 175
Foucault current   183
fractional density   4
fractionation   60
fullerenes   49, 155
FUors   23

g

galaxy   40
– elliptical   40f.
– irregular   40
– spiral   40f.

Galilean moon   1, 186ff.
gamma (γ)-process   31
Ganymede   186ff.
gas giant planet   99, 181
gas planet   101
gas-grain surface reaction   174
gas-phase reaction   81, 92
GEMS (glass with embedded metal and sulfide)   175
germane   183
giant planet   100, 180, 203
– moon   180
GJ (Gliese+Jahreiss) catalog   206
GJ 581   274
GJ 1214b   211
Gl (Gliese) catalog   206
Gl 581e   274
Gliese 581   207, 274
Gliese 581d (Gl 581d)   207ff., 274
globular clusters   41
globulars   41
glucose-6-phosphate   224, 255
gluon   8
glycine   76, 89ff., 161, 249, 264
glycolaldehyde   77, 250
glycyl–glycine   259
goethite   130
grain   36, 80, 100ff.
– chemical composition   82
– chemical reaction   82
– chemistry   80
– ice   164
– ice mantle   88
– interplanetary   83
– interstellar   84ff.
– main stream particle   164, 180
– oxidic   164
– presolar   162
– silica-rich   166
– spectroscopic features   84
– structure   82
– surface formation   93
– surface reaction   81
– X-type   33, 164, 180
graphene   154
graphite   154ff.
Great Andromeda Nebula   40
Great Oxygen Event   239
guanine   251

h

hadron epoch   8
Hale-Bopp comet   168ff.
Halley's comet   168ff., 264

Halo 41
halo star 31
haloalkaliphiles 232
Halomonas 233
haloperoxidase 272
Haumea 103ff., 178
Hayashi tracks 21
HD (Henry Draper) catalog 206
HD 21389 68
HD 124314 73
HD 189733b 204f.
HD 209458b 205
HD 216956 208
heavy bombardment 213
HED (Howardites, Eucrites, and Diogenites) meteorites 150
helium burning 22ff.
helium flash 22
hematite 126ff.
Henry Draper (HD) classification 203
Hertzsprung–Russel (HR) diagram 19, 42, 208
heterocyclic polyaromatic compound 195
HI region 63
hibonite 163
homolytic fission 93
hornblendes 120
hot core 47, 88ff.
hot Jupiters 20, 204ff.
hot molecular cores 45
HR (Harvard revised) number 206
HR 8799 208
HR 8799d 207f.
Hubble constant 11
Hyakutake comet 173
hydrazine 185
hydride 81
hydride ion 78
hydro-superoxyl radical 190
hydrocarbon 184
hydrocarbon-nitrile aerosol 195
hydrocyanic acid/isohydrocyanic acid 172
hydrogen 62ff., 77ff., 193, 237ff.
– burning 21ff., 43
– fusion 21ff.
– ionized 94
– isotope 9f.
– neutral 94
– problem 81
– species 62
hydrogen cyanide 57, 248
– protonated 195
hydrogen fluoride 228
hydrogen isocyanide 57

hydrogen peroxide 142, 189, 268
hydrogenation 80
– carbon monoxide 93
– catalytic 173
hydrothermal conversion
– olivine 134
hydroxyl radical 68
hydroxylapatite 224
hydroxyvinyl-cyanide 252
hyperthermophiles 238ff.

i

ice giant 99, 181
ilmenite 126
impact parameter 55
infrared dwarf 23
insertion reaction 53f.
insoluble organic matter (IOM) 156ff.
interaction energy 55
intergalactic coronal gas 47
interplanetary dust 33, 146
interplanetary dust particle 87, 162
interplanetary medium 103
interstellar cloud 68, 87
– chemistry 50
– hydrogen species 62
– reaction type 50
interstellar dust 59
interstellar gas 65
interstellar grain 84ff.
– spectroscopic feature 84
interstellar ion 67
interstellar matter 46
interstellar medium 45ff.
interstellar reddening 80
interstellar species 61
– detection 61
inverse micelle 229
inversion 75
inversion barrier 75
inversion transition 75
Io 183ff.
ion–molecule reaction 51, 174ff., 194
ion–neutral reaction 51, 263
ionized diffuse cloud 45
IRC+10216 78
iron sulfide 87, 234
iron sulfide cell 243
iron sulfide mineral 267
iron–nickel hydrogenases 239
iron–sulfur cluster 244
iron–sulfur world 220, 242
isocyanides
– metal 38

L-isoleucine   257
isotope abundances   154
isotopic fractionation   166
isotopic self-shielding   166
isovaline   158f., 180
Ivuna meteorites   161
Ixion   104ff., 178

j
J-coupling   64
jarosite   130
Jeans' mass   18
Jovian planet   100, 146
Jovian stratosphere   184
Jupiter   106, 176ff.
– atmosphere   183
– body and orbital characteristics   102
– characteristics   181
– ionosphere   183
– magnetic field   183
– troposphere   183, 210
Jupiter trojans   103

k
kaolinite   179
Keplerate   229
Kepler's laws   1
kieserite   138
kinetic isotope effect   71
KREEP-rich magma   108
Kuiper belt   103, 176
Kuiper belt object   100, 176, 193

l
L class star   203ff.
L subdwarf   210
L-dwarf atmosphere   210
labile organic matter   156
lambda-doubling   69
large Magellanic Cloud   41
Large Molecular Heimat   76, 263
last uniform common ancestor (LUCA)   216ff., 236
late heavy bombardment   100, 186
lazulite   254
Leo   78
lepton   9
life
– extreme condition   230
– origin   213ff.
– water   226
limestone   137
limonite   130
liquid surface   145

lithoautotrophs   235
lonsdalite   154f.
LUCA, see last uniform common ancestor
Lulin comet   168
luminosity   20f.
luminous red novae   24
Lunar cataclysm   213
Lunar crust   108
Lyman band   65
Lyman series   63

m
M class star   208
M dwarf   203ff.
magma
– KREEP-rich   108
magnesite   118, 137
magnesium   29, 148
magnesium hydroxides   134
magnesium silicates   134, 210
magnetic buoyancy instability   36
magnetite   126ff., 150, 269
– crystal   270f.
magnetosomes   244
magnetotactic bacteria   244
Makemake   103ff., 178f.
malachite   249
marcasite   242
maria   108
Mars   58, 106, 126ff., 267ff.
– atmosphere   140f.
– body   102
– carbonates   137
– chemistry in the atmosphere   140
– dichotomy   130
– geological feature   129
– Moons   127
– meteorites   129ff.
– methane   133
– nightglow   144
– orbital feature   102, 127f.
– Phobos and Deimos   107, 127ff.
– radicals   143
– sulfates   137
– surface chemistry   129
– trojans   103, 127
– water   137
medium-energy Solar neutrinos   28
membrane   229
mercaptomethane   237
Mercury   104ff.
– body and orbital characteristics   102
– exosphere   113ff.

messenger RNA (mRNA)   217
metabolism   214f.
metallicity   17, 35, 203f.
meteorite   39, 103, 146ff., 255ff.
– achondrites (S, stony)   151
– chondrites (C)   150ff.
– CI-type   167
– E-type (enstatite)   152
– Mars   131f.
– metal (M)   152
– stony (S)   151
– stony-iron   152
meteoroide   103
methane   93, 248, 269
– clathrates   227
– Mars   133f.
– photolysis   184
– Titan   192
methane-metabolizing archaea   232
methane-oxidizing
  chemolithoautotrophs   232
methanofuran   136
methanogenesis   134ff.
methanogenic bacteria   237
methanogens   237ff.
methanol   93
methanotrophs   237ff., 270
methyl cation   57
methyl-cobalamine   244
S-methyl-ethanesulfonate   239
1-methyladeninium   260, 261
methylamine   75, 263
(S)-2-methylbutyraldehyde   257
methylmethanofuran   136
2-methylpyrimidine   258
2-methylpyrimidine-5-carbaldehyde   258
methylsulfide   237
methylthiol   237ff.
methylyne radicals   68
micas   120
micelle   229
micrometeorites   162
migratory insertion   245
Milanković cycles   127
Milky Way   40, 47
Miller–Urey experiment   241ff., 274
mineral
– comet   175
minihalo   11
mitochondria   219
mixing ratio   119
Mizar   21
moissanite   156
molality   4

molar concentration   4
mole fraction   4, 119
molecular anion   78
molecular cloud   45ff., 94
molecular hydrogen   77
molecular hydrogen fraction   69
molecular mass   5
molecule   65
molecule–molecule reaction   182
molybdenum   230ff.
montmorillonite   219ff., 259ff.
Moon (Earth Moon)   107
moon   101
– giant planet   180
moonlets   101
multicarbon molecule   72
Murchison meteorite   152ff., 166, 255ff.
Murray meteorite   265

n
naked singularity   25
Nakhlites   131
nanobacteria   133, 217
naphthalene   195
neon   29
neon-dependent process   29
Neptune   106, 176ff.
– atmosphere   183
– body and orbital characteristics   102
– characteristics   181
– trojans   103
neutral exchange   53, 173
neutral–neutral reaction   53
neutralino   11
neutrino   9, 27f.
neutron   9, 31f.
neutron capture   30f.
neutron star   23ff.
Nice model   100
nickel   239
nicotineadenine-dinucleotidephosphate
  (NADPH)   224
nightglow   144
nitrate   239
nitriles   92
nitrite   239
nitrogen family   54ff.
nitrogen oxide   143f.
nitrogenase   246
non-carbon life form   220
nonaqueous life form   220
nova   24
nuclear fusion sequence   28
nucleation model   181

nucleobase 250
nucleosynthesis 9f.
number density 4, 119

o

OGLE (optical gravitational lensing experiment) 206
oligomerization 260
oligophosphate 254
olivine 110, 126ff., 138, 150, 163
– Fe-rich 175
– hydrothermal conversion 134
– magnesium-rich 151
olivine-rich basalt 109
Oort cloud 76, 103
open clusters 42
ζ Ophiuchi 73
Orcus 103ff., 178
organic matter
– insoluble (IOM) 156f.
– labile 156
– refractory 156
Orgueil meteorite 161, 265
Orion 63
orthosilicate 126
oxygen 29, 77, 190, 248
oxygen burning 29
oxygen family 54f.
ozone 142

p

π bonding 220f., 240
p-process 31
panspermia 213, 275
paraffin 86
Parker instability 36
Paschen series 63
pentlandite 242
peptide 259
peptide nucleic acid (PNA) 255
perchlorates 131
perihelion 104
ζ Persei 68
Phobos 107, 127ff.
phosphabenzene 224
phosphane 183, 223
phosphate
– biological role 220
– ester 224f.
– mobile 254
phosphines 223
phospholipids 225
phosphorine 224

photochemistry 184
photodetachment 53
photodisintegration 30f.
photodissociation 38, 54, 121f., 166ff., 188
– carbon dioxide 121, 142
photolysis 143
– ammonia 185
– methane 184
photon
– high-energy 11
photosynthesis 239
phyllosilicate 138, 150, 176ff., 222
physical sputtering 114
piezophiles 235
pigeonite 151
pion 11
pioneer organism 220, 242f., 274
Piscis Austrinus 208
pL-planet 210
plagioclase feldspar 108f.
Planck epoch 8
planet
– definition 101
planetary nebula (PN) 21f., 43, 71
planetisimals 99f.
Pleiades 42
Pluto 103ff., 177
Pluto–Charon system 177
pM-planet 210
polyaromatic hydrocarbon 270
polycyclic aromatic hydrocarbon (PAH) 53, 68, 85f., 131, 163
– AGB star 158
polymict ureilite 151
polyoxidometalate (POM) 229
polyoxidomolybdate 229
polyphosphazene 223
polyynes 49ff.
population I star 12, 29ff.
population II star 12
population III star 11f.
population III.1 star 11f.
population III.2 star 11f.
porphyrinogenic ligand system 238
positron 9, 31
postasymptotic giant star 71
power law 70
pp (proton–proton) chain 25
pre-supernova
– explosive collapse 33
preplanetary nebula (PPN) 35ff., 50
presolar grain 162
primeval atom 7

propionitrile 91
propyne 184
protein 219
proton 9, 31
proton affinity 66
proton capture 30
protoplanetary disc 50
protoplanetary nebula 50
protostar 25
Proxima Centauri 104
pulsar 23f.
– radio 24
– γ-ray 24
purine 250ff.
pyrimidine 250ff.
pyrite 123ff., 234ff., 251
pyroclastic glasses 108
pyroxene 131, 151, 163, 175
pyrrhotite 126, 242

q
QB1os (cubewanos) 177
Quaoar 103ff., 178
quark 8
quark–gluon plasma 8
quartz 222f.
quasar 23
quasicrystalline 226

r
r (rapid)-process 30ff.
radiative association 53
radiative attachment 79
radiative combination 92
radiative decay 65
radiative dissociation 79
radiative electron attachment 78
radiative recombination 53, 143
radio pulsar 24
radiolysis 187ff.
rapid-process, see *r*-process
γ-ray 11
γ-ray pulsar 24
re-equilibration 174
reaction network 54ff.
red dwarf 23, 207
red giant 24, 71
red supergiant 46
red-shifted radiation 7
reflection nebula 45, 94
refractory organic matter 156
regolith 109, 138
reproduction 215

resonance charge exchange 53
retrograde rotation 116
Rhizobium 246
ribose 250
RNA (ribonucleic acid) 216ff.
RNA polymerase 217
RNA world 219
rotational quantum number 64
rp-process 30ff.
rutile 142, 163

s
s (slow)-process 30ff., 166
salt-induced peptide formation 249, 275
Saturn 101ff., 180ff.
– atmosphere 183
– body and orbital characteristics 102
– characteristics 181
– magnetic field 183
– moon 58
Scattered Disk 103, 176
Schwarzschild limit 25
Schwarzschild radius 25
Schwasmann-Wachmann 3 170
secondary electron 93
Sedna 104ff., 176ff.
self-condensation 57
self-dissociation 226
self-replication 215
self-shielding 71, 166, 264
serine 76, 92
serpentines 134
serpentinisation 134f., 193
Shergottites 131
Shoemaker-Levy 171
siderite 118, 137
silica 134, 163, 222, 266
silica hydrates 221
silicate
– biological role 220
silicon 59
– biochemistry 221
– burning 30
– family 59
– SiCN 59
– SiNC 59
silicon carbide 59, 156ff.
silicon nitride 163
silicon-carbon compound 222
singularity 25
slow (s)-process, see *s*-process
small Magellanic Cloud 41
small solar system bodies 146

Soai reaction 258
Solar neutrinos
– medium-energy 28
Solar System 29, 71, 99ff.
Solar wind 99ff., 115ff., 145ff., 168ff., 183ff.
solfataras 139
Solomon process 65
spectral class 19
spin–orbit coupling 63f.
spinels 163
sputtering
– chemical 114
– physical 114
star
– classification 17ff., 206
– evolution 17ff.
– formation 17
– massive 30
– metallicity 35
star designation
– HD catalog 206
stars of the Medici (Medicea Sidera) 1
stellar cemetery 22
stellar class 19f.
stellar wind 25ff.
stony meteorites 151
stratosphere 184
stromatolite fossils 213
subdwarf 208
sugar 57, 77
sulfate 239
sulfate reduction 232
sulfate-reducing bacteria 232, 267
sulfide-oxidizing bacteria 267
sulfite–sulfate shuttle 235
sulfur
– allotropes 195
– chemistry 59
– elemental 187
– species 173
sulfur dioxide 187f.
sulfur-oxidizing chemolithoautotrophs 232
sulfur-oxidizing microorganism 232
sulfuric acid 121
sulfuryl chloride 188
Sun 21, 101
super-Earths 205ff.
super-Jupiters 20, 203ff.
superbubble 24
supernova 21ff.
– type I 24
– type II 23f.
– X-type grain 164

supernova remnants 24
superoxide 268
supersymmetric particle 11
surface reaction 80f.
– grain 81

t

T class star 203
T-Tauri star 25
T-Tauri variable 21ff.
Tagish Lake meteorite 156ff.
Taurus 78
Temple 1 comet 168ff.
terrestrial planet 99
thermal decomposition 118
thermophiles 235ff.
thermophilic methanogens 237
thioacetic acid 245
– *S*-methyl ester 242
Thioalkalivibrio 233
thiocyanate 234
thiomolybdenum cluster 229, 272
Tholins 178
three-atomic system 66
tidal heating 186
Titan 58, 191ff.
titanium oxides 209
TMC-1 78
total angular momentum quantum number 63f.
transfer RNA (tRNA) 216
translucent cloud 45
tremolite 266
tricarbon 72
triple-alpha process 28
Triton 185ff.
troilite 242f.
Trojan 100ff.
Trojan asteroid 127
tungsten 238
Tycho Brahe's supernova 24

u

universe
– development 7
– origin 7
uracil 252
Uranus 106, 180ff.
– atmosphere 183
– body and orbital characteristics 102
– characteristics 181
urea 252ff.
ureilite 151

v

V342 Pegasi   207
L-valine   249
vanadium oxides   121, 209
Varuna   103ff., 178f.
Vega   21
Venus   58, 101ff., 115ff.
– atmosphere   118ff., 266ff.
– body and orbital characteristics   102
– chemical reactions   121
– geological feature   115
– mineral   125
– orbit feature   115
Vesta meteorites (V meteorites, HED (Howardites, Eucrites, and Diogenites) meteorites)   150
Vestoids   151
vinylcyanide   76
virus   216
volcanism   186
volume(tric) density   4
vortices   36
vp process   31

w

Wächtershäuser iron–sulfur world   241, 274
Wächterhäuser pioneer organisms   267

WASP (wide area search for planets)   206
water   93, 142, 266
– biological role   220
– life   219ff.
– *ortho*- and *para*-H_2O 174
– radiolysis   193
water-ice 143
water-ice clathrates   182
wavellite   254
Werner band   65
white dwarf   21ff.
white smoker   236
Wild 2 comet   168ff., 264
Wolf-Rayet (WR) star   23f.

x

X-ray absorption near edge structure (XANES)   87
X-ray fluorescence (XRF) analysis   125
Xena   103
xenon-HL   167

z

zeolites   229
zero age   21, 42